This book is dedicated to our wives:
Vera S. Bettelheim and Lucy G. Landesberg
whose help, understanding and patience
enabled us to write this book.

Preface

In preparing the third edition of this Laboratory Manual, we wish to thank our colleagues who made this new edition possible by adopting our Manual for their courses. This third edition coincides with the publication of the fifth edition of the textbook: "Introduction to General, Organic and Biochemistry" by Bettelheim and March, which shares the outline and the pedagogical philosophy with this book. As in previous editions, the emphasis has been for the clearest possible writing in the procedures of meaningful, reliable experiments. Throughout the years feedback from different Colleges and Universities made us aware that we have managed to achieve a manual that eases the student's task in performing experiments. In our new edition we strive to maintain this standard and improve upon it where possible.

The major changes in this new edition are four fold: (1) We increased the number of experiments by four, thereby providing an even wider selection of experiments. The new experiments demonstrate the manifestation of entropy, the extended use of molecular models in studying the structures of organic compounds, the nature of the secondary structure of DNA, and the analysis of vitamin A. (2) We changed significantly the procedures in five experiments, especially in those which required multistep isolation procedures. It seems that those extended procedures, in a few hands, resulted in loss of products. In such cases the subsequent characterization of products was impossible, creating a degree of frustration. In this new edition, when the characterization of the product is the main pedagogical aim, for example, enzyme activity, we simply recommend the purchase of the enzyme from biochemical supply houses. (3) We further improved our previous aims to minimize the use of hazardous chemicals and to work on a semimicroscale. (4) We changed about 50% of the Pre-Lab and Post-Lab questions.

As in the previous editions three basic goals were followed in all the experiments: (a) the experiments should illustrate the concepts learned in the classroom; (b) it should be clearly and concisely written so that students will understand easily the task at hand and will be able to perform the experiments in a 2 1/2 hr. laboratory period; (c) the experiments should not only be simple demonstrations, but also should contain a sense of discovery.

It did not escape our attention that in adopting this text of Laboratory Experiments, the instructor must pay attention to budgetary constraints. All experiments in this manual require only inexpensive equipment, if any. A few spectrophotometers and pH meters are necessary in a number of experiments. A few experiments may require more specialized, albeit inexpensive equipment, for example, a few viscometers.

The 50 experiments in this book will provide suitable choice for the instructor to select about 25 experiments for a 2 semester or 3 quarter course. The following are the principle features of this book:

1. Twenty two experiments illustrate the principles of general chemistry, 11 that of organic chemistry, and 17 of biochemistry. Whenever it was feasible, health and nutritional aspects were emphasized.

2. Each experiment starts out with background information that goes beyond the textbook material. All the relevant principles and their applications are reviewed in this background section.

3. The procedure part provides a step-by-step description of the experiments. Clarity of writing in this section is of utmost importance for successful execution of the experiments. **Caution!** signs alert the students when dealing with dangerous chemicals, such as strong acids or bases.

4. Pre-Lab questions are provided to familiarize the students with the concepts and procedures before they start the experiments. By requiring the students to answer these questions and by grading their answers, we accomplish the task of preparing the students for the experiments.

5. In the Report Sheet we not only ask for the registration of the raw data, but we also request some calculations to yield secondary data.

6. The Post-Lab questions are designed so that the student should be able to reflect upon the results, interpret them, and relate their significance.

7. At the end of the book in Appendix 3, we provide the Stockroom Personnel with detailed instructions on preparation of solutions and other chemicals for each experiment. We also give detailed instructions as to how much material is needed for a class of 25 students.

An Instructor's Manual that accompanies this book is **solely for the use of the Instructor.** It helps in the grading process by providing ranges of the experimental results we obtained from class use. In addition it alerts the instructor to some of the difficulties that may be encountered in certain experiments. The disposal of waste material is discussed for each experiment.

We hope that you will find our book of Laboratory Experiments helpful in instructing your students. We anticipate that students will like the book and find it inspiring in studying different aspects of chemistry.

Garden City, NY
April 1997

Frederick A. Bettelheim
Joseph M. Landesberg

Acknowledgments

These experiments have been used by our colleagues over the years and their criticism and expertise were instrumental in refinement of the experiments. We thank Stephen Goldberg, Robert Halliday, Cathy Ireland, Mahadevappa Kumbar, Jerry March, Sung Moon, Donald Opalecky, Reuben Rudman, Charles Shopsis, Kevin Terrance, and Stanley Windwer for their advice and helpful comments. We acknowledge the contributions of Dr. Jessie Lee, Community College of Philadelphia. We extend our appreciation to the entire staff at Saunders College Publishing, especially to John Vondeling, Vice President/Publisher, and Beth Rosato, Developmental Editor, for their encouragement and excellent efforts in producing this book.

Practice Safe Laboratory

A few precautions can make the laboratory experience relatively hazard free and safe. These experiments are on a small-scale and as such, many of the dangers found in the chemistry laboratory have been minimized. In addition to specific regulations that you may have for your laboratory, the following DO and DON'T RULES should be observed at all times.

DO RULES

❏ **Do wear approved safety glasses or goggles at all times.**

The first thing you should do after you enter the laboratory is to put on your safety eyewear. The last thing you should do before you leave the laboratory is to remove them. Contact lens wearers must wear additional safety goggles; prescription glasses can be used instead.

❏ **Do wear protective clothing.**

Wear sensible clothing in the laboratory: e.g., no shorts, no tank tops, no sandals. Be covered from the neck to the feet. Laboratory coats or aprons are recommended. Tie back long hair, out of the way of flames.

❏ **Do know the location and use of all safety equipment.**

This includes eyewash facilities, fire extinguishers, fire showers, and fire blankets. In case of fire, do not panic, clear out of the immediate area, and call your instructor for help.

❏ **Do use proper techniques and procedures.**

Closely follow the instructions given in this laboratory manual. These experiments have been student tested; however, accidents do occur but can be avoided if the steps for an experiment are followed. Pay heed to the **CAUTION!** signs in a procedure.

❏ **Do discard waste material properly.**

Organic chemical waste should be collected in appropriate waste containers and *not flushed down sink drains*. Dilute, nontoxic solutions may be washed down the sink with plenty of water. Insoluble and toxic waste chemicals should be collected in properly labeled waste containers. Follow the directions of your instructor for alternative or special procedures.

❏ **Do be alert, serious, and responsible.**

❑ **Do not eat or drink in the laboratory.**

❑ **Do not smoke in the laboratory.**

❑ **Do not taste any chemical or breathe any vapors given off by a reaction.**

If there is a need to smell a chemical, you will be shown how to do it safely.

❑ **Do not get any chemicals on your skin.**

Wash off the exposed area with plenty of water should this happen. Notify your instructor at once.

❑ **Do not clutter your work area.**

Your laboratory manual and the necessary chemicals, glassware, and hardware are all that should be on your bench top. This will avoid spilling chemicals and breaking glassware.

❑ **Do not enter the chemical storage area or remove chemicals from the supply area.**

❑ **Do not perform unauthorized experiments.**

❑ **Do not take unnecessary risks.**

These DO and DON'T RULES for a safe laboratory are not an exhaustive list, but are a minimum list of precautions that will make the laboratory a safe and fun activity. Should you have any questions about a hazard, ask your instructor *first* - not your laboratory partner. Finally, if you wish to know about the dangers of any chemical you work with, read the Material Safety Data Sheet (MSDS). These sheets should be on file in the chemistry department office.

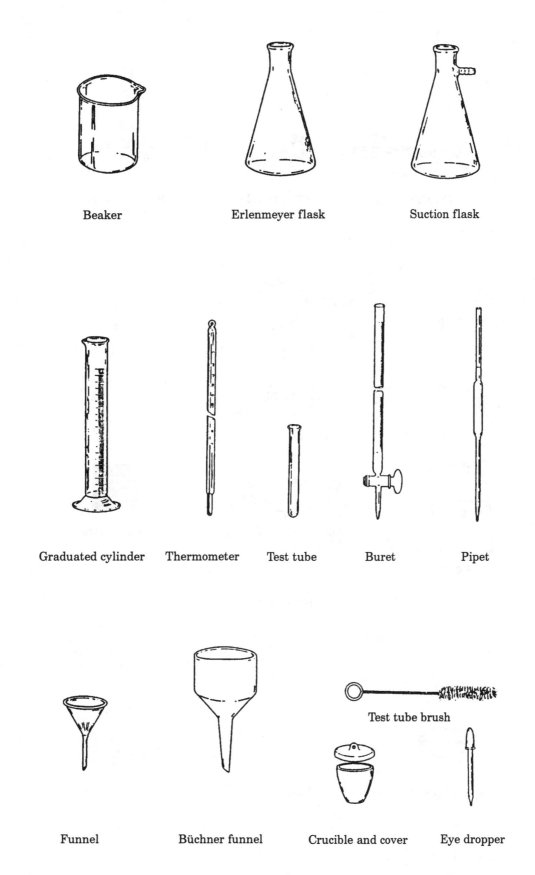

Beaker

Erlenmeyer flask

Suction flask

Graduated cylinder

Thermometer

Test tube

Buret

Pipet

Funnel

Büchner funnel

Crucible and cover

Eye dropper

Test tube brush

Figure 1 • Common laboratory equipment (From Weiner, S.A., and Peters, E.I.: Introduction to Chemical Principles. W.B. Saunders, Philadelphia, 1980.)

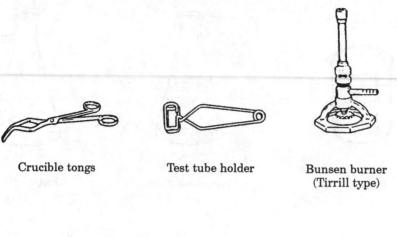

Crucible tongs

Test tube holder

Bunsen burner
(Tirrill type)

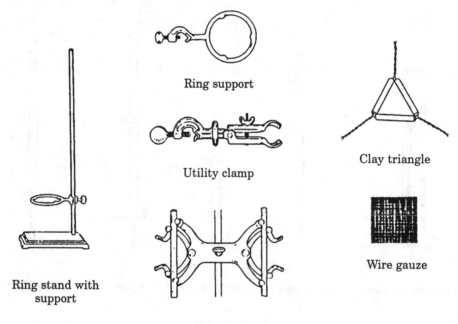

Ring support

Utility clamp

Clay triangle

Wire gauze

Ring stand with
support

Buret clamp

Evaporating dish

Watch glass

Tripod

Figure 1 • (continued)

Table of Contents

Laboratory techniques: use of the laboratory gas burner; basic glassworking

The Laboratory Gas Burner

Tirrill or Bunsen burners provide a ready source of heat in the chemistry laboratory. In general, since chemical reactions proceed faster at elevated temperatures, the use of heat enables the experimenter to accomplish many experiments more quickly than would be possible at room temperature. The burner illustrated in Fig. 1.1 is typical of the burners used in most general chemistry laboratories.

Figure 1.1
The Bunsen burner.

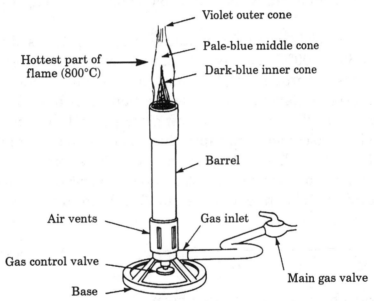

A burner is designed to allow gas and air to mix in a controlled manner. The gas often used is "natural gas," mostly the highly flammable and odorless hydrocarbon methane, CH_4. When ignited, the flame's temperature can be adjusted by altering the various proportions of gas and air. The gas flow can be controlled either at the main gas valve or at the gas control valve at the base of the burner. Manipulation of the air vents at the bottom of the barrel allows air to enter and mix with the gas. The hottest flame has a violet outer cone, a pale-blue middle cone, and a dark blue inner cone; the air vents, in this case, are opened sufficiently to assure complete combustion of the gas. Lack of air produces a cooler, luminous yellow flame. This flame lacks the inner cone and most likely is smoky, and often deposits soot on objects it contacts. Too much air blows out the flame.

Basic Glassworking

In the chemistry laboratory, it is often necessary to modify apparatus made from glass or to connect pieces of equipment with glass tubing. Following correct procedures for working with glass, especially glass tubing, is important.

Glass is a super-cooled liquid. Unlike crystalline solids which have sharp melting points, glass softens when heated, flows, and thus can be worked. Bending, molding, and blowing are standard operations in glassworking.

Not all glass is the same; there are different grades and compositions. Most laboratory glassware is made from borosilicate glass (containing silica and borax compounds). Commercially, this type of glass is known as *Pyrex* (made by Corning Glass) or *Kimax* (made by Kimble glass). This glass does not soften very much below 800°C and, therefore, requires a very hot flame in order to work it. A Bunsen burner flame provides a hot enough temperature for general glassworking. In addition, borosilicate glass has a low thermal coefficient of expansion. This refers to the material's change in volume per degree change in temperature. Borosilicate glass expands or contracts slowly when heated or cooled. Thus, glassware composed of this material can withstand rapid changes in temperature and can resist cracking.

Soft glass consists primarily of silica sand, SiO_2. Glass of this type softens in the region of 300–400°C, and because of this low softening temperature is not suitable for most laboratory work. It has another unfortunate property that makes it a poor material for laboratory glassware. Soft glass has a high thermal coefficient of expansion. This means that soft glass expands or contracts very rapidly when heated or cooled; sudden, rapid changes in temperature introduce too much stress into the material, and the glass cracks. While soft glass can be worked easily using a Bunsen burner, care must be taken to prevent breakage; with annealing, by first mildly reheating and then uniformly, gradually cooling, stresses and strains can be controlled.

Objectives

1. To learn how to use a Bunsen burner.
2. To learn basic glassworking by bending and fire-polishing glass tubing.

Procedure

The Laboratory Gas Burner; Use of the Bunsen Burner

1. Before connecting the Bunsen burner to the gas source, examine the burner and compare it to Fig. 1.1. Be sure to locate the gas control valve and the air vents and see how they work.

2. Connect the gas inlet of your burner to the main gas valve by means of a short piece of thin-walled rubber tubing. Be sure the tubing is long enough to provide some slack for movement on the bench top. Close the gas control valve. If your

burner has a screw-needle valve, turn the knob clockwise. Close the air vents. This can be done by rotating the barrel of the burner (or sliding the ring over the air vents if your burner is built this way).

3. Turn the main gas valve to the open position. Slowly open the gas control valve counterclockwise until you hear the hiss of gas. Quickly strike a match or use a gas striker to light the burner. With a lighted match, hold the flame to the top of the barrel. The gas should light. How would you describe the color of the flame? Hold a Pyrex test tube in this flame. What do you observe?

4. Carefully turn the gas control valve, first clockwise and then counterclockwise. What happens to the flame size? (If the flame should go out, or if the flame did not light initially, shut off the main gas valve and start over, as described above.)

5. With the flame on, adjust the air vents by rotating the barrel (or sliding the ring). What happens to the flame as the air vents open? Adjust the gas control valve and the air vents until you obtain a flame about 3 or 4 in. high, with an inner cone of blue (Fig. 1.1). The tip of the pale blue inner cone is the hottest part of the flame.

6. Too much air will blow out the flame. Should this occur, close the main gas valve immediately. Relight following the procedure in step no. 3.

7. Too much gas pressure will cause the flame to rise away from the burner and "roar" (Fig. 1.2). If this happens, reduce the gas flow by closing the gas control valve until a proper flame results.

Figure 1.2
The flame rises away
from the burner.

8. "Flashback" sometimes occurs. If so, the burner will have a flame at the bottom of the barrel. Quickly close the main gas valve. Allow the barrel to cool. Relight following the procedures in step no. 3.

Basic Glassworking; Working with Glass Tubing

Cutting glass tubing

1. Obtain a length of glass tubing (5–6 mm in diameter). Place the tubing flat on the bench top, and with a grease pencil mark off a length of 30 cm. Grasp a triangular file with one hand, placing your index finger on a flat side of the file. With your other hand, hold the tubing firmly in place against the bench top. At

the mark, press the edge of the file down firmly on the glass, and in one continuous motion scratch the glass (Fig. 1.3).

Figure 1.3
Cutting glass tubing with
a triangular file.

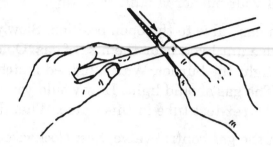

2. Place a drop of water on the scratch (this seems to help the glass break). Wrap the tubing with cloth or paper towels and grasp with both hands, as shown in Fig. 1.4. Place your thumbs on the unscratched side of the tubing, one thumb on either side of the scratch. Position the scratch away from your body and face. Snap the tubing by simultaneously pushing with both thumbs and pulling with both hands toward your body. The tubing should break cleanly where the glass was scratched. Should the tubing not break, repeat the procedure described above.

Figure 1.4
Breaking glass tubing.

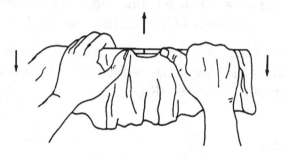

Glass bends

1. Turn off the Bunsen burner and place a wing top on the barrel. The wing top will spread out the flame so that a longer section of glass will be heated to softness. Relight the burner and adjust the flame until the blue inner cone appears along the width of the wing top (Fig. 1.5).

Figure 1.5
Wing top on the Bunsen burner.

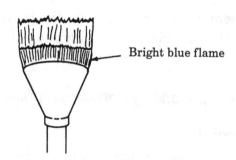

Bright blue flame

2. Hold the midsection of the newly-cut glass tubing in the flame. Keep the tubing in the hottest part of the flame, just above the spread-out blue cone (Fig. 1.6). Rotate the tubing continuously to obtain uniform heating. As the glass gets hot,

the flame should become yellow; this color is due to sodium ions, which are present in the glass.

Figure 1.6

Holding the glass tubing in the flame.

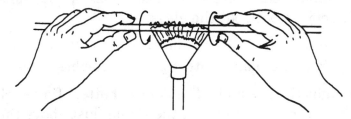

When the glass gets soft and feels as if it is going to sag, remove the glass from the flame. Hold it steady without twisting or pulling (Fig. 1.7), and quickly, but gently, bend it to the desired angle (Fig. 1.8).

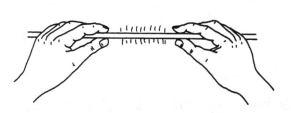

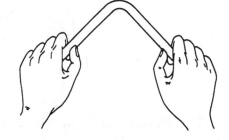

Figure 1.7 • Hold before bending.

Figure 1.8 • Quickly bend.

A good bend has a smooth curve with no constrictions (Fig. 1.9).

Figure 1.9

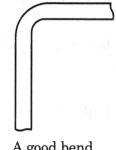

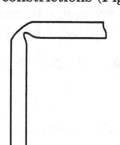

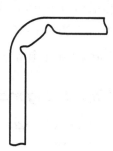

A good bend.

Poor bends.

CAUTION!

Hot glass looks like cold glass. When finished with a piece of hot glass, place it out of the way on your bench top, on a piece of wire gauze. Glass cools slowly, so do not attempt to pick up any piece until you test it. Hold your hand above the glass without touching; you will be able to sense any heat. If your fingers get burnt by touching hot glass, immediately cool them with cold water and notify your instructor.

Fire polishing

1. To remove sharp edges from cut glass, a hot flame is needed to melt and thereby smooth out the glass.

2. If the wing top is on the burner, turn off the gas and carefully remove the wing top from the barrel with a pair of crucible tongs. The wing top may be hot.

3. Relight the gas and adjust to the hottest flame. Hold one end of the cooled tubing in the hottest part of the flame (just above the blue inner cone). Slowly rotate the tube (Fig. 1.10).

Figure 1.10
Fire polishing.

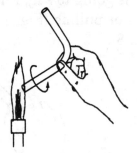

The flame above the glass tubing should become yellow as the glass gets hot and melts. Be careful not to overmelt the glass, in order to prevent the end from closing. After a short time (approx. 1 min.), remove the glass from the flame and examine the end; fire polishing will round the edges. Reheat if necessary to complete the polishing. When the end is completely smooth, lay the hot glass on a piece of wire gauze to cool. Be sure the glass is completely cooled before you attempt to polish the other end.

4. Show your instructor your glass bend with the ends completely fire polished.

Making stirring rods

Cut some solid glass rods (supplied by the instructor) into 20 cm lengths. Fire polish the ends.

Drawing capillary tubes

1. Cut a piece of glass tubing about 20 cm in length.

2. Heat the middle of the glass tubing in the flame just above the inner blue cone. Don't use a wing top. Rotate the tube in the flame until it softens (Fig. 1.11 A).

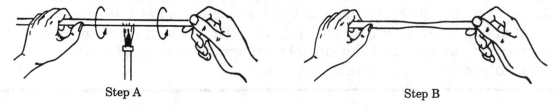

Step A   Step B

Figure 1.11 • Techniques for drawing capillary tubes.

3. As the glass sags, remove the tubing from the flame. Gently pull on each end, as straight as possible, until the capillary is as small as desired (Fig. 1.11 B).

4. Carefully place the tubing on the bench top and allow the glass to cool.

5. With a triangular file, carefully cut a piece of the drawn-out capillary tube (approx. 10 cm). Seal one end by placing it in the flame. Show your instructor your sealed capillary tube.

Chemicals and Equipment

1. Glass tubing (6-mm and 8-mm OD)
2. Glass rod (6-mm OD)
3. Bunsen burner
4. Wing top
5. Wire gauze
6. Crucible tongs

Experiment 1

PRE-LAB QUESTIONS

1. How can you control the amount of air mixed with the gas in the Bunsen burner?

2. At what temperature does soft glass soften?

3. At what temperature does Pyrex glass soften?

4. What is the purpose of annealing glass?

5. Why is laboratory glassware made of borosilicate glass?

Experiment 1

REPORT SHEET

Bunsen burner

1. What is the color of the flame when the air vents are closed?

2. What happened to the Pyrex test tube in this flame?

3. What happens to the flame when the gas control valve is turned?

4. Describe the effect on the flame as the air vents were opened.

Glassworking

Let the instructor comment on your glass experiments.

1. 90° angle bend:

2. Fire polishing:

3. Glass stirring rod:

4. Capillary tube:

POST-LAB QUESTIONS

1. Why must glass tubing be wrapped with a cloth or paper towel before breaking?

2. As glass heats and begins to melt, the flame becomes yellow. What causes the color?

3. What is the purpose of fire polishing glass?

4. When attempting to make a capillary tube, the glass tube collapsed before it could be drawn out. What may have gone wrong in the procedure?

Laboratory measurements

Units of measurement

The metric system of weights and measures is used by scientists of all fields, including chemists. This system uses the base 10 for measurements; for conversions, measurements may be multiplied or divided by 10. Table 2.1 lists the most frequently used factors in the laboratory which are based on powers of 10.

Table **2.1**	**Frequently Used Factors**		
Prefix	**Power of 10**	**Decimal Equivalent**	**Abbreviation**
Micro	10^{-6}	0.000001	μ
Milli	10^{-3}	0.001	m
Centi	10^{-2}	0.01	c
Kilo	10^{3}	1000	k

The measures of length, volume, mass, energy, and temperature are used to evaluate our physical and chemical environment. Table 2.2 compares the metric system with the more recently accepted SI system (International System of Units). The laboratory equipment associated with obtaining these measures is also listed.

Table **2.2**	**Units and Equipment**		
Measure	**SI Unit**	**Metric Unit**	**Equipment**
Length	Meter (m)	Meter (m)	Meterstick
Volume	Cubic meter (m^3)	Liter (L)	Pipet, graduated cylinder, Erlenmeyer flask, beaker
Mass	Kilogram (kg)	Gram (g)	Balance
Energy	Joule (J)	Calorie (cal)	Calorimeter
Temperature	Kelvin (K)	Degree Celsius (°C)	Thermometer

Accuracy, precision, meaning, and significant figures

Chemistry is a science that depends on experience and observation for data. It is an empirical science. An experiment that yields data requires the appropriate measuring devices in order to get accurate measurements. Once the data is in hand, calculations are done with the numbers obtained. How good the calculations are depends on a number of factors: (1) how careful you are in taking the measurements (laboratory techniques), (2) how good your measuring device is in getting a true measure (accuracy), and (3) how reproducible the measurement is (precision).

The measuring device usually contains a scale. The scale, with its subdivisions or graduations, tells the limits of the device's accuracy. You cannot expect to obtain a measurement better than your instrument is capable of reading. Consider the portion of the ruler shown in Fig. 2.1.

Figure 2.1 • Reading a metric ruler.

There are major divisions labeled at intervals of 1 cm and subdivisions of 0.1 cm or 1 mm. The accuracy of the ruler is to 0.1 cm (or 1 mm); that is the measurement that is known for certain. However, it is possible to estimate to 0.01 cm (or 0.1 mm) by reading in between the subdivisions; this number is less accurate and of course, is less certain. In general, you should be able to record the measured value to one more place than the scale is marked. For example, in Fig. 2.1 there is a reading marked on the ruler. This value is 8.75 cm: two numbers are known with certainty, *8.7*, and one number, *0.05*, is uncertain since it is the *best estimate* of the fractional part of the subdivision. The number recorded, 8.75, contains 3 significant figures, 2 certain plus 1 uncertain. When dealing with *significant figures*, remember: (1) the uncertainty is in the last recorded digit, and (2) the number of significant figures contains the number of digits definitely known, plus one more that is estimated.

The manipulation of significant figures in multiplication, division, addition, and subtraction is important. It is particularly important when using electronic calculators which give many more digits than are useful or significant. If you keep in mind the principle that the final answer can be no more accurate than the least accurate measurement, you should not go wrong. A few examples will demonstrate this.

EXAMPLE 1

Divide 9.3 by 4.05. If this calculation is done by a calculator, the answer found is 2.296296296. However, a division should have as an answer the same number of significant figures as the least accurately known (fewest significant figures) of the numbers being divided. One of the numbers, 9.3, contains only 2 significant figures. Therefore, the answer can only have 2 significant figures, i.e. 2.3 (rounded off).

EXAMPLE 2

Multiply 0.31 by 2.563. Using a calculator, the answer is 0.79453. As in division, a multiplication can have as an answer the same number of significant figures as the least accurately known (fewest significant figures) of the numbers being multiplied. The number 0.31 has 2 significant figures (the zero fixes the decimal point), therefore, the answer can only have 2 significant figures, i.e., 0.79 (rounded off).

EXAMPLE 3

Add 3.56 + 4.321 + 5.9436. A calculator gives 13.8246. With addition (or subtraction), the answer is significant to the least number of decimal places of the numbers added (or subtracted). The least accurate number is 3.56, measured only to the hundredth's place. The answer should be to this accuracy, i.e., 13.82 (rounded off to the hundredth's place).

Finally, how do precision and accuracy compare? *Precision* is a determination of the reproducibility of a measurement. It tells you how closely several measurements agree with one another. Several measurements of the same quantity showing high precision will cluster together with little or no variation in value; however, if the measurements show a wide variation, the precision is low. *Random errors* are errors which lead to differences in successive values of a measurement and affect precision; some values will be off in one direction or another. One can estimate the precision for a set of values for a given quantity as follows: estimate = $\pm \Delta/2$, where Δ is the difference between the highest and lowest value.

Accuracy is a measure of how closely the value determined agrees with a known or accepted value. Accuracy is subject to *systematic errors*. These errors cause measurements to vary from the known value and will be off in the same direction, either too high or too low. A consistent error in a measuring device will affect the accuracy, but always in the same direction. It is important to use properly calibrated measuring devices. If a measuring device is not properly calibrated, it may give high precision, but with none of the measurements being accurate. However, a properly calibrated measuring device will be both precise and accurate.

(See Fig. 2.2.) A systematic error is expressed as the difference between the known value and the average of the values obtained by measurement in a number of trials.

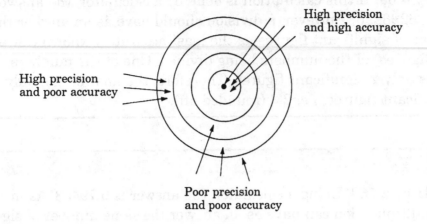

Figure 2.2 • Precision and accuracy illustrated by a target.

Objectives

1. To learn how to use simple, common equipment found in the laboratory.
2. To learn to take measurements.
3. To be able to record these measurements with precision and accuracy using the proper number of significant figures.

Procedure

Length: use of the meterstick (or metric ruler)

1. The meterstick is used to measure length. Examine the meterstick in your kit. You will notice that one side has its divisions in inches (in.) with subdivisions in sixteenths of an inch; the other side is in centimeters (cm) with subdivisions in millimeters (mm). Some useful conversion factors are listed below.

1 km	= 1000 m	1 in. = 2.54 cm
1 m	= 100 cm	1 ft. = 30.48 cm
1 cm	= 10 mm	1 yd. = 91.44 cm
1 m	= 1000 mm	1 mi. = 1.6 km

   The meterstick can normally measure to 0.001 m, 0.1 cm, or 1 mm.

2. With your meterstick (or metric ruler), measure the length and width of this laboratory manual. Express the measurements in inches (to the nearest sixteenth of an inch) and centimeters (to the nearest 0.1 cm). Record your response on the Report Sheet (1).

3. Convert the readings in cm to mm and m (2).

4. Calculate the area of the manual in in², cm², and mm² (3). Be sure to express your answers to the proper number of significant figures.

EXAMPLE 4

A student measured a piece of paper and found it to be 20.3 cm by 29.2 cm. The area was found to be

$$20.3 \text{ cm} \times 29.2 \text{ cm} = 593 \text{ cm}^2$$

Volume: use of a graduated cylinder, an Erlenmeyer flask, and a beaker

1. Volume in the metric system is expressed in liters (L) and milliliters (mL). Another way of expressing milliliters is in cubic centimeters (cm³ or cc). Several conversion factors for volume measurements are listed below.

1 L	= 1000 mL	1 qt.	= 0.96 L
1 mL	= 1 cm³ = 1 cc	1 gal.	= 3.79 L
1 L	= 0.26 gal.	1 fl. oz.	= 29.6 mL

2. The graduated cylinder is a piece of glassware used for measuring the volume of a liquid. Graduated cylinders come in various sizes with different degrees of accuracy. A convenient size for this experiment is the 100-mL graduated cylinder. Note that this cylinder is marked in units of 1 mL; major divisions are of 10 mL and subdivisions are of 1 mL. Estimates can be made to the nearest 0.1 mL. When a liquid is in the graduated cylinder, you will see that the level in the cylinder is curved with the lowest point at the center. This is the *meniscus*, or the dividing line between liquid and air. When reading the meniscus for the volume, be sure to read the *lowest* point on the curve and not the upper edge. To avoid errors in reading the meniscus, the eye's line of sight must be perpendicular to the scale (Fig. 2.3). In steps 3 and 4, use the graduated cylinder to see how well the marks on an Erlenmeyer flask and a beaker measure the indicated volume.

Figure 2.3

Reading the meniscus on a graduated cylinder.

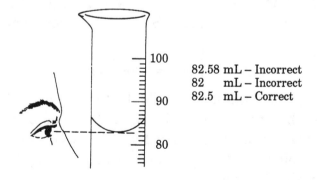

82.58 mL – Incorrect
82   mL – Incorrect
82.5 mL – Correct

3. Take a 50-mL graduated Erlenmeyer flask (Fig. 2.4) and fill with water to the 50 mL mark. Transfer the water, completely and without spilling, to a 100-mL graduated cylinder. Record the volume on the Report Sheet (4) to the nearest 0.1 mL; convert to L.

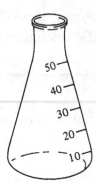

Figure 2.4

A 50-mL graduated
Erlenmeyer flask.

4. Take a 50-mL graduated beaker (Fig. 2.5), and fill with water to the 40 mL mark. Transfer the water, completely and without spilling, to a dry 100-mL graduated cylinder. Record the volume on the Report Sheet (5) to the nearest 0.1 mL; convert to L.

Figure 2.5

A 50-mL graduated beaker.

5. What is the error in mL and in percent for obtaining 50.0 mL for the Erlenmeyer flask and 40.0 mL for the beaker (6)?

6. Which piece of glassware will give you a more accurate measure of liquid: the graduated cylinder, the Erlenmeyer flask, or the beaker (7)?

Mass: use of the laboratory balance

1. Mass measurements of objects are carried out with the laboratory balance. Many types of balances are available for laboratory use. The proper choice of a balance depends upon what degree of accuracy is needed for a measurement. The standard units of mass are the kilogram (kg) in the SI system, and is the gram (g) in the metric system. Some conversion factors are listed below.

   $$1 \text{ kg} = 1000 \text{ g} \qquad 1 \text{ lb.} = 454 \text{ g}$$
   $$1 \text{ g} = 1000 \text{ mg} \qquad 1 \text{ oz.} = 28.35 \text{ g}$$

   Three types of balances are illustrated in Figs. 2.6, 2.8, and 2.10. A platform triple beam balance is shown in Fig. 2.6. This balance can weigh objects up to 2610 g. Since the scale is marked in 0.1-g divisions, it is mostly used for rough weighing; weights to 0.01 g can be estimated. Fig. 2.7 illustrates how to take a reading on this balance.

Figure 2.6

A platform triple beam balance.

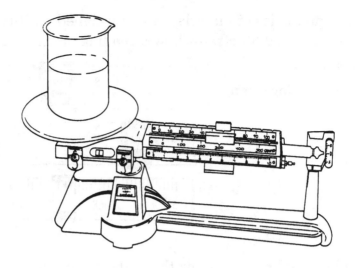

Figure 2.7

Reading on a platform triple beam balance.

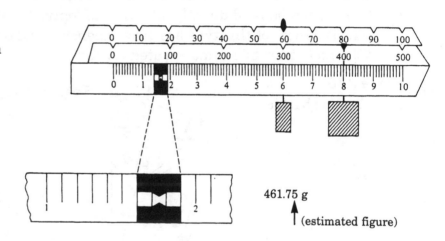

461.75 g
(estimated figure)

The single pan, triple beam (or Centigram) balance is shown in Fig. 2.8. This balance has a higher degree of accuracy since the divisions are marked in 0.01 g (estimates can be made to 0.001 g).

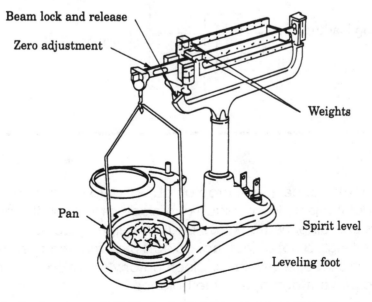

Beam lock and release

Zero adjustment

Weights

Pan

Spirit level

Leveling foot

Figure 2.8 • A single pan, triple beam balance.

Smaller quantities of material can be weighed on this balance (to a maximum of 311 g). Fig. 2.9 illustrates how a reading on this balance would be taken.

Figure 2.9

Reading on a single pan, triple beam balance.

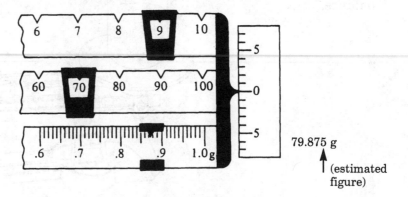

79.875 g

↑ (estimated figure)

Top loading balances show the highest accuracy (Fig. 2.10). Objects can be weighed very rapidly with these balances because the total weight, to the nearest 0.001 g, can be read directly off either an optical scale (Fig. 2.11) or a digital readout. Balances of this type are very expensive and one should be used only after the instructor has demonstrated its use.

Figure 2.10

A top loading balance.

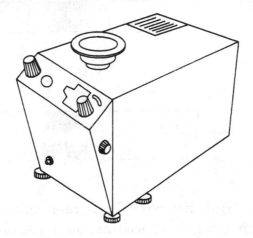

Figure 2.11

Reading on a top loading balance.

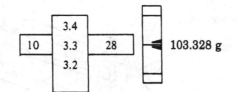

103.328 g

CAUTION!

When using any balance, never drop an object onto the pan; place it gently in the center of the pan. Never place chemicals directly on the pan; use either a glass container (watch glass, beaker, weighing bottle) or weighing paper. Never weigh a hot object; hot objects may mar the pan. Buoyancy effects will cause incorrect weights. Clean up any chemical spills in the balance area to prevent damage to the balance.

2. Weigh a quarter, a test tube (100 × 13 mm), and a 125-mL Erlenmeyer flask. Express each weight to the proper number of significant figures. Use a platform balance, a Centigram balance, and a top loading balance for these measurements. Use the table on the Report Sheet to record each weight.

3. The triple beam balance (Centigram) (Fig. 2.8) is operated in the following way.

   a. Place the balance on a level surface; use the leveling foot to level.

   b. Move all the weights to the zero position at left.

   c. Release the beam lock.

   d. The pointer should swing freely in an equal distance up and down from the zero or center mark on the scale. Use the zero adjustment to make any correction to the swing.

   e. Place the object on the pan (remember the caution).

   f. Move the weight on the middle beam until the pointer drops; make sure the weight falls into the "V" notch. Move the weight back one notch until the pointer swings up. This beam weighs up to 10 g, in 1 g-increments.

   g. Now move the weights on the back beam until the pointer drops; again be sure the weight falls into the "V" notch. Move the weight back one notch until the pointer swings up. This beam weighs up to 1 g, in 0.1 g-increments.

   h. Lastly, move the smallest weight (the cursor) on the front beam until the pointer balances, that is, swings up and down an equal distance from the zero or center mark on the scale. This last beam weighs to 0.1 g in 0.01 g-increments.

   i. The weight of the object on the pan is equal to the weights shown on each of the three beams (Fig. 2.8). Weights to 0.001 g may be estimated.

   j. Repeat the movement of the cursor to check your precision.

   k. When finished, move the weights to the left, back to zero, and arrest the balance with the beam lock.

Temperature: use of the thermometer

1. Routine measurements of temperature are done with a thermometer. Thermometers found in chemistry laboratories may use either mercury or a colored fluid as the liquid, and degrees Celsius (°C) as the units of measurement. The fixed reference points on this scale are the freezing point of water, 0°C, and the boiling point of water, 100°C. Between these two reference points, the scale is divided into 100 units, with each unit equal to 1°C. Temperature can be estimated to 0.1°C. Other thermometers use either the Fahrenheit (°F) or the Kelvin (K) temperature scale and use the same reference points, that is, the freezing and boiling points of water. Conversion between the scales can be accomplished using the formulas below.

$$\text{°F} = \tfrac{9}{5}\text{°C} + 32.0 \qquad \text{°C} = \tfrac{5}{9}(\text{°F} - 32.0) \qquad K = \text{°C} + 273.15$$

EXAMPLE 5

Convert 37.0°C to °F and K.

$$°F = \frac{9}{5}(37.0°C) + 32.0 = 98.6°F$$

$$K = 37.0°C + 273.15 = 310.2 \text{ K}$$

2. Use the thermometer in your kit and record to the nearest 0.1°C the temperature of the laboratory at *room temperature*. Use the Report Sheet to record your results.

3. Record the temperature of boiling water. Set up a 250-mL beaker containing 100 mL water, and heat on a hot plate until boiling. Hold the thermometer in the boiling water for 1 min. before reading the temperature (*be sure not to touch the sides of the beaker*). Using the Report Sheet, record your results to the nearest 0.1°C.

4. Record the temperature of ice water. Into a 250-mL beaker, add enough crushed ice to fill halfway. Add distilled water to the level of the ice. Stir the ice water gently with the thermometer for 1 min. (*use extreme caution; be careful not to hit the walls of the beaker*) and read the thermometer to the nearest 0.1°C. Record your results on the Report Sheet.

CAUTION!

When reading the thermometer, do not hold the thermometer by the bulb. Body temperature will give an incorrect reading. If you are using a mercury thermometer and the thermometer should break accidentally, call the instructor for proper disposal of the mercury. Mercury is toxic and very hazardous to your health. Do not handle the liquid or breathe its vapor.

5. Convert your answers to questions 2, 3, and 4 into °F and K .

Chemicals and Equipment

1. 50-mL graduated beaker
2. 50-mL graduated Erlenmeyer flask
3. 100-mL graduated cylinder
4. Meterstick or ruler
5. Quarter
6. Balances
7. Hot plates

Experiment 2

PRE-LAB QUESTIONS

1. Define the accuracy of a measurement.

2. Define the precision of a measurement.

3. Solve the following problems and record the answers to the proper number of significant figures.

    a. $35.2 \times 43.21 =$

    b. $6.51 \div 1.2 =$

    c. $3.05 + 9.506 + 3.001 =$

    d. $9.07 - 6.1 =$

4. Using the SI system, give the basic units of measurement for the following:

    a. mass

    b. volume

    c. density

    d. temperature

NAME _____   SECTION _____   DATE _____

PARTNER _____   GRADE _____

Experiment 2

REPORT SHEET

Length

1. Length _____ in.   _____ cm

   Width _____ in.   _____ cm

2. Length _____ mm   _____ m

   Width _____ mm   _____ m

3. Area _____ in^2 _____ cm^2 _____ mm^2
   (Show calculations)

Volume

4. Erlenmeyer flask _____ mL   _____ L

5. Beaker _____ mL   _____ L

6. Error in volume:
   Erlenmeyer flask _____ mL   _____ %

   Beaker _____ mL   _____ %

$$\% \text{ Error} = \frac{\text{Error in volume}}{\text{Total volume}} \times 100$$

7.

Mass

OBJECT	BALANCE					
	Platform		Centigram		Top Loading	
	g	mg	g	mg	g	mg
Quarter						
Test tube						
125-mL Erlenmeyer						

Temperature

	°C	°F	K
Room Temperature			
Ice Water			
Boiling Water			

How well do your thermometer readings agree with the accepted values for the freezing point and boiling point of water? Express any discrepancy as a deviation in degrees.

POST-LAB QUESTIONS

1. A student needs 0.032 g of sample. Which balance will provide the most accurate weight?

2. Student A measured the volume of a liquid once in a graduated cylinder; student B repeated the same measurement three times and reported the average volume. Which student had greater precision in the measurement? Which student had greater accuracy?

3. A 351 mg sample was placed on a watch glass weighing 8.732 g. What is the combined weight of the watch glass and sample in grams?

4. Phoenix, Arizona, had a temperature of 105°F one day. On the same day, Davis, California, reported a temperature of 42°C. Which city was hotter?

Experiment 3

Density determination

Background

Samples of matter can be identified by using characteristic physical properties. A substance may have a unique color, odor, melting point, or boiling point. These properties do not depend on the quantity of the substance and are called *intensive properties*. Density also is an intensive property and may serve as a means for identification.

The *density* of a substance is the *ratio of its mass per unit volume*. Density can be found mathematically by dividing the mass of a substance by its volume. The formula is $\mathbf{d} = \frac{\mathbf{m}}{\mathbf{V}}$, where $\mathbf{d}$ is density, $\mathbf{m}$ is mass, and $\mathbf{V}$ is volume. While mass and volume do depend on the quantity of a substance (these are *extensive properties*), the ratio is constant at a given temperature. The units of density, reported in standard references, is in terms of g/mL (or g/cc or g/cm^3) at 20°C. The temperature is reported since the volume of a sample will change with temperature and, thus, so does the density.

EXAMPLE

A bank received a yellow bar, marked gold, of mass 453.6 g, and volume 23.5 cm^3. Is it gold? (Density of gold = 19.3 g/cm^3 at 20°C.)

$$d = \frac{m}{V} = \frac{453.6 \text{ g}}{23.5 \text{ cm}^3} = 19.3 \text{ g/cm}^3$$

Yes, it is gold.

Objectives

1. To determine the densities of regular- and irregular-shaped objects and use them as a means of identification.
2. To determine the density of water.
3. To determine the density of a small irregular-shaped object by flotation technique.

Density of a Regular-Shaped Object

1. Obtain a solid block from the instructor. Record the code number.

2. Using your metric ruler, determine the dimensions of the block (length, width, height) and record the values to the nearest 0.01 cm (1). Calculate the volume of the block (2). Repeat the measurements for a second trial.

3. Using a triple beam balance (Centigram) or a top loading balance (if available), determine the mass of the block (3). Record the mass to the nearest 0.001 g. Calculate the density of the block (4). Repeat the measurements for a second trial.

Density of an Irregular-Shaped Object

1. Obtain a sample of unknown metal from your instructor. Record the code number.

2. Obtain a mass of the sample of approximately 5 g. Be sure to record the exact quantity to the nearest 0.001 g (5).

3. Fill a 10-mL graduated cylinder approximately halfway with water. Record the exact volume to the nearest 0.1 mL (6).

4. Place the metal sample into the graduated cylinder. (If the pieces of metal are too large for the opening of the 10-mL graduated cylinder, use a larger graduated cylinder.) Be sure all of the metal is below the water line. Gently tap the sides of the cylinder with your fingers to ensure that no air bubbles are trapped in the metal. Read the new level of the water in the graduated cylinder to the nearest 0.1 mL (7). Assuming that the metal does not dissolve or react with the water, the difference between the two levels represents the volume of the metal sample (8) (Fig. 3.1).

Figure 3.1

Measurement of volume of an irregular-shaped object.

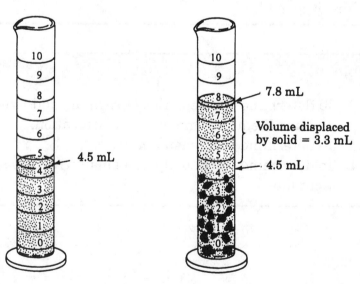

4.5 mL

7.8 mL

Volume displaced by solid = 3.3 mL

4.5 mL

5. Carefully recover the metal sample and dry it with a paper towel. Repeat the experiment.

6. Calculate the density of the metal sample from your data (9). Determine the average density from your trials, reporting to the proper number of significant figures.

7. Determine the identity of your metal sample by comparing its density to the densities listed in Table 3.1 (10).

Table 3.1	Densities of Selected Metals	
Sample	**Formula**	**Density (g/mL)**
Aluminum	Al	2.70
Iron	Fe	7.86
Tin (white)	Sn	7.29
Zinc	Zn	7.13
Lead	Pb	11.30

8. Recover your metal sample and return it as directed by your instructor.

Use of the Spectroline Pipet Filler

1. Examine the Spectroline pipet filler and locate the valves marked "A," "S," and "E" (Fig. 3.2). These operate by pressing the flat surfaces between the thumb and forefinger.

2. Squeeze the bulb with one hand while you press valve "A" with two fingers of the other hand. The bulb flattens as air is expelled. If you release your fingers when the bulb is flattened, the bulb remains collapsed.

Figure 3.2
The Spectroline pipet filler.

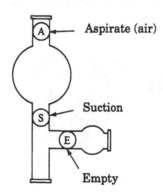

3. Carefully insert the pipet end into the Spectroline pipet filler (Fig. 3.3). The end should insert easily and not be forced.

4. Place the tip of the pipet into the liquid to be pipetted. Make sure that the tip is below the surface of the liquid at all times.

Figure 3.3

Using the Spectroline
pipet filler to pipet.

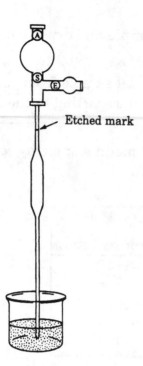

Etched mark

5. With your thumb and forefinger, press valve "S." Liquid will be drawn up into
   the pipet. By varying the pressure applied by your fingers, the rise of the liquid
   into the pipet can be controlled. Allow the liquid to fill the pipet to a level
   slightly above the etched mark on the stem. Release the valve; the liquid
   should remain in the pipet.

6. Withdraw the pipet from the liquid. Draw the tip of the pipet lightly along the
   wall of the beaker to remove excess water.

7. Adjust the level of the meniscus of the liquid by carefully pressing valve "E."
   The level should lower until the curved meniscus touches the etched mark
   (Fig. 3.4). Carefully draw the tip of the pipet lightly along the wall of the
   beaker to remove excess water.

Figure 3.4

Adjusting the curved meniscus
of the liquid to the etched mark.

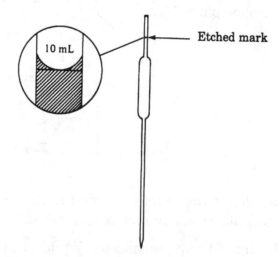

Etched mark

10 mL

8. Drain the liquid from the pipet into a collection flask by pressing valve "E."
   Remove any drops on the tip by touching the tip of the pipet against the inside
   walls of the collection flask. Water should remain inside the tip; the pipet is
   calibrated with this water in the tip.

Density of Water

1. Obtain approximately 50 mL of distilled water from your instructor. Record the temperature of the water (11).

2. Take a clean, dry 50-mL beaker; weigh to the nearest 0.001 g (12).

3. With a 10-mL volumetric pipet, transfer 10.00 mL of distilled water into the preweighed beaker using a Spectroline pipet filler (Fig. 3.3). Immediately weigh the beaker and water and record the weight to the nearest 0.001 g (13). Calculate the weight of the water by subtraction (14). Calculate the density of the water at the temperature recorded (15).

CAUTION!

Never use your mouth when pipetting.

---

4. Repeat step 3 for a second trial. Be sure all the glassware used is clean and dry.

5. Calculate the average density (16). Compare your average value at the recorded temperature to the value reported for that temperature in a standard reference.

Density of a Small Irregular-Shaped Object by Flotation Technique

1. Obtain two small (2 mm) plastic chips from your instructor.

2. Place a 50-mL graduated cylinder containing a small magnetic spin-bar on a magnetic stirrer. Add 30 mL of acetone and begin to stir the liquid slowly. Add the plastic chips to the liquid. Stop the stirring and note that the chips will sink to the bottom.

3. With slow intermittent stirring, add 3–4 mL of water dropwise. Watch the plastic chips as you add the water; see if they rise or stay on the bottom. If they stay on the bottom, keep adding more drops of water until the chips float in the middle of the liquid. At this point, *the liquid has the same density as that of the plastic chips.*

4. Weigh a clean and dry 50-mL beaker to the nearest 0.001 g. Record the weight on your Report Sheet (17).

5. Using a Spectroline pipet filler (Fig. 3.3), transfer exactly 10.00 mL of liquid from the graduated cylinder to the beaker. Weigh to the nearest 0.001 g (18), and by subtraction determine the weight of the liquid. Record it on your Report Sheet (19).

6. Repeat step 5 for a second trial. Be sure all the glassware used is clean and dry.

7. Calculate the density of the liquid, and hence the density of the plastic chips (20). Determine the average density of the plastic chips.

Chemicals and Equipment

1. Magnetic spin-bar
2. Magnetic stirrer
3. Spectroline pipet filler
4. 10-mL volumetric pipet
5. Solid wood block
6. Aluminum
7. Lead
8. Tin
9. Zinc
10. Polyethylene plastic chips
11. Acetone

Experiment 3

PRE-LAB QUESTIONS

1. The density of tin is 5.75 g/mL. What is its density in the SI units of kg/m³?

2. Is density an intensive property or an extensive property? Explain.

3. Is the density of hexane (d = 0.659 g/mL at 20°C) the same, higher, or lower at 20°C than at 40°C?

4. A diver discovered a cache of yellow ingots believed to be pure gold. The ingots had a mass of 1500 g and a volume of 132.7 cm³ at 20°C. Was this gold? How do you know? (*Hint:* read the **Background** section again.)

EXPERIMENT 3

REPORT SHEET

Report all measurements and calculations to the correct number of significant figures.

Density of a regular-shaped object	Trial 1	Trial 2
Unknown code number _____		
1. Length	_____ cm	_____ cm
Width	_____ cm	_____ cm
Height	_____ cm	_____ cm
2. Volume (L × W × H)	_____ cm^3	_____ cm^3
3. Mass	_____ g	_____ g
4. Density: (3)/(2)	_____ g/cm^3	_____ g/cm^3
Average density of block		_____ g/cm^3

Density of an irregular-shaped object	Trial 1	Trial 2
Unknown code number _____		
5. Mass of metal sample	_____ g	_____ g
6. Initial volume of water	_____ mL	_____ mL
7. Final volume of water	_____ mL	_____ mL
8. Volume of metal: (7) − (6)	_____ mL	_____ mL
9. Density of metal: (5)/(8)	_____ g/mL	_____ g/mL
Average density of metal		_____ g/mL
10. Identity of unknown metal	_____	

Density of water

	Trial 1	Trial 2
11. Temperature of water	_____ °C	_____ °C
12. Weight of 50-mL beaker	_____ g	_____ g
Volume of water	10.00 mL	10.00 mL
13. Weight of beaker and water	_____ g	_____ g
14. Weight of water: (13) − (12)	_____ g	_____ g
15. Density of water: (14)/10.00 mL	_____ g/mL	_____ g/mL
16. Average density of water		_____ g/mL
Density found in literature		_____ g/mL

Density by flotation technique

	Trial 1	Trial 2
17. Weight of 50-mL beaker	_____ g	_____ g
Volume of liquid	10.00 mL	10.00 mL
18. Weight of beaker and liquid	_____ g	_____ g
19. Weight of liquid: (18) − (17)	_____ g	_____ g
20. Density of liquid: (19)/10.00 mL	_____ g/mL	_____ g/mL
Average density of plastic chips		_____ g/mL

POST-LAB QUESTIONS

1. The density of a liquid is 0.695 g/mL. How many mL are needed to have 25.0 g of liquid?

2. In determining the density of a liquid, a student left air bubbles trapped in the volumetric pipet. Did this give a density *less* than expected or *greater* than expected? Why?

3. Explain how the density of an irregular-shaped wooden block is affected if

   a. the center is hollow;

   b. the block is not completely submerged when determining the volume.

4. Assume that the plastic chips in your flotation experiment were floating on top of the acetone. Could you still use water as a second liquid to bring the chips to the middle of the liquid? Explain.

Experiment 4

The separation of the components of a mixture

Mixtures are not unique to chemistry; we use and consume them on a daily basis . The beverages we drink each morning, the fuel we use in our automobiles, and the ground we walk on are mixtures. Very few materials we encounter are pure. Any material made up of two or more substances that are not chemically combined is a mixture.

The isolation of pure components of a mixture requires the separation of one component from another. Chemists have developed techniques for doing this. These methods take advantage of the differences in physical properties of the components. The techniques to be demonstrated in this laboratory are the following:

1. *Sublimation*. This involves heating a solid until it passes directly from the solid phase into the gaseous phase. The reverse process, when the vapor goes back to the solid phase without a liquid state in between, is called condensation or deposition. Some solids which sublime are iodine, caffeine, and *para*-dichlorobenzene (mothballs).

2. *Extraction*. This uses a solvent to selectively dissolve one component of the solid mixture. With this technique, a soluble solid can be separated from an insoluble solid.

3. *Decantation*. This separates a liquid from an insoluble solid sediment by carefully pouring the liquid from the solid without disturbing the solid (Fig. 4.1).

Figure 4.1
Decantation.

4. *Filtration*. This separates a solid from a liquid through the use of a porous material as a filter. Paper, charcoal, or sand can serve as a filter. These materials allow the liquid to pass through but not the solid (see Fig. 4.4 in the **Procedure** section).

5. *Evaporation*. This is the process of heating a mixture in order to drive off, in the form of vapor, a volatile liquid, so as to make the remaining component dry.

The mixture that will be separated in this experiment contains three components: naphthalene, $C_{10}H_8$, common table salt, NaCl, and sea sand, SiO_2. The separation will be done according to the scheme in Fig. 4.2 by

1. heating the mixture to sublime the naphthalene;

2. dissolving the table salt with water to extract; and

3. evaporating water to recover dry NaCl and sand.

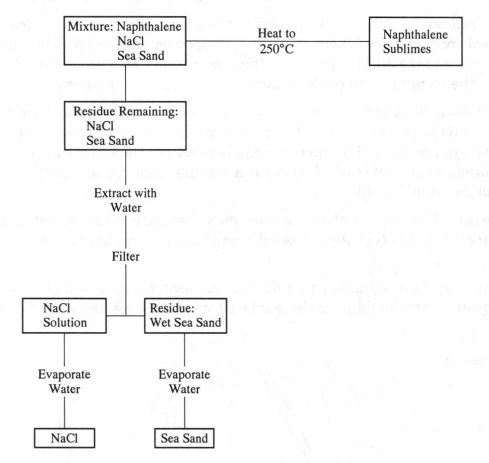

Figure 4.2 • Separation scheme.

Objectives

1. To demonstrate the separation of a mixture.
2. To examine some techniques for separation using physical methods.

1. Obtain a clean, dry 150-mL beaker and carefully weigh to the nearest 0.001 g. Record this weight for beaker 1 on the Report Sheet (1). Obtain a sample of the unknown mixture from your instructor; use a mortar and pestle to grind the mixture into a fine powder. With the beaker still on the balance, carefully transfer approximately 2 g of the unknown mixture into the beaker. Record the weight of the beaker with the contents to the nearest 0.001 g (2). Calculate the exact sample weight by subtraction (3).

2. Place an evaporating dish on top of the beaker containing the mixture. Place the beaker and evaporating dish on a wire gauze with an iron ring and ring stand assembly as shown in Fig. 4.3. Place ice in the evaporating dish, being careful not to get any water on the underside of the evaporating dish or inside the beaker.

Figure 4.3
Assembly for sublimation.

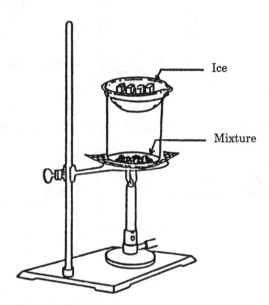

Ice

Mixture

3. Carefully heat the beaker with a Bunsen burner, increasing the intensity of the flame until vapors appear in the beaker. A solid should collect on the underside of the evaporating dish. After 10 min. of heating, remove the Bunsen burner from under the beaker. Carefully remove the evaporating dish from the beaker and collect the solid by scraping it off the dish with a spatula. Drain away any water from the evaporating dish and add ice to it, if necessary. Stir the contents of the beaker with a glass rod. Return the evaporating dish to the beaker and apply the heat again. Continue heating and scraping off solid until no more solid collects. Discard the naphthalene into a special container provided by your instructor.

4. Allow the beaker to cool until it reaches room temperature. Weigh the beaker with the contained solid (4). Calculate the weight of the naphthalene that sublimed (5).

5. Add 25 mL of distilled water to the solid in the beaker. Heat and stir for 5 min.

6. Weigh a second clean, dry 150-mL beaker to the nearest 0.001 g (6).

7. Assemble the apparatus for gravity filtration as shown in Fig. 4.4.

Figure 4.4
Gravity filtration.

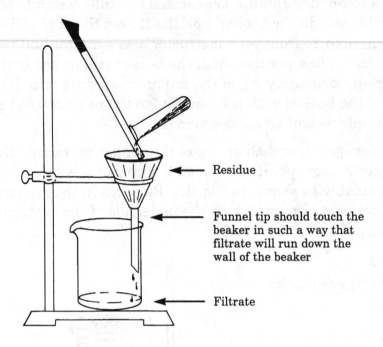

Residue

Funnel tip should touch the
beaker in such a way that
filtrate will run down the
wall of the beaker

Filtrate

8. Fold a piece of filter paper following the technique shown in Fig. 4.5.

Figure 4.5

Steps for folding a filter
paper for gravity filtration.

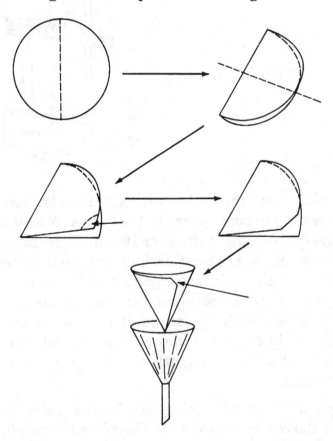

9. Wet the filter paper with water and adjust the paper so that it lies flat on the glass of the funnel.

10. Position the second beaker under the funnel.

11. Pour the mixture through the filter, first decanting most of the liquid into beaker 2, and then carefully transferring the wet solid into the funnel with a rubber policeman. Collect all the liquid (called the filtrate) in beaker 2.

12. Rinse beaker 1 with 5–10 mL of water, pour over the residue in the funnel, and add the liquid to the filtrate; repeat with an additional 5–10 mL of water.

13. Place beaker 2 and its contents on a wire gauze with an iron ring and ring stand assembly as shown in Fig. 4.6a. Add boiling chips (two or three are enough) to the solution and begin to heat gently with a Bunsen burner. Control the flame in order to prevent boiling over. As the volume of liquid is reduced, solid sodium chloride will appear. Reduce the flame to avoid bumping of the solution and spattering of the solid. When all of the liquid is gone, cool the beaker to room temperature. With a spatula or forceps, remove the boiling chips. Weigh the beaker and the solid residue to the nearest 0.001 g (7). Calculate the weight of the recovered NaCl by subtraction (8).

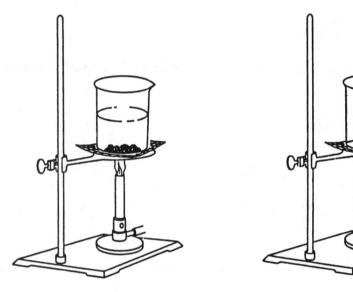

a. Evaporation of a volatile liquid from a solution.   b. Heating a solid to dryness.

Figure 4.6 • Assembly for evaporation.

14. Carefully weigh a third clean, dry 150-mL beaker to the nearest 0.001 g (9). Transfer the sand from the filter paper to beaker 3. Heat the sand to dryness in the beaker with a burner using the ring stand and assembly shown in Fig. 4.6b (or use an oven at T = 90–100°C, if available). Heat carefully to avoid spattering; when dry, the sand should be freely flowing. Allow the sand to cool to room temperature. Weigh the beaker and the sand to the nearest 0.001 g (10). Calculate the weight of the recovered sand by subtraction (11).

15. Calculate

    a. Percentage yield using the formula:

$$\% \text{ yield} = \frac{\text{grams of solid recovered}}{\text{grams of initial sample}} \times 100$$

    b. Percentage of each component in the mixture by using the formula:

$$\% \text{ component} = \frac{\text{grams of component isolated}}{\text{grams of initial sample}} \times 100$$

EXAMPLE

A student isolated the following from a sample of 1.132 g:

    0.170 g of naphthalene
    0.443 g of NaCl
    <u>0.499 g of sand</u>
    1.112 g solid recovered

The student calculated the percentage yield and percentage of each component as follows:

$$\% \text{ yield} = \frac{1.112}{1.132} \times 100 = 98.2\%$$

$$\% \text{ C}_{10}\text{H}_8 = \frac{0.170}{1.132} \times 100 = 15.0\%$$

$$\% \text{ NaCl} = \frac{0.443}{1.132} \times 100 = 39.1\%$$

$$\% \text{ sand} = \frac{0.499}{1.132} \times 100 = 44.1\%$$

Chemicals and Equipment

    1. Unknown mixture
    2. Balances
    3. Boiling Chips
    4. Evaporating dish, 6 cm
    5. Filter paper, 15 cm
    6. Mortar and pestle
    7. Oven (if available)
    8. Ring stands (3)
    9. Rubber policeman

Experiment 4

PRE-LAB QUESTIONS

1. Of the five methods listed for the separation of the components found in a mixture, which one would you use to separate salt from water in a sample of sea water?

2. Can any of the methods listed in the **Background** section be used to separate the elements found in a compound? Explain.

3. What separation technique(s) is (are) used when making coffee from ground coffee in a percolator?

4. Define sublimation.

Experiment 4

REPORT SHEET

1. Weight of beaker 1 _____ g

2. Weight of beaker 1 and mixture _____ g

3. Weight of mixture: (2) − (1) _____ g

4. Weight of beaker 1 and solid
   after sublimation _____ g

5. Weight of naphthalene: (2) − (4) _____ g

6. Weight of beaker 2 _____ g

7. Weight of beaker 2 and NaCl _____ g

8. Weight of NaCl: (7) − (6) _____ g

9. Weight of beaker 3 _____ g

10. Weight of beaker 3 and sand _____ g

11. Weight of sand: (10) − (9) _____ g

Calculations

12. Weight of recovered solids:

    (5) + (8) + (11) _____ g

13. Percentage yield (percentage of
    solids recovered):

    (12)/(3) × 100 _____ %

14. Percentage of naphthalene:

    (5)/(3) × 100 _____ %

15. Percentage of NaCl:

    (8)/(3) × 100   _____%

16. Percentage of sand:

    (11)/(3) × 100   _____%

POST-LAB QUESTIONS

1. A student started this experiment with a mixture weighing 2.345 g. After separating the components, a total of 2.765 g of material was recovered. Assuming that all the weighings and calculations were done correctly, what was the most likely source of error in the experiment?

2. How would results be affected if the evaporating dish on top of the beaker, containing the ice, did not fit well and did leave openings? Propose a simple test to support your assumptions.

3. *para*-Dichlorobenzene is sold commercially as mothballs. Suggest why it can be used successfully in a closed garment bag to prevent damage to clothes by moth larvae.

4. A sample of milk weighing 10.0 g was extracted with the volatile organic solvent diethyl ether. After evaporating the diethyl ether, 0.35 g of fat was recovered. What was the percent of fat content in the milk?

5. Dry cleaners remove oil and grease spots from clothing by using an organic solvent called perchloroethylene. What method of separation does the cleaner use?

Experiment 5

● Resolution of a mixture by distillation

Background

Distillation is one of the most common methods of purifying a liquid. It is a very simple method: a liquid is brought to a boil, the liquid becomes a gas, the gas condenses and returns to the liquid state, and the liquid is collected.

Everyone has had an opportunity to heat water to a boil. As heat is applied, water molecules increase their kinetic energy. Some molecules acquire sufficient energy to escape from the liquid phase and enter into the vapor phase. The vapor above the liquid exerts a pressure, called the vapor pressure. As more and more molecules obtain enough energy to escape into the vapor phase, the vapor pressure of these molecules increases. Eventually the vapor pressure equals the pressure exerted externally on the liquid (this external pressure usually is caused by the atmosphere). Boiling occurs when this condition is met, and the temperature where this occurs is called the boiling point.

In distillation, the process described is carried out in an enclosed system, such as is illustrated in Fig. 5.2. The liquid in the boiling flask is heated to a boil, and the vapor rises through tubing. The vapor then travels into a tube cooled by water, which serves as a condenser, where the vapor returns to the liquid state. If the mixture has a low-boiling component (a volatile substance with a high vapor pressure), it will distill over first and can be collected. Higher-boiling and nonvolatile components (substances with low vapor pressure) remain in the boiling flask. Only by applying more heat will the higher-boiling component be distilled. Nonvolatile substances will not distill.

Normal distillations, procedures carried out at atmospheric pressure, require "normal" boiling points. However, when boiling takes place in a closed system, it is possible to change the boiling point of the liquid by changing the pressure in the closed system. If the external pressure is reduced, usually by using a vacuum pump or a water aspirator, the boiling point of the liquid is reduced. Thus, heat sensitive liquids, some of which decompose when boiled at atmospheric pressure, distill with minimum decomposition at reduced pressure and temperature. The relation of temperature to vapor pressure for the organic compound aniline can be shown by the curve in Fig. 5.1. The organic liquid aniline, $C_6H_5NH_2$, can be distilled at 184°C (760 mm Hg) or at 68°C (10 mm Hg).

Figure 5.1

Temperature–vapor pressure
curve for aniline.

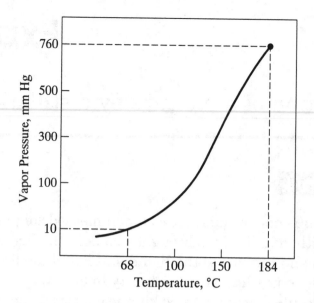

Objectives

1. To use distillation to separate a mixture.
2. To show that distillation can purify a liquid.

Procedure

1. In this experiment a salt–water mixture will be separated by distillation. The volatile water will be separated from the nonvolatile salt (sodium chloride, NaCl). The purity of the collected distilled water will be demonstrated by chemical tests specific for sodium ions (Na^+) and chloride ions (Cl^-).

2. Assemble an apparatus as illustrated in Fig. 5.2. A kit containing the necessary glassware can be obtained from your instructor. The glassware contains standard taper joints, which allow for quick assembly and disassembly. Before fitting the pieces together, apply a light coating of silicone grease to each joint to prevent the joints from sticking.

3. Use 100-mL round bottom flasks for the boiling flask and the receiving flask. Fill the boiling flask with 50 mL of the prepared salt–water mixture. Add two boiling chips to the boiling flask to ensure smooth boiling of the mixture and to prevent bumping. Be sure that the rubber tubing to the condenser enters the lower opening and empties out of the upper opening. Turn on the water faucet and allow the water to fill the jacket of the condenser slowly, so as not to trap air. Take care not to provide too much flow, otherwise the hoses will disconnect from the condenser. Adjust the bulb of the thermometer to below the junction of the condenser and the distillation column. Be sure that the opening of the vacuum adapter is open to the atmosphere.

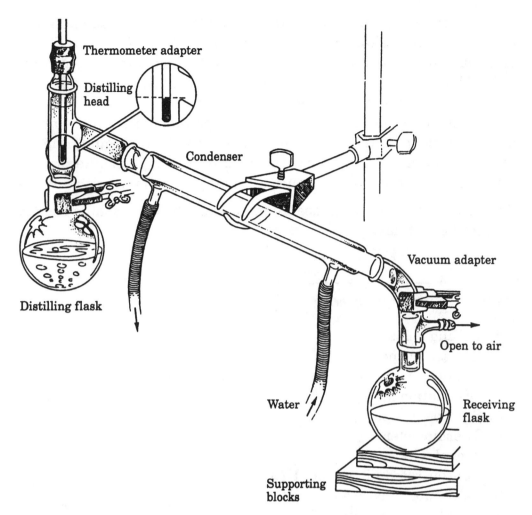

Figure 5.2 • A distillation apparatus.

4. Gently heat the boiling flask with a Bunsen burner. Eventually the liquid will boil, vapors will rise and enter the condenser, and liquid will recondense and be collected in the receiving flask.

5. Discard the first 1 mL of water collected. Record the temperature of the vapors as soon as the 1 mL of water has been collected. Continue collection of the distilled water until approximately one-half of the mixture has distilled. Record the temperature of the vapors at this point. Turn off the Bunsen burner and allow the system to return to room temperature.

6. The distilled water and the liquid in the boiling flask will be tested.

7. Place in separate clean, dry test tubes (100 × 13 mm) 2 mL of distilled water and 2 mL of the residue liquid from the boiling flask. Add to each sample 5 drops of silver nitrate solution. Look for the appearance of a white precipitate. Record your observations. Silver ions combine with chloride ions to form a white precipitate of silver chloride.

$$Ag^+ + Cl^- \longrightarrow AgCl(s) \text{ (White precipitate)}$$

8. Place in separate clean, dry test tubes (100 × 13 mm) 2 mL of distilled water and 2 mL of the residue liquid from the boiling flask. Obtain a clean nickel wire from your instructor. In the hood, dip the wire into concentrated nitric acid and hold the wire in a Bunsen burner flame until the yellow color in the flame disappears. Dip the wire into the distilled water sample. Put the wire into the Bunsen burner flame. Record the color of the flame. Repeat the above procedure, cleaning the wire, dipping the wire into the liquid from the boiling flask, and observing the color of the Bunsen burner flame. Record your observations. Sodium ions produce a bright yellow flame with a Bunsen burner.

9. Make sure you wipe the grease from the joints before washing the glassware used in the distillation.

Chemicals and Equipment

1. Boiling chips
2. Bunsen burner
3. Clamps
4. Distillation kit
5. Silicone grease
6. Thermometer
7. Nickel wire
8. Concentrated nitric acid, HNO_3
9. Salt-water mixture
10. 0.5 M silver nitrate, 0.5 M $AgNO_3$

Experiment 5

PRE-LAB QUESTIONS

1. When does boiling take place?

2. In Denver, Colorado, the atmospheric pressure is 740 torr. Will the boiling point of water be higher or lower than 100°C?

3. Why do we need to add boiling chips to the boiling flask when carrying out distillation?

4. A student has a mixture of two liquids. Liquid A boils at 112°C, and liquid B boils at 145°C. In distillation, which liquid will distill first? Why?

NAME _____   SECTION _____   DATE _____

PARTNER _____   GRADE _____

Experiment 5

REPORT SHEET

1. Barometric pressure　　　　　　　　　　　　_____

2. Boiling point of water at measured pressure　　_____

3. Temperature of vapor after collecting 1 mL　　_____

4. Temperature of vapor at end of distillation　　_____

Solution	Observation with 0.5 M AgNO$_3$	Color in Flame Test
Distilled water		
Liquid in boiling flask		

POST-LAB QUESTIONS

1. Why did the distilled water give a negative test for Cl$^-$ anion?

2. If you examine Fig. 5.2, the distilling flask appears to be completely closed. However, a student reports the observed temperature for the boiling liquid as the "normal" boiling point. Why?

3. You have a liquid with a boiling point listed as 450°C. Your heating source under the distilling flask can provide only enough heat to bring the liquid to 350°C. How would you be able to distill the liquid?

Determination of the formula
of a metal oxide

Through the use of chemical symbols and numerical subscripts, the formula of a compound can be written. The simplest formula that may be written is the *empirical formula*. In this formula, the subscripts are in the form of the simplest whole number ratio of the atoms in a molecule or of the ions in a formula unit. The *molecular formula*, however, represents the actual number of atoms in a molecule. For example, although CH_2O represents the empirical formula of the sugar, glucose, $C_6H_{12}O_6$ represents the molecular formula. For water, H_2O, and carbon dioxide, CO_2, the empirical and the molecular formulas are the same. Ionic compounds are generally written as empirical formulas only; for example, common table salt is NaCl.

The formation of a compound from pure components is independent of the source of the material or of the method of preparation. If elements chemically react to form a compound, they always combine in definite proportions by weight. This concept is known as the *Law of Definite Proportion*.

If the weight of each element that combines in an experiment is known, then the number of moles of each element can be determined. The empirical formula of the compound formed is the ratio between the number of moles of elements in the compound. This can be illustrated by the following example. If 32.06 grams of sulfur is burned in the presence of 32.00 grams of oxygen, then 64.06 grams of sulfur dioxide results. Thus

$$\frac{32.06 \text{ g S}}{32.06 \text{ g/mole S}} = 1 \text{ mole of Sulfur}$$

$$\frac{32.00 \text{ g O}}{16.00 \text{ g/mole O}} = 2 \text{ moles of Oxygen}$$

and the mole ratio of sulfur : oxygen is 1 : 2. The empirical formula of sulfur dioxide is SO_2. This also is the molecular formula.

In this experiment, the moderately reactive metal, magnesium, is combined with oxygen. The oxide, magnesium oxide, is formed. The equation for this reaction, based on the known chemical behavior is

$$2Mg(s) + O_2(g) \xrightarrow{\text{heat}} 2MgO(s)$$

If the mass of the magnesium is known and the mass of the oxide is found in the experiment, the mass of the oxygen in the oxide can be calculated:

$$\frac{\text{mass of magnesium oxide}}{\text{mass of oxygen}}$$

As soon as the masses are known, the moles of each component can be calculated. The moles can then be expressed in a simple whole number ratio and an empirical formula written.

EXAMPLE

When 2.43 g of magnesium was burned in oxygen, 4.03 g of magnesium oxide was produced.

$$\frac{\text{mass of magnesium oxide} \quad = \quad 4.03 \text{ g}}{- \text{ mass of magnesium} \qquad\quad = \quad 2.43 \text{ g}}$$
$$\text{mass of oxygen} \qquad\qquad = \quad 1.60 \text{ g}$$

$$\text{No. of moles of magnesium} \ = \ \frac{2.43 \text{ g}}{24.3 \text{ g/mole}} \ = \ 0.100 \text{ moles}$$

$$\text{No. of moles of oxygen} \ = \ \frac{1.60 \text{ g}}{16.0 \text{ g/mole}} \ = \ 0.100 \text{ moles}$$

The molar ratio is $0.100 : 0.100 \ = \ 1 : 1$
The empirical formula is $Mg_1 O_1$ or MgO.

$$\% \ Mg \ = \ \frac{2.43 \text{ g}}{4.03 \text{ g}} \ \times \ 100 \ = \ 60.3\%$$

In the present experiment, magnesium metal is heated in air. Air is composed of approximately 78% nitrogen and 21% oxygen. A side reaction occurs between some of the magnesium and the nitrogen gas:

$$3Mg(s) \ + \ N_2(g) \ \xrightarrow{\text{heat}} \ Mg_3N_2(s)$$

Not all of the magnesium is converted into magnesium oxide; some becomes magnesium nitride. However, the magnesium nitride can be converted to magnesium oxide by the addition of water:

$$Mg_3N_2(s) \ + \ 3H_2O(l) \ \xrightarrow{\text{heat}} \ 3MgO(s) \ + \ 2NH_3(g)$$

As a result, all of the magnesium is transformed into magnesium oxide.

1. To prepare a metal oxide.
2. To determine the empirical formula of a compound.
3. To demonstrate the Law of Definite Proportion.

Procedure

CAUTION!

A hot crucible can cause severe burns if handled improperly. Be sure to allow the crucible to cool sufficiently before handling. Always handle a hot crucible with crucible tongs.

Cleaning the Crucible

1. Obtain a porcelain crucible and cover. Carefully clean the crucible *in the hood* by adding 10 mL of 6 M HCl to the crucible; allow the crucible to stand for 5 min. with the acid. With crucible tongs, pick up the crucible, discard the HCl, and rinse the crucible with distilled water from a plastic squeeze bottle.

2. Place the crucible in a clay triangle, which is mounted on an iron ring and attached to a ring stand. Be sure the crucible is firmly in place in the triangle. Place the crucible cover on the crucible slightly ajar (Fig. 6.1a).

Figure 6.1
(a) Heating the crucible;
(b) picking up the crucible with crucible tongs.

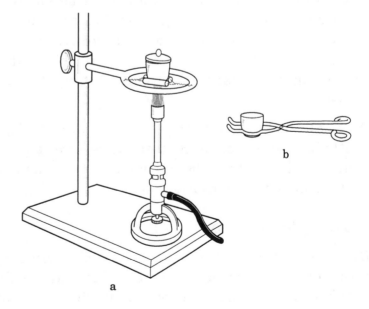

a

b

3. Begin to heat the crucible with the aid of a Bunsen burner in order to evaporate water. Increase the heat, and, with the most intense flame (the tip of the inner blue cone), heat the crucible and cover for 5 min.; a cherry red color should appear when the bottom is heated strongly. Remove the flame. With tongs, remove the crucible to a heat resistant surface and allow the crucible and cover to reach room temperature.

4. When cool, weigh the crucible and cover to 0.001 g (1). (Be sure to handle with tongs since fingerprints leave a residue.)

5. Place the crucible and cover in the clay triangle again. Reheat to the cherry red color for 5 min. Allow the crucible and cover to cool to room temperature. Reweigh when cool (2). Compare weight (1) and weight (2). If the weight differs by more than 0.005 g, heat the crucible and cover again for 5 min. and reweigh when cool. Continue heating, cooling, and weighing until the weight of the crucible and cover are constant to within 0.005 g.

Forming the Oxide

1. Using forceps to handle the magnesium ribbon, cut a piece approximately 12 cm in length and fold the metal into a ball; transfer to the crucible. Weigh the crucible, cover, and magnesium to 0.001 g (3). Determine the weight of magnesium metal (4) by subtraction.

2. Transfer the crucible to the clay triangle; the cover should be slightly ajar .(Fig. 6.1a).

3. Using a small flame, gently apply heat to the crucible. Should fumes begin to appear, remove the heat and cover the crucible immediately. Again place the cover ajar and continue to gently heat for 10 min. (If fumes appear, cover as before.) Remove the flame and allow the assembly to cool for 2 min. With tongs, remove the cover. If the magnesium has been fully oxidized, the contents should be a dull gray. Shiny metal means there is still free metal present. The cover should be replaced as before and the crucible heated for an additional 5 min. Reexamine the metal and continue heating until no shiny metal surfaces are present.

4. When all the metal appears as the dull gray oxide, half-cover the crucible and gently heat with a small Bunsen flame. Over a period of 5 min., gradually adjust the intensity of the flame until it is at its hottest, then heat the crucible to the cherry red color for 5 min.

Completing the Reaction

1. Discontinue heating and allow the crucible assembly to cool to room temperature. Remove the cover and, with a glass rod, *carefully* break up the solid in the crucible. With 0.5 mL (10 drops) of distilled water dispensed from an eye dropper, wash the glass rod, adding the water to the crucible.

2. Set the cover ajar on the crucible and *gently* heat with a small Bunsen flame to evaporate the water. (Be careful to avoid spattering while heating; if spattering occurs, remove the heat and quickly cover the crucible completely.)

3. When all the water has been evaporated, half-cover the crucible and gradually increase to the hottest flame. Heat the crucible and the contents with the hottest flame for 10 min.

4. Allow the crucible assembly to cool to room temperature. Weigh the cool crucible, cover, and magnesium oxide to 0.001 g (5).

5. Return the crucible, cover, and magnesium oxide to the clay triangle. Heat at full heat of the Bunsen flame for 5 min. Allow to cool and then reweigh (6). The two weights, (5) and (6), must agree to within 0.005 g; if not, the crucible assembly must be heated for 5 min., cooled, and reweighed until two successive weights are within 0.005 g.

Calculations

1. Determine the weight of magnesium oxide (7) by subtraction.

2. Determine the weight of oxygen (8) by subtraction.

3. From the data obtained in the experiment, calculate the empirical formula of magnesium oxide.

Chemicals and Equipment

1. Clay triangle
2. Porcelain crucible and cover
3. Crucible tongs
4. Magnesium ribbon
5. Eye dropper
6. 6 M HCl

NAME

SECTION

DATE

PARTNER

GRADE

Experiment 6

PRE-LAB QUESTIONS

1. Define the following terms:

a. empirical formula

b. molecular formula

2. The molecular formula of cyclohexane, an organic solvent, is C_6H_{12}. Write the empirical formula. Calculate the mass of a mole of material in grams.

3. Calculate the percentage, by weight, of each of the elements (C, H) in cyclohexane.

Experiment 6

REPORT SHEET

1. Weight of crucible and cover (1)   _____ g

2. Weight of crucible and cover (2)   _____ g

3. Weight of crucible, cover, and Mg (3)   _____ g

4. Weight of Mg metal (4): (3) − (2)   _____ g

5. Weight of crucible, cover, and oxide (5)   _____ g

6. Weight of crucible, cover, and oxide (6)   _____ g

7. Weight of magnesium oxide (7): (6) − (2)   _____ g

8. Weight of oxygen (8): (7) − (4)   _____ g

9. Number of moles of magnesium
   (4)/24.30 g/mole   _____ moles

10. Number of moles of oxygen
    (8)/16.00 g/mole   _____ moles

11. Simplest whole number ratio
    of Mg atoms to O atoms   ____ : _____

12. Empirical formula for magnesium oxide   _____

13. % Mg in the oxide from data
    [(4)/(7)] × 100   _____ %

14. % Mg calculated from the formula MgO
    [24.30 g/40.30 g] × 100   _____ %

15. Error
    $$\frac{(14) - (13)}{(14)} \times 100$$ _____ %

POST-LAB QUESTIONS

1. Write the balanced equation for the formation of carbon monoxide (CO), from the elements of carbon, C, and oxygen, O_2.

2. What substance would be present in the crucible along with the magnesium oxide and magnesium nitride if the crucible were not sufficiently heated?

3. What error in calculation would result if spattering occurred when water was evaporated from the magnesium oxide?

4. Calculate the percentage of sulfur, by weight, in sulfur trioxide, SO_3.

Experiment 7

Classes of chemical reactions

Background

The Periodic Table shows over 100 elements. The chemical literature describes millions of compounds that are known—some isolated from natural sources, some synthesized by laboratory workers. The combination of chemicals, in the natural environment or the laboratory setting, involves chemical reactions. The change in the way that matter is composed is a *chemical reaction*, a process wherein reactants (or starting materials) are converted into products. The new products often have properties and characteristics that are entirely different from that of the starting materials.

Four ways in which chemical reactions may be classified are combination, decomposition, single replacement (substitution), or double replacement (metathesis).

Two elements reacting to form a compound is a *combination reaction*. This process may be described by the general formula:

$$A + B \longrightarrow AB$$

The rusting of iron or the combination of iron and sulfur are good examples.

$$4Fe(s) + 3O_2(g) \longrightarrow 2Fe_2O_3(s) \text{ (rust)}$$

$$Fe(s) + S(s) \longrightarrow FeS(s)$$

A compound which breaks down into elements or simpler components typifies the *decomposition reaction*. This reaction has the general formula:

$$AB \longrightarrow A + B$$

Some examples of this type of reaction are the electrolysis of water into hydrogen and oxygen:

$$2H_2O(l) \longrightarrow 2H_2(g) + O_2(g)$$

and the decomposition of potassium iodate into potassium iodide and oxygen:

$$2KIO_3(s) \longrightarrow 2KI(s) + 3\,O_2(g)$$

The replacement of one component in a compound by another describes the *single replacement* (or *substitution*) reaction. This reaction has the general formula:

$$AB + C \longrightarrow CB + A$$

Processes which involve oxidation (the loss of electrons or the gain of relative positive charge) and reduction (the gain of electrons or the loss of relative positive charge) are typical of these reactions. Use of Table 7.1, the activity series of common metals, enables chemists to predict which oxidation-reduction reactions are possible. A more active metal, one higher in the table, is able to displace a less active metal, one listed lower in the table, from its aqueous salt. Thus aluminum metal displaces copper metal from an aqueous solution of copper(II) chloride; but copper metal will not displace aluminum from an aqueous solution of aluminum(III) chloride.

$$2Al(s) + 3CuCl_2(aq) \longrightarrow 3Cu(s) + 2AlCl_3(aq)$$

$$Cu(s) + AlCl_3(aq) \longrightarrow \text{No Reaction}$$

(Note that Al is oxidized to Al^{3+} and Cu^{2+} is reduced to Cu.)

Hydrogen may be displaced from water by a very active metal. Alkali metals are particularly reactive with water, and the reaction of sodium with water often is exothermic enough to ignite the hydrogen gas released.

$$2Na(s) + 2HOH(l) \longrightarrow 2NaOH(aq) + H_2(g) + heat$$

(Note that Na is oxidized to Na^+ and H^+ is reduced to H_2.)

Active metals, those above hydrogen in the series, are capable of displacing hydrogen from aqueous mineral acids such as HCl or H_2SO_4; however, metals below hydrogen will not replace hydrogen. Thus zinc reacts with aqueous solutions of HCl and H_2SO_4 to release hydrogen gas, but copper will not.

$$Zn(s) + 2HCl(aq) \longrightarrow ZnCl_2(aq) + H_2(g)$$

$$Cu(s) + H_2SO_4(aq) \longrightarrow \text{No reaction}$$

Table 7.1	Activity Series of Common Metals	
K	(potassium)	Most active
Na	(sodium)	
Ca	(calcium)	
Mg	(magnesium)	
Al	(aluminum)	
Zn	(zinc)	
Fe	(iron)	Activity increases
Pb	(lead)	
H_2	(hydrogen)	
Cu	(copper)	
Hg	(mercury)	
Ag	(silver)	
Pt	(platinum)	
Au	(gold)	Least active

Two compounds reacting with each other to form two different compounds describes *double replacement* (or *metathesis*). This process has the general formula:

$$AB + CD \longrightarrow AD + CB$$

There are two replacements in the sense that A replaces C in CD and C replaces A in AB. This type of reaction generally involves ions which form in solution either from the dissociation of ionic compounds or the ionization of molecular compounds. The reaction of an aqueous solution of silver nitrate with an aqueous solution of sodium chloride is a good example. The products are sodium nitrate and silver chloride. We know a reaction has taken place since the insoluble precipitate silver chloride forms and separates from solution.

$$AgNO_3(aq) + NaCl(aq) \longrightarrow NaNO_3(aq) + AgCl(s)$$
(White precipitate)

In general, a double replacement results if one combination of ions leads to a precipitate, a gas or an un-ionized or very slightly ionized species such as water. In all of these reaction classes, it is very often possible to use your physical senses to observe whether a chemical reaction has occurred. The qualitative criteria may involve the formation of a gaseous product, the formation of a precipitate, a change in color, or a transfer of energy.

Objectives

1. To demonstrate the different types of chemical reactions.
2. To be able to observe whether a chemical reaction has taken place.
3. To use chemical equations to describe a chemical reaction.

Procedure

Combination Reactions

1. Obtain a piece of aluminum foil approximately 2 × 0.5 in. Hold the foil at one end with a pair of forceps or crucible tongs and hold the other end in the hottest part of the flame of a Bunsen burner. Observe what happens to the foil. Record your observation and complete a balanced equation if you see that a reaction has occurred (1). Place the foil on a wire gauze to cool.

2. Obtain a piece of copper foil approximately 2 × 0.5 in. (A copper penny, one minted before 1982, may be substituted.) Hold the foil at one end with a pair of forceps or crucible tongs and hold the other end in the hottest part of the flame of a Bunsen burner. Observe what happens to the metal. Record your observation and complete a balanced equation if you see that a reaction has occurred (2). Place the foil on a wire gauze to cool.

3. Scrape some of the gray solid from the surface of the aluminum into a test tube (100 × 13 mm). Add 1 mL of water and shake the test tube. Is the solid soluble? Record your observation (3).

4. Scrape some of the black solid from the surface of the copper into a test tube (100 × 13 mm). Add 1 mL of water and shake the test tube. Is the solid soluble? Record your observation (4).

Decomposition Reactions

1. *Decomposition of ammonium carbonate.* Place 0.5 g of ammonium carbonate into a clean, dry test tube (100 × 13 mm). Gently heat the test tube in the flame of a Bunsen burner (Fig. 7.1). As you heat, hold a piece of wet red litmus paper at the mouth of the test tube. What happens to the solid? Are any gases produced? What happens to the color of the litmus paper? Ammonia gas acts as a base and turns moist red litmus paper blue. Record your observations and complete a balanced equation if you see that a reaction has occurred (5).

CAUTION!

When heating the contents of a solid in a test tube, do not point the open end towards anyone.

Figure 7.1
Position for holding a test tube in a Bunsen burner flame.

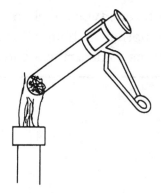

2. *Decomposition of potassium iodate.*

   a. Obtain three clean, dry test tubes (100 × 13 mm). Label them and add 0.5 g of compound according to the table below.

Test Tube No.	Compound
1	KIO_3
2	KIO_3
3	KI

b. Heat test tube no. 1 with the hottest flame of the Bunsen burner as shown in Fig. 7.2. Keep the test tube holder at the upper end of the test tube. While test tube no. 1 is being heated, hold a glowing wooden splint just inside the opening of the test tube. (The splint should not be flaming but should be glowing with sparks after the flame has been blown out. *Do not drop the glowing splint into the hot KIO_3.*) Oxygen supports combustion. The glowing splint should glow brighter or may burst into flame in the presence of oxygen. Record what happens to the glowing splint and complete a balanced equation for the decomposition reaction (6).

Figure 7.2
Testing for oxygen gas.

c. Remove the test tube from the flame and set it aside to cool.

d. Add 10 mL of distilled water to each of the three test tubes and mix thoroughly to ensure that the solids are completely dissolved. Add 10 drops of 0.1 M $AgNO_3$ solution to each test tube. Observe what happens to each solution. Record the colors of the precipitates and complete balanced equations for these reactions (7). (The KIO_3 and KI solids can be distinguished by the test results with $AgNO_3$: AgI is a yellow precipitate; $AgIO_3$ is a white precipitate.) What compound is present in test tube no. 1 after heating (8)?

Single Replacement Reactions

1. In a test tube rack, set up labeled test tubes (100 × 13 mm) numbered from 1 through 9. Place 1 mL (approx. 20 drops) of the appropriate solution in the test tube with a small piece of metal as outlined in the table below.

Test Tube No.	Solution	Metal
1	H_2O	Ca
2	H_2O	Fe
3	H_2O	Cu
4	3 M HCl	Zn
5	6 M HCl	Pb
6	6 M HCl	Cu
7	0.1 M $NaNO_3$	Al
8	0.1 M $CuCl_2$	Al
9	0.1 M $AgNO_3$	Cu

2. Observe the mixtures over a 20 min. period of time. Note any color changes, any evolution of gases, any formation of precipitates, or any energy changes (hold each test tube in your hand and note whether the solution becomes warmer or colder) that occur during each reaction; record your observations in the appropriate spaces on the Report Sheet (9). Write a complete and balanced equation for each reaction that occurred. For those cases where no reaction took place, write "No Reaction."

3. Dispose of the unreacted metals as directed by your instructor. *Do not discard into the sink.*

Double Replacement Reaction

1. Each experiment in this part requires mixing equal volumes of two solutions in a test tube (100 × 13 mm). Use about 10 drops of each solution. Record your observation at the time of mixing (10). When there appears to be no evidence of a reaction, feel the test tube for an energy change (exothermic or endothermic). The solutions to be mixed are outlined in the table below.

Test Tube No.	Solution No. 1	Solution No. 2
1	0.1 M NaCl	0.1 M KNO_3
2	0.1 M NaCl	0.1 M $AgNO_3$
3	0.1 M Na_2CO_3	3 M HCl
4	3 M NaOH	3 M HCl
5	0.1 M $BaCl_2$	3 M H_2SO_4
6	0.1 M $Pb(NO_3)_2$	0.1 M K_2CrO_4
7	0.1 M $Fe(NO_3)_3$	3 M NaOH
8	0.1 M $Cu(NO_3)_2$	3 M NaOH

2. For those cases where a reaction occurred, write a complete and balanced equation. Indicate precipitates, gases, and color changes. Table 7.2 lists some insoluble salts. For those cases where no reaction took place, write "No Reaction."

3. Discard the solutions as directed by your instructor. *Do not discard into the sink.*

Table **7.2**	Some Insoluble Salts
AgCl	Silver chloride (white)
Ag_2CrO_4	Silver chromate (red)
$AgIO_3$	Silver iodate (white)
AgI	Silver iodide (yellow)
$BaSO_4$	Barium sulfate (white)
$Cu(OH)_2$	Copper(II) hydroxide (blue)
$Fe(OH)_3$	Iron(III) hydroxide (red)
$PbCrO_4$	Lead(II) chromate (yellow)
PbI_2	Lead(II) iodide (yellow)
$PbSO_4$	Lead(II) sulfate (white)

Chemicals and Equipment

1. Aluminum foil
2. Aluminum wire
3. Copper foil
4. Copper wire
5. Ammonium carbonate, $(NH_4)_2CO_3$
6. Potassium iodate, KIO_3
7. Potassium iodide, KI
8. Calcium turnings
9. Iron filings
10. Mossy zinc
11. Lead shot
12. 3 M HCl
13. 6 M HCl
14. 3 M H_2SO_4
15. 3 M NaOH
16. 0.1 M $AgNO_3$
17. 0.1 M NaCl
18. 0.1 M $NaNO_3$
19. 0.1 M Na_2CO_3
20. 0.1 M KNO_3
21. 0.1 M K_2CrO_4
22. 0.1 M $BaCl_2$
23. 0.1 M $Cu(NO_3)_2$
24. 0.1 M $CuCl_2$
25. 0.1 M $Pb(NO_3)_2$
26. 0.1 M $Fe(NO_3)_3$

Experiment 7

PRE-LAB QUESTIONS

For each of the reactions below, classify as a combination, decomposition, single replacement, or double replacement.

1. $Mg(s) + Cl_2(g) \longrightarrow MgCl_2(s)$ _____

2. $2Zn(s) + O_2(g) \longrightarrow 2ZnO(s)$ _____

3. $Pb(NO_3)_2(aq) + H_2SO_4(aq) \longrightarrow 2HNO_3(aq) + PbSO_4(s)$ _____

4. $NH_3(aq) + HCl(aq) \longrightarrow NH_4Cl(aq)$ _____

5. $AgNO_3(aq) + NaI(aq) \longrightarrow AgI(s) + NaNO_3(aq)$ _____

6. $AgNO_3(aq) + NaCl(aq) \longrightarrow AgCl(s) + NaNO_3(aq)$ _____

7. $Zn(s) + H_2SO_4(aq) \longrightarrow ZnSO_4(aq) + H_2(g)$ _____

8. $H_2CO_3(aq) \longrightarrow CO_2(g) + H_2O(l)$ _____

9. $2H_2O(l) \longrightarrow 2H_2(g) + O_2(g)$ _____

10. $2K(s) + 2H_2O(l) \longrightarrow 2KOH(s) + H_2(g)$ _____

Experiment 7

REPORT SHEET

Write *complete, balanced equations* for all cases that a reaction takes place. Your observation that a reaction occurred would be by a color change, by the formation of a gas, by the formation of a precipitate, or by an energy change (exothermic or endothermic). Those cases showing no evidence of a reaction, write "No Reaction."

Classes of chemical reactions   *Observation*

Combination reactions

1. _____ Al + _____ O_2 $\longrightarrow$ _____

2. _____ Cu + _____ O_2 $\longrightarrow$ _____

3. Solubility of aluminum oxide   _____

4. Solubility of copper oxide   _____

Decomposition reactions

5. _____ $(NH_4)_2CO_3$ $\longrightarrow$ _____

6. _____ KIO_3 $\longrightarrow$ _____

7. Residue of KIO_3 and $AgNO_3$ solution   _____

_____ KIO_3 + _____ $AgNO_3$ $\longrightarrow$ _____

_____ KI + _____ $AgNO_3$ $\longrightarrow$ _____

8. The residue present after heating KIO_3 _____

Classes of chemical reactions

Single replacement reaction

9. *Test tube no.*

1. ___ Ca + ___ H_2O $\longrightarrow$ _____

2. ___ Fe + ___ H_2O $\longrightarrow$ _____

3. ___ Cu + ___ H_2O $\longrightarrow$ _____

4. ___ Zn + ___ HCl $\longrightarrow$ _____

5. ___ Pb + ___ HCl $\longrightarrow$ _____

6. ___ Cu + ___ HCl $\longrightarrow$ _____

7. ___ Al + ___ $NaNO_3$ $\longrightarrow$ _____

8. ___ Al + ___ $CuCl_2$ $\longrightarrow$ _____

9. ___ Cu + ___ $AgNO_3$ $\longrightarrow$ _____

Double replacement reaction

10. *Test tube no.*

1. ___ NaCl + ___ KNO_3 $\longrightarrow$ _____

2. ___ NaCl + ___ $AgNO_3$ $\longrightarrow$ _____

3. ___ Na_2CO_3 + ___ HCl $\longrightarrow$ _____

4. ___ NaOH + ___ HCl $\longrightarrow$ _____

5. ___ $BaCl_2$ + ___ H_2SO_4 $\longrightarrow$ _____

6. ___ $Pb(NO_3)_2$ + ___ K_2CrO_4 $\longrightarrow$ _____

7. ___ $Fe(NO_3)_3$ + ___ NaOH $\longrightarrow$ _____

8. ___ $Cu(NO_3)_2$ + ___ NaOH $\longrightarrow$ _____

POST-LAB QUESTIONS

1. Calcium metal, Ca, reacts with water, H_2O, but gold metal, Au, does not. Why?

2. From the following list of chemicals, select two combinations that would lead to a double replacement reaction. Write the complete, balanced equations for the reactions.

$$KCl \quad HNO_3 \quad AgNO_3 \quad PbCl_2 \quad Na_2SO_4$$

3. Potassium chlorate, $KClO_3$, decomposes upon heating.

   a. Write a complete balanced equation for the decomposition.

   b. What gas is given off and how may it be detected?

   c. What chemical can be used to detect the salt which remains from the decomposition? Write a complete balanced equation for the reaction of the chemical with the salt.

Experiment 8

Chemical properties of consumer products

Background

Concern for the environment has placed considerable attention on the identification of chemicals that enter our everyday world. Analytical chemistry deals with these concerns in both a quantitative and qualitative sense. In quantitative analysis, the concern is for exact amounts of certain chemicals present in a sample; experiments in this manual will deal with this problem (for example, see Experiments 21, 22, 48, and 49). Qualitative analysis is limited to establishing the presence or absence of certain chemicals in detectable amounts in a sample. This experiment will focus on the qualitative determination of inorganic chemicals. Later experiments in this manual will deal with organic chemicals.

The simplest approach to the detection of inorganic chemicals is to use tests that will identify the ions that make up the inorganic sample. These ions are cations and anions. *Cations* are ions that carry positive charges; Na^+, NH_4^+, Ca^{2+}, Cu^{2+}, and Al^{3+} are representative examples. *Anions* are ions that carry negative charges; Cl^-, HCO_3^-, CO_3^{2-}, SO_4^{2-} and PO_4^{3-} are examples of this type. Since each ion has unique properties, each will give a characteristic reaction or test. By examining an aqueous solution of the chemical, qualitative spot tests often will identify the cation and anion present. The tests used will bring about some chemical change. This change will be seen in the form of a solid precipitate, gas bubbles, or a color change.

This experiment will use chemicals commonly found around the house, so-called consumer chemical products. You may not think of these products as chemicals nor refer to them by their inorganic chemical names. Nevertheless, they are chemicals, and simple qualitative analytical techniques can be used to identify the ions found in their make up.

Table salt, NaCl. Table salt is most commonly used as a flavoring agent. Individuals with high blood pressure (hypertension) are advised to restrict salt intake in order to reduce the amount of sodium ion, Na^+ absorbed. When dissolved in water, table salt releases the sodium cation, Na^+ and the chloride anion, Cl^-. Chloride ion is detected by silver nitrate, $AgNO_3$; a characteristic white precipitate of silver chloride forms.

$$Ag^+(aq) + Cl^-(aq) \longrightarrow AgCl(s)$$
$$\text{White}$$

Sodium ions produce a characteristic bright yellow color in a flame.

Ammonia, NH_3. Ammonia is a gas with a strong irritating odor. The gas dissolves readily in water, giving an aqueous ammonia solution; the solution is commonly referred to as ammonium hydroxide. Aqueous ammonia solutions are used as cleaning agents because of their ability to solubilize grease, oils, and waxes. Ammonia solutions are basic and will change moistened red litmus paper to blue. Ammonium salts (for example, ammonium chloride, NH_4Cl) react with strong bases to form ammonia gas.

$$NH_4^+(aq) + OH^-(aq) \longrightarrow NH_3(g) + H_2O(l)$$

Baking soda, sodium bicarbonate, $NaHCO_3$. Baking soda, sodium bicarbonate, $NaHCO_3$, acts as an antacid in some commercial products (e.g. Alka Seltzer) and as a leavening agent, helping to "raise" a cake. When sodium bicarbonate reacts with acids, carbon dioxide, a colorless, odorless gas, is released.

$$HCO_3^-(aq) + H^+(aq) \longrightarrow CO_2(g) + H_2O(l)$$

The presence of CO_2 can be confirmed with barium hydroxide solution, $Ba(OH)_2$; a white precipitate of barium carbonate results.

$$CO_2(g) + Ba(OH)_2(aq) \longrightarrow \underset{\text{White}}{BaCO_3(s)} + H_2O(l)$$

Epsom salt, $MgSO_4 \cdot 7\,H_2O$. Epsom salt has several uses; it may be taken internally as a laxative or purgative, or it may be used externally as a solution for soaking one's feet. When dissolved in water, epsom salt releases magnesium cations, Mg^{2+} and sulfate anions, SO_4^{2-}. The magnesium cation may be detected by first treating with a strong base, such as NaOH, and then with the organic dye, p-nitrobenzene azoresorcinol. The magnesium hydroxide, $Mg(OH)_2$, which initially forms, combines with the dye to give a blue color. This behavior is specific for the magnesium cation.

$$Mg^{2+}(aq) + 2OH^-(aq) \longrightarrow Mg(OH)_2(s) \xrightarrow{\text{dye}} \text{Blue complex}$$

The sulfate anion, SO_4^{2-}, reacts with barium chloride, $BaCl_2$, to form a white precipitate of barium sulfate, $BaSO_4$.

$$Ba^{2+}(aq) + SO_4^{2-}(aq) \longrightarrow \underset{\text{White}}{BaSO_4(s)}$$

Bleach, sodium hypochlorite, NaOCl. Bleach sold commercially is a dilute solution of sodium hypochlorite, NaOCl, usually 5% in concentration. The active agent is the hypochlorite anion. In solution, it behaves as if free chlorine, Cl_2, were present. Chlorine is an effective oxidizing agent. Thus in the presence of iodide salts, such as potassium iodide, KI, iodide anions are oxidized to free iodine, I_2; chlorine is reduced to chloride anions, Cl^-.

$$Cl_2(aq) + 2I^-(aq) \longrightarrow I_2(aq) + 2Cl^-(aq)$$

The free iodine gives a reddish-brown color to water. However, since iodine is more soluble in organic solvents, such as hexane, C_6H_{14}, the iodine dissolves in the organic solvent. The organic solvent separates from the water, and the iodine colors the organic solvent violet.

Sodium phosphate, Na$_3$PO$_4$. In some communities that use well water for their water supply, dissolved calcium and magnesium salts make the water "hard." Normal soaps do not work well as a result. In order to increase the efficiency of their products, especially in hard water areas, some commercial soap preparations, or detergents, contain sodium phosphate, Na$_3$PO$_4$. The phosphate anion is the active ingredient and keeps the calcium and magnesium ions from interfering with the soap's cleaning action. Other products containing phosphate salts are plant fertilizers; here, ammonium phosphate serves as the source of phosphorus. The presence of the phosphate anion can be detected with ammonium molybdate, (NH$_4$)$_2$MoO$_4$. In acid solution, phosphate anions combine with the molybdate reagent to form a bright yellow precipitate.

$$PO_4^{3-}(ag) + 12MoO_4^{2-}(ag) + 3NH_4^+(ag) + 24H^+(ag) \longrightarrow (NH_4)_3PO_4(MoO_3)_{12}(s) + 12H_2O(l)$$
$$\text{Yellow}$$

Objectives

1. To examine the chemical properties of some common substances found around the house.
2. To use spot tests to learn which inorganic cations and anions are found in these products.

Procedure

CAUTION!

Although we are using chemical substances common to our everyday life, conduct this experiment as you would any other. Wear safety glasses; do not taste anything; mix only those substances as directed.

Analysis of Table Salt, NaCl

1. Place a small amount (covering the tip of a small spatula) of table salt in a test tube (100 × 13 mm). Add 1 mL (approx. 20 drops) of distilled water and mix to dissolve. Add 2 drops of 0.1 M AgNO$_3$. Record your observation (1).

2. Take a small spatula and clean the tip by holding it in a Bunsen burner flame until the yellow color disappears. Allow to cool but do not let the tip touch anything. Place a few crystals of table salt on the clean spatula tip and heat in the flame of the Bunsen burner. Record your observation (2).

Analysis of Household Ammonia, NH_3, and Ammonium Ions, NH_4^+

1. Place 1 mL of household ammonia in a test tube (100 × 13 mm). Hold a piece of dry red litmus paper over the mouth of the test tube (be careful not to touch the glass with the paper). Record your observation (3). Moisten the red litmus paper with distilled water and hold it over the mouth of the test tube. Record your observation (4).

2. Place a small amount (covering the tip of a small spatula) of ammonium chloride, NH_4Cl, in a test tube (100 × 13 mm). Add 0.5 mL (about 10 drops) of 6 M NaOH to the test tube. Hold a moist piece of red litmus in the mouth of the test tube (be careful not to touch the glass with the paper). Does the litmus change color? If the litmus paper does not change color, gently warm the test tube (do not boil the solution). Record your observation (5).

3. Place a small amount (covering the tip of a small spatula) of commercial fertilizer in a test tube (100 × 13 mm). Add 0.5 mL (about 10 drops) of 6 M NaOH to the test tube. Test as above with moist red litmus paper. Record your observation and conclusion (6).

Analysis of Baking Soda, $NaHCO_3$

1. Place a small amount (covering the tip of a small spatula) of baking soda in a test tube (100 × 13 mm). Dissolve the solid in 1 mL of distilled water. Add 5 drops of 6 M H_2SO_4 and tap the test tube to mix. Record your observation (7).

2. Test the escaping gas for CO_2. Make a loop in a wire; the loop should be about 5 mm in diameter. Dip the wire loop into 1% barium hydroxide, $Ba(OH)_2$, solution; a drop of solution should cling to the loop. Carefully lower the wire loop down into the mouth of the test tube. Avoid touching the walls. Record what happens to the drop (8).

Analysis of Epsom Salt, $MgSO_4 \cdot 7 H_2O$

1. Place a small amount (covering the tip of a small spatula) of Epsom salt into a test tube (100 × 13 mm). Dissolve in 1 mL (about 20 drops) of distilled water. Add 5 drops of 6 M NaOH. Then add 5 drops of the "organic dye" solution (0.01% p-nitrobenzene azoresorcinol). Record your observation (9).

2. Place a small amount (covering the tip of a small spatula) of Epsom salt into a test tube (100 × 13 mm). Dissolve in 1 mL (about 20 drops) of distilled water. Add one drop of 3 M HNO_3 followed by 2 drops of 1 M $BaCl_2$ solution. Record your observation (10).

Analysis of Bleach, NaOCl

Place a small amount (covering the tip of a small spatula) of potassium iodide, KI in a test tube (100 × 13 mm). Dissolve in 1 mL (about 20 drops) of distilled water. Add 1 mL of bleach to the solution, followed by 10 drops of hexane, C_6H_{14}. Cork the test tube and shake vigorously. Set aside and allow the layers to separate. Note the color of the upper organic layer and record your observation (11).

Analysis of Sodium Phosphate, Na_3PO_4

Label three clean test tubes no. 1, no. 2, and no. 3. In test tube no. 1, place 2 mL of 0.1 M Na_3PO_4; in test tube no. 2, place a small amount (covering the tip of a small spatula) of a detergent; in test tube no. 3, place a small amount (covering the tip of a small spatula) of a fertilizer. Add 2 mL of distilled water to the solids in test tubes no. 2 and no. 3 and mix. Add 6 M HNO_3 dropwise to all three test tubes until the solutions test acid to litmus paper (blue litmus turns red when treated with acid). Mix each solution well and then add 10 drops of the $(NH_4)_2MoO_4$ reagent to each test tube. Warm the test tube in a water bath maintained between 60–70°C. Compare the three solutions and record your observations (12).

Chemicals and Equipment

1. Bunsen burner
2. Copper wire
3. Litmus paper, blue
4. Litmus paper, red
5. Commercial ammonia solution, NH_3
6. Ammonium chloride, NH_4Cl
7. Commercial baking soda, $NaHCO_3$
8. Commercial bleach, $NaOCl$
9. Detergent, Na_3PO_4
10. Epsom salt, $MgSO_4 \cdot 7\ H_2O$
11. Garden fertilizer, $(NH_4)_3PO_4$
12. Table salt, $NaCl$
13. Ammonium molybdate reagent, $(NH_4)_2MoO_4$
14. 1 M $BaCl_2$
15. 1% $Ba(OH)_2$
16. 3 M HNO_3
17. 6 M HNO_3
18. Potassium iodide, KI
19. 0.1 M $AgNO_3$
20. 6 M $NaOH$
21. 0.1 M Na_3PO_4
22. 6 M H_2SO_4
23. 0.01% p-nitrobenzene azoresorcinol ("organic dye" solution)
24. Hexane, C_6H_{14}

Experiment 8

PRE-LAB QUESTIONS

1. Spot tests can be used qualitatively to identify the cation and anion present in an aqueous solution of a chemical. For each ion below, indicate the chemical used for the spot test and the *observable* chemical change that is expected.

Cl^-

NH_4^+

HCO_3^-

SO_4^{2-}

2. Below is a list of the materials to be analyzed. Complete the table by providing the name and formula of the salt found in each product, the name and formula of the cation, and the name and formula of the anion.

Product	Salt	Cation	Anion
1. Baking soda			
2. Bleach			
3. Detergent			
4. Epsom salt			
5. Fertilizer			
6. Table salt			

NAME ______________________ SECTION ____________ DATE ____________

PARTNER ______________________ GRADE ______________________

Experiment 8

REPORT SHEET

Analysis of table salt, NaCl

1. $AgNO_3 + NaCl$ ____________

2. Color of flame ____________

Analysis of household ammonia, NH_3, and ammonium ions, NH_4^+

3. Color of dry litmus with ammonia fumes ____________

4. Color of wet litmus with ammonia fumes ____________

5. Color of wet litmus with $NH_4Cl + NaOH$ ____________

6. Presence of ammonium ions in fertilizer ____________

Analysis of baking soda, $NaHCO_3$

7. $H_2SO_4 + NaHCO_3$ ____________

8. Presence of CO_2 gas ____________

Analysis of Epsom salt, $MgSO_4 \cdot 7\ H_2O$

9. Presence of magnesium cation ____________

10. Presence of sulfate anion ____________

Analysis of bleach, NaOCl

11. Color of hexane layer ____________

Analysis of sodium phosphate, Na_3PO_4

12. Presence of phosphate no. 1 ____________

no. 2 ____________

no. 3 ____________

POST-LAB QUESTIONS

1. Some commercial table salt, NaCl, is "iodized" by including small quantities of sodium iodide, NaI. How could you test for the presence of the iodide anion?

2. If tap water gives a white precipitate with 1% barium hydroxide solution, what ion in the water is most likely responsible for this effect? Write the equation for the reaction.

3. A bottle of club soda releases a gas when opened. What is the gas and how would you test for it?

4. Baking powder is used to produce a "rising action" in the dough for baked goods. It contains sodium bicarbonate, $NaHCO_3$, and calcium acid phosphate, $CaHPO_4$. Write an equation that shows what happens to these salts when liquid is added in the mixing bowl. What brings about the "rising action" in the dough?

5. Many fertilizers contain ammonium phosphate, $(NH_4)_3PO_4$. How could you test for the presence of ammonium cations, NH_4^+? How could you test for the presence of phosphate anions, PO_4^{3-}?

Calorimetry: the determination of the specific heat of a metal

Background

Any chemical or physical change involves a change in energy. Heat is a form of energy that can be observed as a flow of energy. Heat can pass spontaneously from an object at a high temperature to an object at a lower temperature. Two objects in contact at different temperatures, given enough time, will eventually reach the same temperature. The flow of heat energy can also be either into or out of a system under study.

The amount of heat can be measured in a device called a *calorimeter*. A calorimeter is a container with insulated walls. The insulation prevents a rapid heat exchange between the contents of the calorimeter and the surroundings. In the closed environment of the system, there is no loss or gain of heat. Since the change in temperature of the contents of the calorimeter is used to measure the magnitude of the heat flow, a thermometer is included with the calorimeter.

The specific heat of any substance can be determined in a calorimeter. The *specific heat* is an intensive physical property of a substance and is the quantity of heat (in calories) necessary to raise the temperature of one gram of substance by one degree Celsius. The specific heats for some common substances are listed in Table 9.1. Notice that specific heat has the units calories per gram per degree Celsius. From Table 9.1, the specific heat of water is 1.00 cal/g °C; this means that it would take one calorie to raise the temperature of one gram of water by 1°C. In contrast, iron has a specific heat of 0.11 cal/g °C; it would take only 0.11 calorie to raise the temperature of one gram of iron by 1°C. Just by comparing these two substances, you can see that water is a convenient coolant and explains its use in the internal combustion engine of automobiles. A small quantity of water is capable of absorbing a relatively large amount of heat, yet shows only a modest rise in temperature.

In general, when a given mass of a substance undergoes a temperature change, the heat energy required for the change is given by the equation

$$Q = m \times S \times \Delta T$$

where Q is the change in heat energy, m is the mass of the substance in grams, S is the specific heat of the substance, and ΔT is the change in temperature (the difference between the *final* and *initial* temperature); thus

$$\text{calories} = g \times (cal/g \ °C) \times °C$$

Table 9.1	Specific Heat Values for Some Common Substances		
Substance	Specific heat (cal/g °C)	Substance	Specific heat (cal/g °C)
Lead (Pb)	0.038	Glass	0.12
Tin (Sn)	0.052	Table salt (NaCl)	0.21
Silver (Ag)	0.056	Aluminum (Al)	0.22
Copper (Cu)	0.092	Wood	0.42
Zinc (Zn)	0.093	Ethyl alcohol (C_2H_6O)	0.59
Iron (Fe)	0.11	Water (H_2O)	1.00

EXAMPLE 1

If 20 g of water is heated so that its temperature rises from 20° to 25°C, then we know that 100 cal have been absorbed.

$$Q = m \times S \times \Delta T$$

$$20 \text{ g} \times 1.0 \frac{\text{cal}}{\text{g °C}} \times (25 - 20)°C = 100 \text{ cal}$$

The specific heat of a metal can be found with a water calorimeter. This can be conveniently done by using the Principle of Conservation of Energy: *Energy can neither be created nor destroyed in any process, but can be transferred from one part of a system to another.* Experimentally, the amount of heat absorbed by a known mass of water can be measured when a known mass of hot metal is placed in the water. The temperature of the water will rise as the temperature of the metal falls. Using the known heat capacity of water, the amount of heat added to the water can be calculated, just as in Example 1. This is exactly the amount of heat given up by the metal.

Heat(cal) lost by metal = Heat(cal) gained by water

$$Q_{metal} = Q_{water}$$

$$m_m \times S_m \times (\Delta T)_m = m_w \times S_w \times (\Delta T)_w$$

All the terms in the above equation are either known or can be determined experimentally, except for the value S_m, the specific heat of the metal. The unknown can then be calculated.

$$S_m = \frac{m_w \times S_w \times (\Delta T)_w}{m_m \times (\Delta T)_m}$$

EXAMPLE 2

An unknown hot metal at 100.0°C with a mass of 50.03 g was mixed with 40.11 g of water at a temperature of 21.5°C. A final temperature of 30.6°C was reached. The heat gained by the water is calculated by

$$Q_w = (40.11 \text{ g}) \times 1.00 \frac{\text{cal}}{\text{g °C}} \times (30.6 - 21.5)°C = 365 \text{ cal}$$

The heat lost by the metal is equal to the heat gained by the water.

$$Q_m = Q_w = 365 \text{ cal}$$

The specific heat of the unknown metal is calculated to be

$$S_m = \frac{365 \text{ cal}}{(50.03 \text{ g}) \times (100.0 - 30.6)°C} = 0.105 \frac{\text{cal}}{\text{g °C}}$$

The specific heat of iron is 0.11 cal/g °C; thus, from the value of S_m determined experimentally, the unknown metal is iron.

If the specific heat of the metal is known, an approximate atomic weight can be determined. This can be done using the relationship between the specific heat of solid metallic objects and their atomic weights observed by Pierre Dulong and Alexis Petit in 1819; it is known as the Law of Dulong and Petit.

$$S_m \times \text{Atomic Weight} = 6.3 \text{ cal/mole °C}$$

EXAMPLE 3

The specific heat from Example 2 is 0.11 cal/g °C (to two significant figures). The approximate atomic weight is calculated to be

$$\text{Atomic Weight} = \frac{6.3 \text{ cal/mole °C}}{0.11 \text{ cal/g °C}} = 57 \text{ g/mole}$$

The atomic weight of iron is 56 g/mole (to two significant figures).

The calculations assume no heat is lost from the calorimeter to the surroundings and that the calorimeter absorbs a negligible amount of heat. However, this is not entirely correct. The calorimeter consists of the container, the stirrer, and the thermometer. All three get heated along with the water. As a result, the calorimeter absorbs heat. Therefore, the heat capacity for the calorimeter will be obtained experimentally, and the value derived applied whenever the calorimeter is used.

$$Q_{calorimeter} = C_{calorimeter} \Delta T$$

EXAMPLE 4

The temperature of 50.0 mL of warm water is 36.9°C. The temperature of 50.0 mL of cold water in a calorimeter is 19.9°C. When the two were mixed together in the calorimeter, the temperature after mixing was 28.1°C. The heat capacity of the calorimeter is calculated as follows (assume the density of water is 1.00 g/mL):

The heat lost by the warm water is

$$(28.1 - 36.9)°C \times 50.0 \text{ g} \times 1.00 \text{ cal/g °C} = -440 \text{ cal}$$

The heat gained by the cold water is

$$(28.1 - 19.9)°C \times 50.0 \text{ g} \times 1.00 \text{ cal/g °C} = 410 \text{ cal}$$

The heat lost to the calorimeter is

$$-440 \text{ cal} + 410 \text{ cal} = -30 \text{ cal}$$

The heat capacity of the calorimeter is

$$\frac{30 \text{ cal}}{(28.1 - 19.9)°C} = 3.7 \text{ cal/°C}$$

In this experiment, you also will plot the water temperature in the calorimeter versus time. Since the calorimeter walls and cover are not perfect insulators, some heat will be lost to the surroundings. In fact, when the hot water (or hot metal) is added to the colder water in the calorimeter, some heat will be lost before the maximum temperature is reached. In order to compensate for this loss, the maximum temperature is obtained by extrapolation of the curve as shown in Fig. 9.2. This gives the maximum temperature rise that would have been recorded had there been no heat loss through the calorimeter walls. Once T_{max} is found, then ΔT can be determined.

Objectives

1. To construct a simple calorimeter.
2. To measure the heat capacity of the calorimeter.
3. To measure the specific heat of a metal.

Determination of the Heat Capacity of the Calorimeter

1. Construct a calorimeter as shown in Fig. 9.1. The two dry 8 oz. styrofoam cups are inserted one into the other, supported in a 250-mL beaker. The plastic lid should fit tightly on the cup. With a suitable-sized cork borer, make two holes in the lid; one hole should be near the center for the thermometer and one hole to the side for the stirring wire. In order to keep the thermometer bulb 2 cm above the bottom of the inner cup, fit a rubber ring (cut from latex rubber tubing) around the thermometer and adjust the ring by moving it up or down the thermometer.

Figure 9.1

The styrofoam calorimeter.

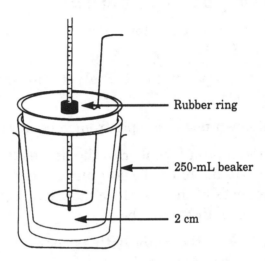

— Rubber ring

— 250-mL beaker

— 2 cm

2. Since the density of water is nearly 1.00 g/mL over the temperature range for this experiment, the amount of water used in the calorimeter will be measured by volume. With a volumetric pipet, place 50.0 mL of cold water in the calorimeter cup; determine and record the mass (1). Cover the cup with the lid-thermometer-stirrer assembly. Stir the water for 5 min., observing the temperature during the time; record the temperature at 1 min. intervals on the Data Sheet. When the system is at equilibrium, record the temperature to the nearest 0.2°C (3).

3. With a volumetric pipet, place 50.0 mL of water in a clean, dry 150-mL beaker; determine and record the mass (2). Heat the water with a low flame until the temperature of the water is about 70°C. Allow the hot water to stand for a few minutes, stirring occasionally during this time period. Quickly record the temperature to the nearest 0.2°C (4) and pour the water completely into the calorimeter that has been assembled and has reached equilibrium (Fig. 9.1).

4. Replace the cover assembly and stir the contents gently. Observe the temperature for 5 min. and record the temperature on the Data Sheet (p. 106) every 30 sec. during that 5 min. period. Plot the temperature as a function of time, as shown in Fig. 9.2. (Use the graph paper on page 107.) Determine from your curve the maximum temperature by extrapolation and record it (5). Determine the ΔT. From the data, calculate the heat capacity of the calorimeter.

Figure 9.2

Plot of temperature vs. time.

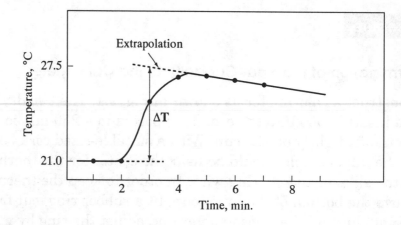

Determination of the Specific Heat of a Metal

1. Dry the styrofoam cups used for the calorimeter calibration. Reassemble the apparatus as in Fig. 9.1.

2. With a volumetric pipet, place 50.0 mL of cold water in the calorimeter cup; record the mass (1).

3. Obtain an unknown metal sample from your instructor.

4. Weigh a clean, dry 50-mL beaker to the nearest 0.01 g (2). Place about 40 g of your unknown sample in the beaker and reweigh to the nearest 0.01 g (3). Determine the mass of the metal by subtraction (4). Pour the sample into a 16 × 150 mm clean, dry test tube.

5. Place the test tube in the water bath as shown in Fig. 9.3. Be sure that all of the metal in the test tube is below the surface of the water. Heat the water to a gentle boil and keep the test tube in the bath for 10 min. Make certain that water does not splash into the test tube.

Figure 9.3

Assembly for heating the metal.

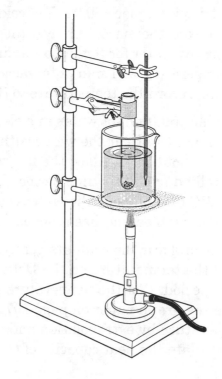

6. While the metal is heating, follow the temperature of the cold water in the calorimeter for 5 min.; record the temperature on the Data Sheet at 1 min. intervals. After 5 min., record the temperature on the Report Sheet of the cold water to the nearest 0.2°C (5).

7. After 10 min. of heating the metal, observe and record the temperature on the Report Sheet of the boiling water in the beaker to the nearest 0.2°C (6).

8. All steps must be done *quickly* and *carefully* at this point. Remove the test tube from the boiling water; dry the outside glass with a paper towel; remove the lid on the calorimeter; add the hot metal to the calorimeter. Be careful no water is added to or lost from the calorimeter on the transfer.

9. Record the calorimeter temperature on the Data Sheet as soon as the apparatus has been reassembled. Note the time when the temperature is determined. Stir the water. Continue to follow the temperature, recording the temperature on the Data Sheet every 30 sec. for the next 4 min.

10. Plot the temperature as a function of time, as shown in Fig. 9.2. (Use the graph paper on page 108.) Determine from your curve the maximum temperature; record the temperature on the Report Sheet (7). Determine the ΔT. From the data, determine the specific heat and the atomic mass of the metal.

Chemicals and Equipment

1. Metal pellets
2. Styrofoam cups (2)
3. Lid for styrofoam cups
4. Metal stirring loop
5. Thermometers, 110°C (2)
6. Latex rubber ring
7. Volumetric pipet, 50-mL

Experiment 9

PRE-LAB QUESTIONS

1. What is the purpose of the insulated walls of the calorimeter?

2. Can heat flow be measured?

3. Why is water a convenient coolant in an automobile radiator?

4. Why can a water calorimeter be used to determine the specific heat of a metal?

5. What are the units of specific heat?

Experiment 9

REPORT SHEET

Determination of the heat capacity of the calorimeter

1. Mass of the cold water
   50.0 mL × 1.00 g/mL   _____ g

2. Mass of the warm water
   50.0 mL × 1.00 g/mL   _____ g

3. Temperature of the equilibrated system:
   cold water and calorimeter   _____ °C

4. Temperature of the warm water   _____ °C

5. Maximum temperature from the graph   _____ °C

6. ΔT of cold water and calorimeter
   (5) − (3)   _____ °C

7. ΔT of warm water
   (5) − (4)   _____ °C

8. Heat lost by warm water
   (2) × 1.00 cal/g °C × (7)   _____ cal

9. Heat gained by cold water and
   the calorimeter: −(8)   _____ cal

10. Heat gained by cold water
    (1) × 1.00 cal/g °C × (6)   _____ cal

11. Heat gained by the calorimeter
    (9) − (10)   _____ cal

12. Heat capacity of calorimeter, C_{cal}
    (11)/(6)   _____ cal/°C

Determination of the specific heat of a metal

1. Mass of cold water
   50.0 mL × 1.00 g/mL   _____ g

2. Mass of 50-mL beaker   _____ g

3. Mass of the beaker plus metal   _____ g

4. Mass of metal: (3) − (2)   _____ g

5. Temperature of the equilibrated system   _____ °C

6. Temperature of hot metal
   (Temperature of boiling water)   _____ °C

7. Maximum temperature from the graph   _____ °C

8. ΔT of cold water and calorimeter
   (7) − (5)   _____ °C

9. Heat gained by the calorimeter and water
   [(1) × 1.00 cal/g °C × (8)] + [C_{cal} × (8)]   _____ cal

10. ΔT of the metal
    (7) − (6)   _____ °C

11. Heat lost by the metal: − (9)   _____ cal

12. Specific heat of the metal
    $$\frac{(11)}{(4) \times (10)}$$ _____ cal/g °C

13. Atomic mass
    $$\frac{6.3 \text{ cal/mole °C}}{(12)}$$ _____ g/mole

POST-LAB QUESTIONS

1. Can a metal container be used in place of a styrofoam cup when constructing a calorimeter?

2. A 50.0 g sample of water has heated from an initial temperature of 20.0°C to a final temperature of 60.5°C. How many calories has the water absorbed?

3. A 60.0 g sample of metal beads was heated to 99.5°C. The metal was added to 50.0 g of water at a temperature of 18.0°C resulting in a temperature rise for the water to 35.5°C.

    a. What was the temperature of the metal at equilibrium?

    b. What was the temperature change of the metal?

    c. What is the specific heat of the metal?

    d. Using Table 9.1, what is most likely the identity of the metal?

Temperature Data Sheet

Calorimeter Calibration		Specific Heat Determination	
Time (min.)	Temp (°C)	Time (min.)	Temp (°C)
0	_____	0	_____
1.0	_____	1.0	_____
2.0	_____	2.0	_____
3.0	_____	3.0	_____
4.0	_____	4.0	_____
5.0	_____	5.0	_____
5.5	_____	5.5	_____
6.0	_____	6.0	_____
6.5	_____	6.5	_____
7.0	_____	7.0	_____
7.5	_____	7.5	_____
8.0	_____	8.0	_____
8.5	_____	8.5	_____
9.0	_____	9.0	_____
9.5	_____	9.5	_____
10.0	_____	10.0	_____

Temperature, °C

Time, min.

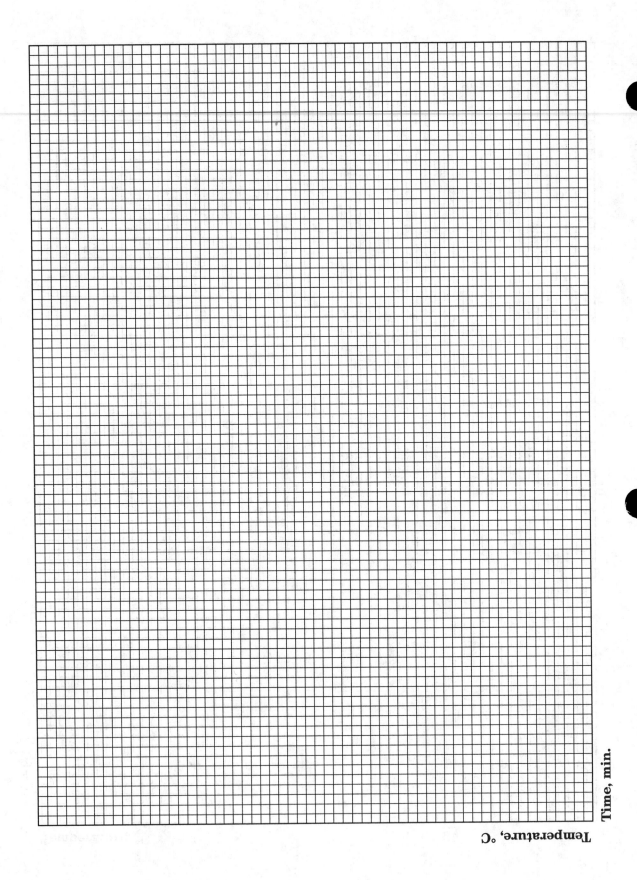

Temperature, °C

Time, min.

Experiment 10

Boyle's Law: the pressure–volume relationship of a gas

Background

The British scientist Robert Boyle made many contributions in the fields of medicine, astronomy, physics, and chemistry. However, he is best known for his work on the behavior of gases. In 1662, Boyle found that when the temperature is held constant, the pressure of a trapped amount of gas (any gas) is inversely proportional to its volume. That is, when the pressure of the gas increases, the volume of the gas decreases; when the pressure of the gas decreases, the volume of the gas increases. Boyle's Law can be written mathematically as follows:

$$V = k \times \frac{1}{P} \quad \text{or} \quad V = \frac{k}{P} \quad \text{or} \quad PV = k$$

where V is the volume of the gas, P is the pressure of the gas, and k is a constant that depends on the temperature and amount of the gas. By looking at these equations, it is easy to see the inverse relationship. For example, if pressure on a sample of trapped gas is *doubled*, the volume of the sample will be reduced by *half* of the value it had been before the increase in pressure. On the other hand, if the pressure is reduced by *half*, the volume will become *doubled*.

If there is a pressure change, for example from P_1 to P_2, then the volume also changes from V_1 to V_2. The relationship between the initial pressure and volume to the new pressure and volume can be expressed as follows.

$$P_1V_1 = P_2V_2 \quad \text{or} \quad \frac{P_1}{P_2} = \frac{V_2}{V_1}$$

Volume may be expressed in a variety of units—liter (L), milliliter (mL), cubic meter (m³), cubic centimeter (cc or cm³). Pressure similarly can be expressed in a variety of units, but the standard unit of pressure is the atmosphere (atm); one atmosphere (1 atm) is defined as the pressure needed to support a column of mercury 760 mm in height at 0°C at sea level. In honor of Evangelista Torricelli, the Italian inventor of the barometer, the unit **torr** is used and is equal to 1 mm Hg. Thus

1 mm Hg = 1 torr
1 atm = 760 mm Hg = 760 torr

In this experiment, a volume of air is trapped in a capillary tube by a column of mercury. The mercury acts as a movable piston. Depending on how the capillary tube is tilted, the mercury column moves and thus causes the volume of trapped air to change. The pressure of the trapped air supports not only the pressure exerted by the atmosphere but also the pressure of the mercury column. The pressure exerted by the mercury column varies depending on the angle of the tilt.

If θ is the angle of the tilt, the pressure of the mercury column can be calculated by the following equation:

$$P_{Hg} = (\text{length of Hg column}) \times \sin \theta$$

The total pressure of the trapped gas is the sum of the atmospheric pressure and the pressure due to the mercury column.

You need only measure the length of the column of air, L_{air}, since the length is directly related to the volume. The column of air geometrically is a regular cylinder. The radius of the cylinder, in this case the capillary tube, remains the same. The volume of a regular cylinder is a constant (πr^2) times the height of the cylinder (L_{air}); the only quantity that varies in this experiment is the value L_{air}.

Objectives

1. To show the validity of Boyle's Law.
2. To measure the volume of a fixed quantity of air as the pressure changes at constant temperature.

Procedure*

CAUTION!

Mercury can be spilled easily. While mercury has a low vapor pressure, its vapor is extremely toxic. Mercury can also be absorbed through the skin. If any mercury is spilled, notify the instructor immediately for proper clean-up.

1. Obtain a Boyle's Law apparatus and a 30°-60°-90° plastic triangle. The Boyle's Law apparatus consists of a piece of glass tubing that contains a column of mercury and is attached to a ruler by means of rubber bands (Fig. 10.1).

2. Record the temperature on the Report Sheet (1).

3. Record the barometric pressure, P_{at}, in mm Hg on the Report Sheet (2).

4. Measure the length of the column of mercury, L_{Hg}, to the nearest 0.5 mm; record this length on the Report Sheet (3).

* Adapted from R. A. Hermens, *J. Chem. Educ.* **60**, (1983), 764.

Figure 10.1

The Boyle's Law apparatus.

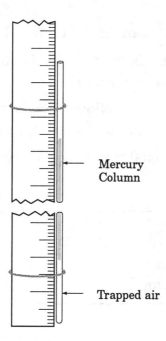

Mercury
Column

Trapped air

5. The length of the column of trapped air is to be measured when the tube is at various angles to the bench top, as outlined in the following table.

Angle of Tube	Position of Open End of Tube
0°	Horizontal
90°	Up
90°	Down
60°	Up
60°	Down
30°	Up
30°	Down

The column length of the trapped air, L_{air}, is measured (from the glass seal to the mercury) to the nearest 0.5 mm and is recorded in the table on the Report Sheet (4). The correct angle can be obtained with the aid of the 30°-60°-90° triangle by placing the Boyle's Law apparatus along the appropriate edge.

CAUTION!

Do not touch the glass tube during the measurements, to avoid any temperature changes. Do not jar the tube at any time. This will avoid separation or displacement of the mercury column when measurements have begun.

6. Calculate the reciprocal, $\frac{1}{L_{air}}$, and enter on the Report Sheet (5).

7. Using the appropriate formula from the table on the Report Sheet (6), calculate the pressure, P, of the column of air (7).

8. Plot the data on graph paper as follows: y-axis, the calculated pressure, P; x-axis, the reciprocal of the length of the trapped air, $\frac{1}{L_{air}}$.

9. Replot the data with P (y-axis) versus L_{air}(x-axis).

Chemicals and Equipment

1. Boyle's Law apparatus
2. 30°-60°-90° plastic triangle

Experiment 10

PRE-LAB QUESTIONS

1. Show a mathematical relationship for Boyle's Law that demonstrates the *inverse* relationship of pressure and volume.

2. Weather reports are always referring to the barometric pressure. What is the cause of this pressure and what is the standard pressure at sea level?

3. Some trapped air underwent a change in volume from 2.0 L to 1.0 L at 20.0°C. What happened to the pressure?

4. Make the following conversions:

2.00 atm = _____ torr

750 mm Hg = _____ atm

700 torr = _____ mm Hg

NAME ______________________ SECTION ____________ DATE ____________

PARTNER ______________________ GRADE ____________

Experiment 10

REPORT SHEET

1. Temperature ____________ °C
2. Barometric pressure, P_{at} ____________ mm Hg
3. Length of mercury column, L_{Hg} ____________ mm

Boyle's Law Data

Angle of Tube (opening*)	(4) Length of Trapped Air, L_{air}, mm	(5) $\frac{1}{L_{air}}$	(6) Pressure (calculation)	(7) Pressure P
0°			P_{at}	
90° (U)			$P_{at} + L_{Hg}$	
90° (D)			$P_{at} - L_{Hg}$	
60° (U)			$P_{at} + (L_{Hg} \sin 60°)$	
60° (D)			$P_{at} - (L_{Hg} \sin 60°)$	
30° (U)			$P_{at} + (L_{Hg} \sin 30°)$	
30° (D)			$P_{at} - (L_{Hg} \sin 30°)$	

*U = up; D = down

POST-LAB QUESTIONS

1. At a fixed temperature, what happens to the volume if the pressure is tripled?

2. In this experiment, the column length of the trapped air was measured rather than the actual volume. Why could this be done as a valid determination of Boyle's Law?

3. In order to demonstrate Boyle's Law, the pressure, P, is plotted versus the inverse of the volume, $\frac{1}{V}$. Explain why the plot is carried out in this way.

4. The pressure on a 150 mL volume of gas is increased from 400 torr to 700 torr while the temperature remains at 25°C. What is the new volume of gas?

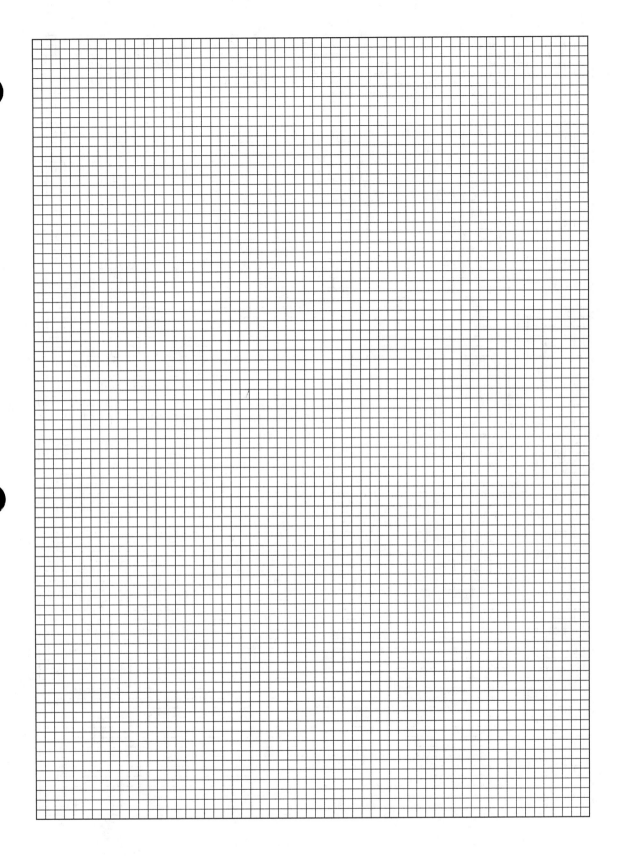

Experiment 11

Charles' Law: the volume–temperature relationship of a gas

Jacques Charles observed that for a fixed quantity of gas, the volume at constant pressure changes when temperature changes: the volume increases when the temperature increases; the volume decreases when the temperature decreases. Although first described by Charles in 1787, it was not until 1802 that Joseph Gay-Lussac expressed the relationship mathematically.

Charles' Law states that when the pressure is held constant, the volume of a fixed mass of ideal gas is in a direct proportion to the temperature in degrees Kelvin. Charles' Law can be written mathematically as follows:

$$V = k \times T \quad \text{or} \quad \frac{V}{T} = k$$

where V is the volume of the gas, T is the temperature in degrees Kelvin, and k is a constant that depends on the pressure and amount of gas. The direct relationship is clear by looking at the equations. If a sample of gas at a fixed pressure has its temperature *doubled*, the volume in turn is *doubled*. Conversely, decreasing the temperature by *one-half* brings about a decrease in volume by *one-half*.

The Law applies, for a given pressure and quantity of gas, at all sets of conditions. Thus for two sets of T and V, the following can be written:

$$\frac{V_1}{T_1} = \frac{V_2}{T_2} \quad \text{or} \quad V_1 T_2 = V_2 T_1 \quad \text{or} \quad \frac{V_1 T_2}{V_2 T_1} = 1$$

where at constant pressure, V_1 and T_1 refer to the set of conditions at the beginning of the experiment, and V_2 and T_2 refer to the set of conditions at the end of the experiment.

This experiment determines the volume of a sample of air when measured at two different temperatures with the pressure held constant.

Objectives

1. To measure the volume of a fixed quantity of air as the temperature changes at constant pressure.
2. To verify Charles' Law.

---

1. Use a clean and dry 250-mL Erlenmeyer flask (Flask no. 1). Fit the flask with a prepared stopper assembly, consisting of a no. 6 one-hole rubber stopper which has a 5 to 8 cm length of glass tubing inserted through the hole. If an assembly needs to be constructed, use the following procedure:

    a. Select a sharpened brass cork borer with a diameter that just allows the glass tubing to pass through it easily.

    b. Lubricate the outside of the cork borer with glycerine and push it through the rubber stopper from the bottom.

    c. Once the cork borer is through the stopper, pass the glass tubing through the cork borer so that the tubing is flush with the bottom of the stopper.

    d. Grasp the tubing and the stopper with one hand to hold these two pieces stationary; with the other hand carefully remove the borer. The glass tubing stays in the stopper. (Check to be certain that the end of the glass tubing is flush with the bottom of the rubber stopper.)

2. Mark the position of the bottom of the rubber stopper on Flask no. 1 with a marking pencil. Connect a 2-ft. piece of latex rubber tubing to the glass tubing.

3. Place 300 mL of water and three (3) boiling stones in an 800-mL beaker. Support the beaker on a ring stand using a ring support and wire gauze, and heat the water with a Bunsen burner to boiling (Fig. 11.1) (or place the beaker on a hot plate and heat to boiling). Keep the water at a gentle boil. Record the temperature of the boiling water on the Report Sheet (1).

Figure 11.1

Equipment to study Charles' Law.

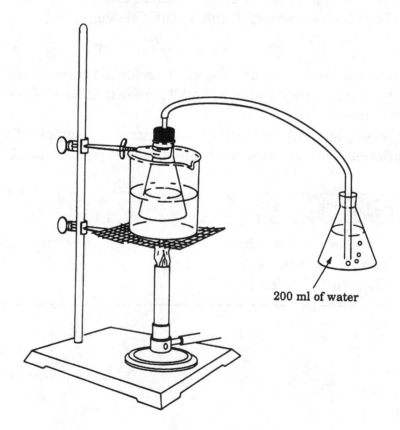

200 ml of water

4. Prepare an ice-water bath using a second 800-mL beaker half-filled with a mixture of ice and water. Record the temperature of the bath on the Report Sheet (3). Set aside for use in step no. 8.

5. Put about 200 mL of water into a second 250-mL Erlenmeyer flask (Flask no. 2) and place the end of the rubber tubing into the water. Make sure that the end of the rubber tubing reaches to the bottom of the flask and stays submerged at all times. (You may wish to hold it in place with a clamp attached to a ring stand.)

6. With a clamp holding the neck of Erlenmeyer Flask no. 1, lower the flask as far as it will go into the boiling water. Secure onto the ring stand (Fig. 11.1)

7. Boil gently for 5 min. Air bubbles should emerge from the rubber tubing submerged in Flask no. 2. Add water to the beaker if boiling causes the water level to go down.

8. When bubbles no longer emerge from the end of the submerged tubing (after 5 min.), carefully lift Flask no. 1 from the boiling water bath and quickly place it into the ice-water bath. Record what you observe happening as Flask no. 1 cools (2).

CAUTION!

The water, the glassware, and the ironware are hot. Be sure to keep the end of the rubber tubing always submerged in the water in Flask no. 2.

---

9. When no more water is drawn into Flask no. 1, raise the flask until the level of water inside the flask is at the same height as the water in the ice-water bath. Then remove the stopper from the Flask no. 1.

10. Take a graduated cylinder and measure the water in Flask no. 1. Record the volume to the nearest 0.1 mL on the Report Sheet (4).

11. Determine the volume of Erlenmeyer Flask no. 1 as follows:

    a. First, fill it with water to the level marked by the marking pencil. Insert the stopper with the glass tubing into the flask to be sure the bottom of the stopper touches the water with no air space present. Adjust the water level if necessary.

    b. Remove the stopper and measure the volume of the water in the flask by pouring it into a graduated cylinder. If a 100-mL graduated cylinder is used, it will be necessary to empty and refill it until all the water from Flask no. 1 has been measured.

    c. The *total* volume of water should be measured to the nearest 0.1 mL. Record this value on the Report Sheet (5).

12. Do the calculations to verify Charles' Law.

Chemicals and Equipment

1. Boiling stones
2. Bunsen burner (or hot plate)
3. 250-mL Erlenmeyer flasks (2)
4. 800-mL beakers (2)
5. Clamps
6. Glass tubing (6 to 8 cm length; 7-mm OD)
7. Marking pencil
8. One-hole rubber stopper (size no. 6)
9. Ring stand
10. Ring support
11. Rubber tubing (2-ft. length)
12. Thermometer, 110°C
13. Wire gauze

Experiment 11

PRE-LAB QUESTIONS

1. How are temperature and pressure related according to Charles' Law?

2. Convert the boiling point of water, 100°C, and the freezing point of water, 0°C, to degrees Kelvin.

3. Complete the following statements about Charles' Law:

    a. As the temperature decreases, the volume _____ .

    b. As the temperature _____ , the volume increases.

4. A gas is contained in a balloon with a volume of 35 L at a temperature of 30°C. What temperature must be reached in order for the balloon to expand to a volume of 50 L, at constant pressure?

Experiment 11

REPORT SHEET

1. Temperature, boiling water (T_2)   _____ °C _____ K

2. Observation as Flask no. 1 cools:

3. Temperature, ice water (T_1)   _____ °C _____ K

4. Volume of water sucked into Flask no. 1 (V_w)   _____ mL

5. Volume of Flask no. 1 (_____ + _____ + _____)   _____ mL

6. Volume of air at the temperature of boiling water (5)   _____ mL
    (V_2)

7. Volume of the air at the temperature of ice-water (V_1)   _____ mL
    ($V_1 = V_2 - V_w$)

8. Verify Charles' Law   _____

$$\frac{V_2 \times T_1}{V_1 \times T_2} =$$

9. Percent deviation from Charles' Law   _____ %

$$\frac{1.00 - (8)}{1.00} \times 100 =$$

POST-LAB QUESTIONS

1. A student assumed the volume of the 250-mL Erlenmeyer flask to be 250 mL without actually measuring the volume. How would this assumption affect the results?

2. Another student allowed all of the water in the beaker to boil away. What effect does this have on the temperature of the gas in the flask?

3. A gas occupies 4.50 L at room temperature, 20°C. If the gas is heated to 100°C, what will be the new volume? (Show your calculations.)

Experiment 12

Properties of gases: determination of the
molecular weight of a volatile liquid

Background

In the world in which we live, the gases with which we are familiar (for example O_2, N_2, CO_2, H_2) possess a molecular volume and show interactions between molecules. When cooled or compressed, these real gases eventually condense into the liquid phase. In the hypothetical world, there are hypothetical gases which we refer to as *ideal* gases. These gases are presumed to possess negligible volume. There are no attractions between molecules of ideal gases, and, as a result, molecules do not stick together upon collision. Such a hypothetical gas does not condense to form a liquid upon cooling. The ideal gas when cooled to a very low temperature should have a volume that is theoretically zero! This low temperature is termed *absolute zero* and is calculated to be −273.15 degrees below zero on the Celsius scale. The temperature that uses absolute zero as the zero point is the Kelvin scale.

The relationship that unites pressure, **P**, volume, **V**, and temperature, **T**, for a given quantity of gas, **n**, in the sample is the ideal gas equation:

$$\mathbf{P} \times \mathbf{V} = \mathbf{n} \times \mathbf{R} \times \mathbf{T}$$

This equation expresses pressure in atmospheres, volume in liters, temperature in degrees Kelvin, and the quantity of gas in moles. These four quantities are related exactly through the use of the ideal gas constant, **R**; the value for R is 0.0821 L atm/K mole. (Notice that the units for R are in terms of V, P, T, and n.)

A container of fixed volume at a given temperature and pressure holds only one possible quantity of gas. This quantity can be calculated by using the ideal gas equation:

$$\mathbf{n} = \frac{\mathbf{P} \times \mathbf{V}}{\mathbf{R} \times \mathbf{T}}$$

The number of moles, **n**, can be determined from the mass of the gas sample, **m**, and the molecular weight of the gas, **M**:

$$\mathbf{n} = \frac{\mathbf{m}}{\mathbf{M}}$$

Substituting this expression into the ideal gas equation gives

$$\mathbf{P} \times \mathbf{V} = \frac{\mathbf{m}}{\mathbf{M}} \times \mathbf{R} \times \mathbf{T}$$

Solving for the molecular weight, **M**:

$$\mathbf{M} = \frac{\mathbf{m} \times \mathbf{R} \times \mathbf{T}}{\mathbf{P} \times \mathbf{V}}$$

This equation gives us the means for determining the molecular weight of a gas sample from the measurements of mass, temperature, pressure, and volume. Most real gases at low pressures (1 atm or less) and high temperatures (300 K or more) behave as an ideal gas and thus, under such conditions, the ideal gas law is applicable for real gases as well.

In this experiment, a small quantity of a volatile liquid will be vaporized in a pre-weighed flask of known volume. Since the boiling point of the liquid will be below that of boiling water when the flask is submerged in a boiling water bath, the liquid will vaporize completely. The gas will drive out the air and fill the flask with the gaseous sample. The gas pressure in the flask is in equilibrium with atmospheric pressure and, therefore, can be determined from a barometer. The temperature of the gas is the same as the temperature of the boiling water. Cooling the flask condenses the vapor. The weight of the vapor can be measured by weighing the liquid, and M can be calculated.

EXAMPLE

A sample of an unknown liquid is vaporized in an Erlenmeyer flask with a volume of 250 mL. At 100°C the vapor exerts a pressure of 0.975 atm. The condensed vapor weighs 0.685 g. Determine the molecular weight of the unknown liquid.

Using the equation

$$M = \frac{m \times R \times T}{P \times V}$$

we can substitute all the known values:

$$M = \frac{(0.685 \text{ g}) (0.0821 \text{ L atm/mole K}) (373 \text{ K})}{(0.975 \text{ atm}) (0.250 \text{ L})} = 86.1 \text{ g/mole}$$

Hexane, with the molecular formula C_6H_{14}, has a molecular weight of 86.17 g/mole.

Objectives

1. Experimentally determine the mass of the vapor of a volatile liquid.
2. Calculate the molecular weight of the liquid by applying the ideal gas equation to the vapor.

Procedure

1. Obtain a sample of unknown liquid from your instructor. Record the unknown number on the Report Sheet (1). The unknown liquid will be one of the liquids found in Table 12.1.

Table 12.1	Volatile Liquid Unknowns		
Liquid	Formula	Molecular Weight	Boiling Point (°C at 1 atm)
Pentane	C_5H_{12}	72.2	36.2
Acetone	C_3H_6O	58.1	56.5
Methanol	CH_4O	32.0	64.7
Hexane	C_6H_{14}	86.2	69.0
Ethanol	C_2H_6O	46.1	78.5
2-Propanol	C_3H_8O	60.1	82.3

2. Weigh together a clean, dry 125-mL Erlenmeyer flask, a 2.5 by 2.5 in. square of aluminum foil, and a 4 in. piece of copper wire. Record the total weight to the nearest 0.001 g (2).

3. Pour approximately 3 mL of the liquid into the flask. Cover the mouth of the flask with the aluminum foil square and crimp the edges tightly over the neck of the flask. Secure the foil by wrapping the wire around the neck and twisting the ends by hand.

4. With a larger square of aluminum foil (3 × 3 in.), secure a second cover on the mouth of the flask with a rubber band.

5. Carefully punch a *single, small* hole in the foil covers with a needle or pin. The assembly is now prepared for heating (Fig. 12.1).

Figure 12.1

Assembly for vaporization of the liquid.

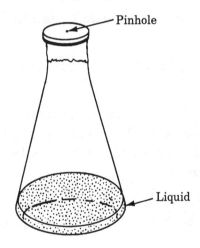

Pinhole

Liquid

6. Take a 1000-mL beaker and add 300 mL of water and a few boiling chips. Heat the water to boiling using a hot plate. Regulate the heat so the boiling does not cause the water to splash.

CAUTION!

Use a hot plate and *not* a Bunsen burner, since the liquids listed in Table 12.1 are flammable.

7. Immerse the flask containing the volatile unknown liquid in the boiling water so that most of the flask is beneath the hot water as shown in Fig. 12.2. (You may need to weigh down the flask with a lead sinker.)

Figure 12.2

Assembly for the determination of molecular weight.

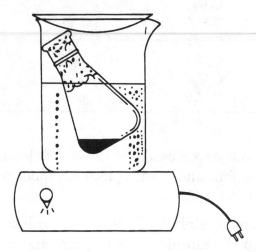

8. Observe the unknown liquid. There is more liquid than is required to fill the flask with vapor. As the liquid evaporates, the level will decrease and excess vapor will escape through the pin hole. When all the liquid appears to be gone, continue to heat for an additional 5 min.

9. Record the temperature of the boiling water (5). Record the temperature of the vapor in the flask (6).

10. Using tongs, carefully remove the flask from the water and set it on the laboratory bench.

11. While the flask cools to room temperature, record the barometric pressure in the laboratory (7).

12. When the flask has cooled to room temperature, wipe dry the outside of the flask with a paper towel. Carefully remove only the second foil cover; blot dry the first foil cover with a paper towel. Take a look inside the flask. Are there droplets of liquid present?

13. Weigh the flask, foil cover, wire, and the condensed liquid. Record to the nearest 0.001 g on the Report Sheet (3). Determine the weight of the condensed liquid (4).

14. Remove the cover from the flask; do not discard the aluminum foil and wire. Rinse the flask with water and refill with water to the rim. Weigh the flask with the water, foil, and wire to the nearest 0.01 g (8).

15. Determine the weight of the water (9). Determine the temperature of the water in the flask (10). Look up the density of water at that temperature, and, using this value, calculate the volume of water in the flask (11). (You may check the volume in the flask by measuring the water with a graduated cylinder.)

16. Calculate the molecular weight of the unknown (12). Identify the unknown (13).

Chemicals and Equipment

1. Aluminum foil
2. Tongs
3. Hot plate
4. Copper wire
5. Rubber bands
6. Boiling chips
7. Absorbent paper towels
8. Unknown liquid chosen from Table 12.1
9. Lead sinkers

NAME _____   SECTION _____   DATE _____

PARTNER _____   GRADE _____

Experiment 12

PRE-LAB QUESTIONS

1. Why would an ideal gas have zero volume at zero degrees Kelvin?

2. The universal gas constant, R, is expressed in terms of what units?

3. An unknown liquid with a weight of 11.5 g fills a flask of 5.0 L at a temperature of 108°C and a pressure of 1.0 atm. What is the molecular weight of the unknown? Using Table 12.1, identify the liquid.

Experiment 12

REPORT SHEET

1. Unknown number   _____

2. Weight of 125-mL Erlenmeyer flask,
   aluminum foil, and copper wire   _____ g

3. Weight of cooled flask, foil, wire, and
   condensed liquid   _____ g

4. Weight of condensed liquid: (3) − (2)   _____ g

5. Temperature of boiling water   _____ °C _____ K

6. Temperature of vapor in flask   _____ °C _____ K

7. Barometric pressure   _____ atm

8. Weight of 125-mL Erlenmeyer flask,
   aluminum foil, copper wire, and water   _____ g

9. Weight of water: (8) − (2)   _____ g

10. Temperature of water   _____ °C

11. Volume of the flask:
    (9)/(density of water)   _____ mL _____ L

12. Molecular Weight of the unknown:
    $$M = \frac{(4) \times (0.0821 \text{ L atm/K mole}) \times (6)}{(7) \times (11)}$$ _____ g/mole

13. The unknown is   _____

POST-LAB QUESTIONS

1. If 1.710 g of a gas occupies a 500-mL flask at 30°C and 750 mm Hg of pressure, what is the molecular weight of the gas? (Remember to use proper units: V in liters, T in degrees Kelvin, P in atmospheres.)

2. Can a real gas ever behave as an ideal gas?

3. Instead of measuring the volume of the Erlenmeyer flask as directed in step no. 15, a student recorded the volume written on the flask. Explain how this would affect the calculation for the molecular weight.

4. Explain how each of the following procedural errors would affect the results in the experiment.

a. The heating was stopped before all of the liquid had evaporated (step no. 8).

b. Water remained on the outside of the flask and on the foil when the final weighing was determined (step no. 13).

Physical properties of chemicals: melting point, sublimation, and boiling point

Background

If you were asked to describe a friend, most likely you would start by identifying particular physical characteristics. You might begin by giving your friend's height, weight, hair color, eye color, or facial features. These characteristics would allow you to single out the individual from a group.

Chemicals also possess distinguishing physical properties which enable their identification. In many circumstances, a thorough determination of the physical properties of a given chemical can be used for its identification. If faced with an unknown sample, a chemist may compare the physical properties of the unknown to properties of known substances that are tabulated in the chemical literature; if a match can be made, an identification can be assumed (unless chemical evidence suggests otherwise).

The physical properties most commonly listed in handbooks of chemical data are color, crystal form (if a solid), refractive index (if a liquid), density (discussed in Experiment 3), solubility in various solvents (discussed in Experiment 15), melting point, sublimation characteristics, and boiling point. When a new compound is isolated or synthesized, these properties almost always accompany the report in the literature.

The transition of a substance from a solid to a liquid to a gas, and the reversal, represent physical changes. This means that there is a change in the form or the state of the substance without any alteration in the chemical composition. Water undergoes state changes from ice to liquid water to steam; however, the composition of molecules in all three states remains H_2O.

$$H_2O(s) \rightleftharpoons H_2O(l) \rightleftharpoons H_2O(g)$$

Ice   Liquid   Steam

The *melting* or *freezing point* of a substance refers to the temperature at which the solid and liquid states are in equilibrium. The terms are interchangeable and correspond to the same temperature; how the terms are applied depends upon the state the substance is in originally. The melting point is the temperature at equilibrium when starting in the solid state and going to the liquid state. The freezing point is the temperature at equilibrium when starting in the liquid state and going to the solid state.

Melting points of pure substances occur over a very narrow range and are usually quite sharp. The criteria for purity of a solid is the narrowness of the melting point range and the correspondence to the value found in the literature. Impurities will lower the melting point and cause a broadening of the range. For example, pure benzoic acid has a reported melting point of 122.13°C; benzoic acid with a melting point range of 121–122°C is considered to be quite pure.

The *boiling point* or *condensation point* of a liquid refers to the temperature when its vapor pressure is equal to the external pressure. If a beaker of liquid is brought to a boil in your laboratory, bubbles of vapor form throughout the liquid. These bubbles rise rapidly to the surface, burst, and release vapor to the space above the liquid. In this case, the liquid is in contact with the atmosphere; the normal boiling point of the liquid will be the temperature when the pressure of the vapor is equal to the atmospheric pressure (1 atm or 760 mm Hg). Should the external pressure vary, so will the boiling point. A liquid will boil at a higher temperature when the external pressure is higher and will boil at a lower temperature when the external pressure is reduced. The change in state from a gas to a liquid represents condensation and is the reverse of boiling. The temperature for this change of state is the same as the boiling temperature but is concerned with the approach from the gas phase.

Just as a solid has a characteristic melting point, a liquid has a characteristic boiling point. At one atmosphere, pure water boils at 100°C, pure ethanol (ethyl alcohol) boils at 78.5°C and pure diethyl ether boils at 34.6°C. The vapor pressure curves shown in Fig. 13.1 illustrate the variation of the vapor pressure of these liquids with temperature. One can use these curves to predict the boiling point at a reduced pressure. For example, diethyl ether has a vapor pressure of 422 mm Hg at 20°C. If the external pressure were reduced to 422 mm Hg, diethyl ether would boil at 20°C.

Sublimation is a process that involves the direct conversion of a solid to a gas and its reversal gas to a solid, without passing through the liquid state, when heated. Relatively few solids do this. Some examples are the solid compounds naphthalene (moth balls), caffeine, iodine, and solid carbon dioxide (commercial Dry Ice). Sublimation temperatures are not as easily obtained as melting points or boiling points.

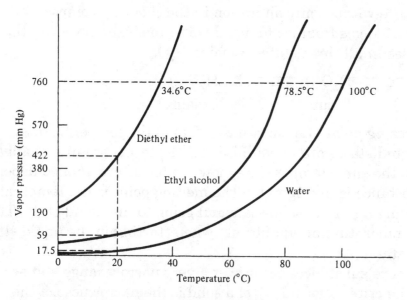

Figure 13.1 • Diethyl ether, ethyl alcohol (ethanol), and water vapor pressure curves.

Objectives

1. To use melting points and boiling points in identifying substances.
2. To use sublimation as a means of purification.

Procedure

Melting Point Determination

1. Unknowns are provided by the instructor. Obtain approximately 0.1 g of unknown solid and place it on a small watch glass. Record the number of the unknown on the Report Sheet (1). (The instructor will weigh out a 0.1 g sample; take approximately that amount with your spatula.) Carefully crush the solid into a powder with the flat portion of a spatula.

2. Obtain a melting point capillary tube. One end of the tube will be sealed. The tube is packed with solid in the following way:

   Step A   Press the open end of the capillary tube vertically into the solid sample (Fig. 13.2 A). A small amount of sample will be forced into the open end of the capillary tube.

   Step B   Invert the capillary tube so that the closed end is pointing toward the bench top. Gently tap the end of the tube against the lab bench top (Fig. 13.2 B). Continue tapping until the solid is forced down to the closed end. A sample depth of 5–10 mm is sufficient.

Figure 13.2

Packing a capillary tube.

Step A

A. Forcing solid into the capillary tube.

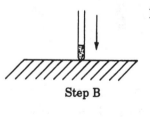

Step B

B. Tapping to force down solid.

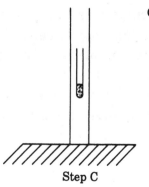

Step C

C. Alternative method for bringing the solid down.

Step C   An alternative method for bringing the solid sample to the closed end uses a piece of glass tubing of approximately 20 to 30 cm. Hold the capillary tube, closed end down, at the top of the glass tubing, held vertically; let the capillary tube drop through the tubing so that it hits the lab bench top. The capillary tube will bounce and bring the solid down. Repeat if necessary (Fig. 13.2 C).

3. The melting point may be determined using either a commercial melting point apparatus or a Thiele tube.

   a. A commercial melting point apparatus will be demonstrated by your instructor.

   b. The use of the Thiele tube is as follows:
   • Attach the melting point capillary tube to the thermometer by means of a rubber ring. Align the mercury bulb of the thermometer so that the tip of the melting point capillary containing the solid is next to it (Fig. 13.3).

Figure 13.3

Proper alignment of the capillary tube and the mercury bulb.

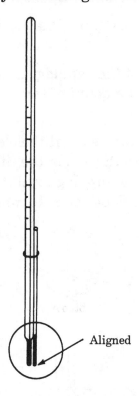

Aligned

   • Use an extension clamp to support the Thiele tube on a ring stand. Add mineral oil or silicone oil to the Thiele tube, and fill to a level above the top of the side arm. Use a thermometer clamp to support the thermometer with the attached melting point capillary tube in the oil. The bulb and capillary tube should be immersed in the oil; keep the rubber ring and open end of the capillary tube out of the oil (Fig. 13.4).
   • Heat the arm of the Thiele tube very slowly with a Bunsen burner flame. Use a small flame and gently move the burner along the arm of the Thiele tube.

- You should position yourself so that you can follow the rise of the mercury in the thermometer as well as observe the solid in the capillary tube. Record the temperature when the solid begins to liquefy (2) (the solid will appear to shrink). Record the temperature when the solid is completely a liquid (3). Express these readings as a melting point range (4).
- Identify the solid by comparing the melting point with the solids listed in Table 13.1 (5).

Figure 13.4

Thiele tube apparatus.

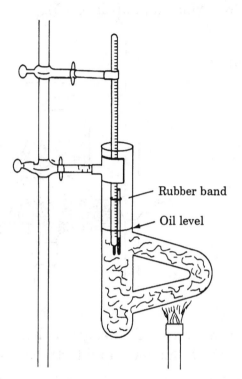

Table **13.1**	Melting Point of Selected Solids
Solids	**Melting Point (°C)**
Acetamide	82
Acetanilide	114
Benzophenone	48
Benzoic acid	122
Biphenyl	70
Lauric acid	43
Naphthalene	80
Stearic acid	70

4. Do as many melting point determinations as your instructor may require. Just remember to use a new melting point capillary tube for each melting point determination.

5. Dispose of the solids as directed by your instructor.

Purification of Naphthalene by Sublimation

1. Place approximately 0.5 g of impure naphthalene into a 100-mL beaker. (Your instructor will weigh out 0.5 g of sample; with a spatula take an amount which approximates this quantity.)

2. Into the 100-mL beaker place a smaller 50-mL beaker. Fill the smaller one halfway with ice cubes or ice chips. Place the assembled beakers on a wire gauze supported by a ring clamp (Fig. 13.5).

Figure 13.5

Set-up for sublimation of naphthalene.

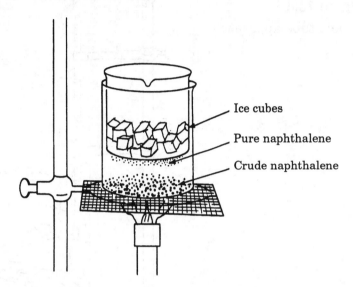

Ice cubes

Pure naphthalene

Crude naphthalene

3. Using a small Bunsen burner flame, gently heat the bottom of the 100-mL beaker by passing the flame back and forth beneath the beaker.

4. You will see solid flakes of naphthalene collect on the bottom of the 50-mL beaker. When a sufficient amount of solid has been collected, turn off the burner.

5. Pour off the ice water from the 50-mL beaker and carefully scrape the flakes of naphthalene onto a piece of filter paper with a spatula.

6. Take the melting point of the pure naphthalene and compare it to the value listed in Table 13.1 (6).

7. Dispose of the crude and pure naphthalene as directed by your instructor.

Boiling Point Determination

> **CAUTION!**
>
> The chemicals used for boiling point determinations are flammable. Be sure all Bunsen burner flames are extinguished before starting this part of the experiment.

1. Obtain an unknown liquid from your instructor and record its number on the Report Sheet (7).

2. Clamp a clean, dry test tube (100 × 13 mm) onto a ring stand. Add to the test tube approximately 3 mL of the unknown liquid and two small boiling chips. Lower the test tube into a 250-mL beaker which contains 100 mL of water and two boiling chips. Adjust the depth of the test tube so that the unknown liquid is below the water level of the water bath (Fig. 13.6).

Figure 13.6

Set-up for determining the boiling point.

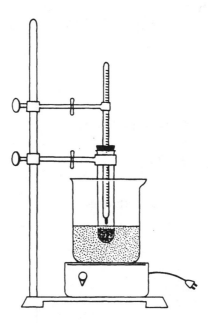

3. With a thermometer clamp, securely clamp a thermometer and lower it into the test tube through a neoprene adapter. Adjust the thermometer so that it is approximately 1 cm above the surface of the unknown liquid.

4. A piece of aluminum foil can be used to cover the mouth of the test tube. (Be certain that the test tube mouth has an opening; the system should not be closed.)

5. Gradually heat the water in the beaker with a hot plate and watch for changes in temperature. As the liquid begins to boil, the temperature above the liquid will rise. When the temperature no longer rises but remains constant, record the temperature to the nearest 0.1°C (8). This is the observed boiling point. From the list in Table 13.2, identify your unknown liquid by matching your observed boiling point with the compound whose boiling point best corresponds (9).

Table **13.2** Boiling Points of Selected Liquids	
Liquid	**Boiling Point (°C at 1 atm)**
Acetone	56
Cyclohexane	81
Ethyl acetate	77
Hexane	69
2-Methyl-1-propanol (Isopropyl alcohol)	83
Methanol (Methyl alcohol)	65
1-Propanol	97

6. Do as many boiling point determinations as required by your instructor.

7. Dispose of the liquid as directed by your instructor.

Chemicals and Equipment

1. Aluminum foil
2. Boiling chips
3. Bunsen burner
4. Hot plate
5. Commercial melting point apparatus (if available)
6. Melting point capillary tubes
7. Rubber rings
8. Thiele tube melting point apparatus
9. Thermometer clamp
10. Glass tubing
11. Acetamide
12. Acetanilide
13. Acetone
14. Benzophenone
15. Benzoic acid
16. Biphenyl
17. Cyclohexane
18. Ethyl acetate
19. Hexane
20. 2-Methyl-1-propanol (Isopropyl alcohol)
21. Lauric acid
22. Methanol (Methyl alcohol)
23. Naphthalene (pure)
24. Naphthalene (impure)
25. 1-Propanol
26. Stearic acid

Experiment 13

PRE-LAB QUESTIONS

1. For a given substance, are the melting point of the solid and the freezing point of the liquid the same temperature? Explain.

2. Refer to Fig. 13.1. Estimate the boiling point of ethanol (ethyl alcohol) at 570 mm Hg.

3. How does one characterize a solid?

4. Water at the summit of Mt. Everest boils at a lower temperature than at sea level. Explain why this is observed.

Experiment 13

REPORT SHEET

Melting point determination

	Trial No. 1	Trial No. 2
1. Unknown number	_____	_____
2. Temperature melting begins	_____ °C	_____ °C
3. Temperature melting ends	_____ °C	_____ °C
4. Melting point range	_____ °C	_____ °C
5. Identification of unknown	_____	_____

Purification of naphthalene by sublimation

6. Melting point range	_____ °C	_____ °C

Boiling point determination

7. Unknown number	_____	_____
8. Observed boiling point	_____ °C	_____ °C
9. Identification of unknown	_____	_____

POST-LAB QUESTIONS

1. A student did a melting point determination for a sample of benzoic acid and found a melting point of 113–120°C. What conclusion can the student draw about the sample?

2. A student in New York City carried out a boiling point determination for cyclohexane (b.p. 81°C) according to the procedure in this laboratory manual. Will this student's observed boiling point be the same as the value obtained by another student at sea level in London, England? Will it be lower or higher? Explain your conclusion.

3. A bottle containing solid iodine crystals fills with a violet haze upon standing at room temperature. What is the violet haze and why does it form?

4. Cocaine is a white solid which melts at 98°C when pure. A forensic chemist working for the New York Police Department has a white solid believed to be cocaine. What can the chemist do to determine quickly whether the sample is cocaine and whether it is pure or a mixture?

Entropy: a measure of disorder

Background

According to the kinetic theory of gases, molecules move randomly, and the higher the temperature, the faster they move. Random motion, in general, means disorder, and as molecules slow down by cooling, more and more order is apparent. When all the molecules of a system become motionless and line up perfectly, we achieve the greatest possible order. There is a measure of such order, called **entropy**. At perfect order, the entropy of the system is zero. When the system has the slightest impurity, the perfect order is broken and the entropy increases. When molecules rotate or move from one place to the next, the disorder increases and so does the entropy. Thus when a crystal melts, one observes an increase in entropy; when a liquid vaporizes, the disorder, hence the entropy, increases. When one combines two pure substances and a mixture is formed, disorder increases and so does the entropy. When the temperature increases, the entropy of a system increases—not just in the gas phase but also in most liquids and solids. This is because an increase in temperature means increased molecular motions. Conversely, when a system is cooled, its entropy mostly decreases. We have learned of the absolute (Kelvin) temperature scale. Absolute zero or −273.15°C is the lowest possible temperature. At this temperature, all molecular motions cease and perfect order reigns. No such perfect order can be reached in reality. But scientists try to get close to systems with very little entropy and have reached temperatures of 0.001 K.

Entropy can also be perceived as part of the energy (heat) that is used up in creating disorder and, therefore, is not available to do useful work. The symbol of entropy is **S**, and the change in entropy going from one state of a system to another is

$$\Delta S = Q/T$$

where Δ is the change, **Q** is the energy (heat) used to create disorder, and **T** is the temperature in Kelvin. Thus the unit of entropy is measured in cal/degree K.

The demonstration of change in order/disorder, that is, a change in entropy, is present in everyday life. For example, the liquid mercury in our thermometer is made of small compact molecules (Fig. 14.1). When it is heated the molecules move faster; they push their neighboring molecules away, the volume of the liquid expands and the disorder, hence the entropy, increases. When the thermometer is cooled the opposite happens. Since there is relatively little interaction between the mercury molecules, the process is completely reversible. Not every material behaves this way.

Figure 14.1

Schematic diagrams of the position of molecules in a liquid (A) before heating and (B) after heating.

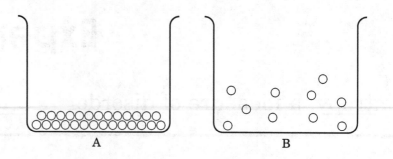

In a plastic material, such as a polypropylene wrapping sheet, giant molecules or polymers are entangled with each other in a random fashion. When we hang a weight on such a sheet, we exert a stress and as a consequence the sheet will elongate slightly. During elongation, the giant molecules align themselves along the stress, in essence, decreasing the entropy and increasing the order (Fig. 14.2 A, B). When such a stressed sheet is heated, the molecular motions increase, just as in mercury, but because of the entanglements, first only the segments of the molecules increase their motion, and only later can we see that whole giant molecules move away from each other. In each case the entropy increases (Fig. 14.2 B, C). When we cool such an amorphous system, the molecular motions decrease and so does the entropy. However, because of the entanglements, upon cooling we do not get back exactly the same disorder as before; we say the heating-cooling is not reversible (Fig. 14.2 C, D).

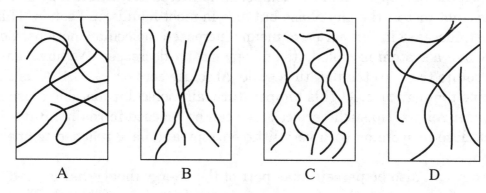

Figure 14.2 • Schematic diagram of molecular segment alignments in a plastic sheet: (A) at rest; (B) under load; (C) under load and heating; (D) under load and cooling.

In rubber-like materials giant molecules are cross-linked, that is, certain interatomic distances are fixed (Fig. 14.3 A). When we hang a load on a rubber band, it will elongate. The molecular segments will align in the direction of the stress. This creates greater order, a decrease in entropy. The elongation of the rubber band is reflected in the increase in the distance between the neighboring cross links (Fig. 14.3 B). When such a system is heated, segments of the molecules between crosslinks move more vigorously than before. Such an increase in segmental motions brings the crosslinks closer together, in essence, the rubber contracts, its entropy decreases (Fig. 14.3 C). When the rubber is cooled again, it will extend to its previous length, an increase in relative disorder and, hence, in entropy (Fig. 14.3 D). This heating and cooling cycle is reversible, as long as we do not break chemical bonds.

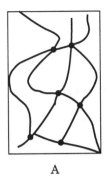 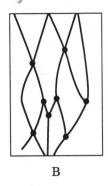 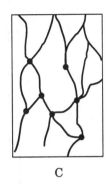 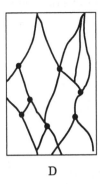

A   B   C   D

Figure 14.3 • Schematic diagram of the segmental movements in rubber: (A) at rest; (B) under load; (C) under load and heating; (D) under load and cooling.

In the following experiments, the changes in the entropy of the systems under investigation will be observed by noting the changes in the behavior of the systems upon heating and cooling.

Objectives

1. Demonstrations on the effect of entropy changes.
2. Investigation of the entropy changes in different systems.

Procedure

1. Take a mercury thermometer and read it at room temperature. Next, boil 50 mL of water in a 125-mL beaker and note the temperature of the water by immersing just the tip of the thermometer in the liquid. Remove the thermometer and let it cool to room temperature. Read the thermometer. Repeat the cycle two more times. Report your observations on the Report Sheet (1).

2. Take a 10 × 2 cm strip of polypropylene sheet. Make two marks 5.0 cm apart with a marker pen in the middle of the strip. Fold about 5 mm of the sheet at one end and place it in a bulldog clip. Make sure the clip holds the strip firmly. Repeat the procedure at the other end of the strip with another bulldog clip. With a paper clip unfolded into an "S" hook, hang the strip assembly on a ring stand. Measure the distance between the two marks with a ruler and record it to the nearest mm on your Report Sheet (2a). With the aid of another paper clip unfolded into an "S" hook, hang a weight of approximately 300 g on the bottom of the strip (Fig. 14.4). Wait a few minutes to allow the elongation to stop. Measure the distance between the two marks to the nearest mm and record it on your Report Sheet (2b).

Figure 14.4

Assembly of elongated
polypropylene sheet.

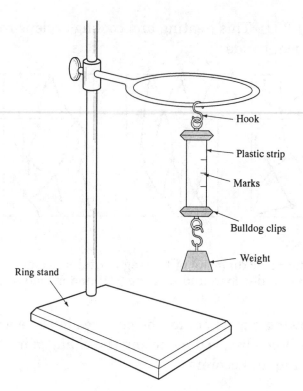

Hook

Plastic strip

Marks

Bulldog clips

Weight

Ring stand

Take a heat gun and turn it on. **Carefully** direct the heat onto the plastic sheet. When the strip starts to elongate, quickly turn the heat gun away and turn it off. (If you heat the strip too much, the plastic will break under the load, and you will need to repeat the experiment.) Allow the strip assembly to come to room temperature. Measure the distance between the marks to the nearest mm and record it on your Report Sheet (2c). Remove the weight from the bottom of the plastic strip. Measure the distance between the marks and record it on your Report Sheet (2d).

3. Take a rubber band that is 3 mm wide and approximately 90 mm long. With the aid of two unfolded paper clips (made into "S" hooks), hang it on a ring stand. Read the length of the rubber band (the distance between the two paper clips) with a ruler and record it to the nearest mm on your Report Sheet (3a). Hang a weight of approximately 300 g on the lower paper clip and allow the rubber band to elongate. Measure the length of your rubber band under the load and record it to the nearest mm on your Report Sheet (3b). Turn on your heat gun. Direct it to the middle of the rubber band and heat both sides. Turn off the heat gun. While the rubber band is still warm, quickly measure its length with a ruler. Record it to the nearest mm on your Report Sheet (3c). Allow a few minutes for the rubber band to come to room temperature. Measure and record the length of the rubber band to the nearest mm on your Report Sheet (3d). Remove the weight from the assembly. Measure and record the length of the rubber band to the nearest mm on your Report Sheet (3e). Repeat the cycle once more and record the corresponding lengths to the nearest mm on your Report Sheet (3f, 3g, 3h, and 3i).

4. Read this entire section of the Procedure before carrying out the experiment so you understand what to do. *The sensation of warmer or cooler temperature can be ascertained if you do the following experiment **rapidly***. Take a new rubber band. While holding it with your fingers between your two hands, let the rubber band touch your upper lip. Move the band away from your lip. Quickly extend the rubber band to about twice its length and touch your upper lip with the extended rubber band. Does the rubber band feel warmer or cooler upon extension? Record your observation on your Report Sheet (4a). Move the rubber band away from your lip. Allow it to contract to its original length. Touch it to your upper lip. Does the contracted (relaxed) rubber band feel warmer or cooler than the extended rubber band? Record your observation on your Report Sheet (4b). Repeat the procedure. Report the results for the second cycle on your Report Sheet (4c, 4d).

Chemicals and Equipment

1. Polypropylene sheet;
   10 × 2 cm strips
2. Rubber bands,
   (3 mm wide × 90 mm long)
3. Ruler (metric)
4. Bulldog clips
5. Paper clips
6. Thermometer
7. Heat gun
8. Weights (300 g)

Experiment 14

PRE-LAB QUESTIONS

1. Which has greater order—liquid mercury at room temperature or at 100°C?

2. Assume that upon elongation of a plastic sheet at 25°C, 2 cal heat was released, i.e. $Q = -2$ cal. Calculate the entropy change, ΔS, in the process. Does the entropy change indicate a greater order or disorder after elongation?

3. Two liquids, alcohol and water, are mixed. Does the resulting mixture have greater or lesser order than the two pure liquids? Does the entropy increase or decrease upon mixing?

Experiment 14

REPORT SHEET

1. **(a)** Reading of thermometer at room temperature _____ °C

   Reading the temperature of boiling water _____ °C

   (b) Reading the room temperature second time _____ °C

   Temperature of boiling water second time _____ °C

   (c) Reading the room temperature third time _____ °C

   Temperature of boiling water third time _____ °C

2. **(a)** Distance between the marks initially _____ mm

   (b) Distance between the marks under load _____ mm

   (c) Distance between the marks after heating
   and cooling under load _____ mm

   (d) Distance between the markings after heating
   and cooling with no load _____ mm

3. **(a)** Length of the rubber band with no load _____ mm

   (b) Length of the rubber band under load _____ mm

   (c) Length of the rubber band after
   heating under load _____ mm

   (d) Length of the rubber band after cooling to room
   temperature under load _____ mm

   (e) Length of the rubber band at room temperature
   after removal of the load _____ mm

(f) Second cycle: length under load _____ mm

(g) Second cycle: length after heating under load _____ mm

(h) Second cycle: length after cooling under load _____ mm

(i) Second cycle: length after cooling at room temperature after removal of the load _____ mm

(j) Are the lengths in (3a), (3e), and (3i) equal?

(k) Are the lengths in (3b) and (3f) equal?

(l) Are the lengths in (3c) and (3g) equal?

(m) Are the lengths in (3d) and (3h) equal?

4. (a) Does the rubber band feel warmer or cooler upon extension?

(b) Does the rubber band feel warmer or cooler upon contraction?

(c) Second cycle: Does the rubber band feel warmer or cooler upon extension?

(d) Second cycle: Does the rubber band feel warmer or cooler upon contraction?

POST-LAB QUESTIONS

1. Were the changes in the expansion of mercury in the thermometer reproducible? Were they reversible?

2. In the experiment on the plastic, polypropylene sheet, what happened to the order/disorder and, hence, to the entropy:
   (a) by putting a weight (load) on the strip;

   (b) by heating the strip under load;

   (c) by cooling the strip under load?

3. In the experiment on the rubber band, what happened to the order/disorder and, hence, to the entropy:
   (a) by putting a weight (load) on the rubber band;

   (b) by heating it under load;

(c) by cooling it to room temperature under load;

(d) by removing the load?

4. Were the measurements with the rubber band reproducible?

5. Judging from the temperature sensation on your upper lip, did the extension of the rubber band absorb or release heat?

Did it create order or disorder?

Did the entropy of the rubber band increase or decrease as a result of extension?

● Solubility and solution

Most materials encountered in every day life are *mixtures*. This means that more than one component is found together in a system. Think back to your morning breakfast beverage; orange juice, coffee, tea, and milk are examples of mixtures.

Some mixtures have special characteristics. A mixture that is uniform throughout, with no phase boundaries, is called a *homogeneous mixture*. If you were to sample any part of the system, the same components in the same proportions would be found in each sample. The most familiar of these homogeneous mixtures is the liquid *solution*; here a *solute* (either a solid or a liquid) is thoroughly and uniformly dispersed into a *solvent* (a liquid). If the solution were allowed to remain standing, the components would not separate, no matter how much time was allowed to pass.

There are limits as to how much solute may be dispersed or dissolved in a given amount of solvent. This limit is the *solubility* and is defined as the *maximum weight of solute that dissolves in 100 g of a given solvent at a given temperature*. For example, sucrose (or table sugar) is soluble to the extent of 203.9 g per 100 g of water at 20°C. This means that if you have 100 g of water, you can dissolve up to 203.9 g of table sugar, but no more, in that quantity of water at 20°C. If more is added, the extra amount sinks to the bottom undissolved. A solution in this state is referred to as *saturated*. A solution with less than the maximum at the same temperature is called *unsaturated*. Solubility also varies with temperature (Fig. 15.1).

Liquids dissolved in liquids similarly may form homogeneous solutions. Some liquids have limited solubility in water. Diethyl ether, $CH_3CH_2OCH_2CH_3$, (an organic liquid) is soluble to the extent of 4 g per 100 g of water at 25°C; an excess of the diethyl ether will result in a separation of phases with the less dense organic liquid floating on the water. Some liquids mix in all proportions; these liquids are completely *miscible*. The mixture of commercial antifreeze, ethylene glycol, $HOCH_2CH_2OH$, and water, used as a coolant in automobile radiators, is such a solution.

The solubility of a given solute in a particular solvent depends on a number of factors. One generalization which can be used for determining solubility is "like dissolves like." This means that the more similar a solute is to the polarity of the solvent, the more likely the two will form a homogeneous solution. A polar solvent, such as water, will dissolve a polar compound: an ionic salt like common table salt, NaCl, will dissolve in water; a polar covalent solid like table sugar, sucrose will dissolve in water. Nonpolar solvents such as naphtha or turpentine will dissolve nonpolar material, such as grease or oil. On the other hand, oil and water do not mix because of their different polar characteristics.

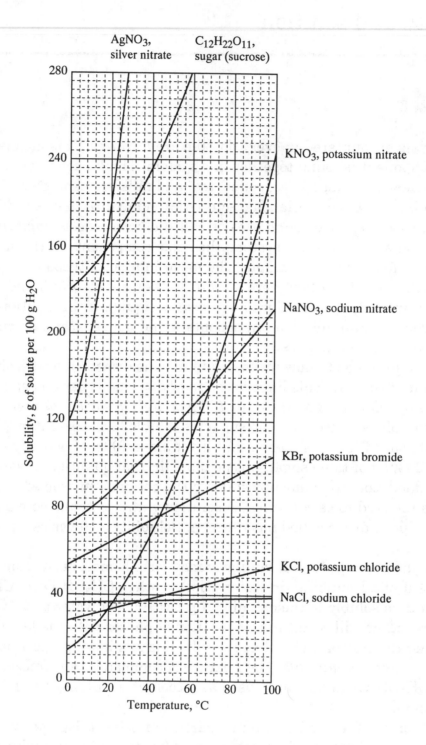

Figure 15.1

The effect of temperature on the solubility of some solutes in water.

When ionic salts dissolve in water, the individual ions separate. These positively and negatively charged particles in the water medium are mobile and can move from one part of a solution to another. Because of this movement, solutions of ions can conduct electricity. *Electrolytes* are substances which can form ions when dissolved in water and can conduct an electric current. These substances are also capable of conducting an electric current in the molten state. *Nonelectrolytes* are substances which do not conduct an electric current. Electrolytes may be further characterized as either strong or weak. A strong electrolyte dissociates almost completely when in a water solution; it is a good conductor of electricity. A weak electrolyte has only a small fraction of its particles dissociated into ions in water; it is a poor conductor of electricity. Table 15.1 lists examples of compounds behaving as electrolytes or nonelectrolytes in a water solution.

Table 15.1	Selected Electrolytes and Nonelectrolytes	
Strong Electrolytes	**Weak Electrolytes**	**Nonelectrolytes**
Sodium chloride, $NaCl$	Acetic acid, CH_3CO_2H	Methanol, CH_3OH
Sulfuric acid, H_2SO_4	Carbonic acid, H_2CO_3	Benzene, C_6H_6
Hydrochloric acid, HCl	Ammonia, NH_3	Acetone, $(CH_3)_2CO$
Sodium hydroxide, $NaOH$		Sucrose, $C_{12}H_{22}O_{11}$

Objectives

1. To show how temperature affects solubility.
2. To demonstrate the difference between electrolytes and nonelectrolytes.
3. To show how the nature of the solute and the solvent affects solubility.

Procedure

Saturated Solutions

1. Place 10 mL of distilled water into a 50-mL beaker; record the temperature of the water (1).

2. While stirring with a glass rod, add sucrose to the water in 2 g portions; keep adding until no more sucrose dissolves. The solution should be saturated. Record the mass of sucrose added (2).

3. Heat the solution on a hot plate to 50°C; maintain this temperature. Again add to the solution, while stirring, sucrose in 2 g portions until no more sucrose dissolves. Record the mass of sucrose added (3).

4. Heat the solution above 50°C until all of the solid dissolves. With beaker tongs, remove the beaker from the hot plate and set it on the bench-top, out of the way. Place an applicator stick, or suspend a string, into the solution and allow the solution to cool. Continue with the next part of this experiment and return to this part after the solution cools to room temperature.

5. Observe what happened to the solution when it cooled to room temperature. Offer an explanation for what has taken place (5). (If no crystals have formed, drop into the solution a single crystal of sucrose or stir the solution with a stirring rod.)

Electrical Conductivity

This part of the experiment can be done in pairs. Obtain and set-up a conductivity apparatus (Fig. 15.2). It consists of two terminals connected to a light bulb and a plug for connection to a 110-volt electrical wall outlet.

Figure 15.2
Conductivity apparatus.

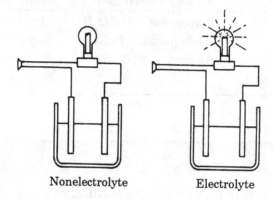

Nonelectrolyte   Electrolyte

CAUTION!

To avoid a shock, do not touch the terminals when the apparatus is plugged in. Be sure to unplug the apparatus between tests and while rinsing and drying. Do not let the terminals touch each other.

The following solutions are to be tested with the conductivity apparatus:
- **a.** distilled water
- **b.** tap water
- **c.** 1 M NaCl
- **d.** 0.1 M NaCl
- **e.** 1 M sucrose, $C_{12}H_{22}O_{11}$
- **f.** 0.1 M sucrose, $C_{12}H_{22}O_{11}$
- **g.** 1 M HCl
- **h.** 0.1 M HCl
- **i.** glacial acetic acid, CH_3CO_2H
- **j.** 0.1 M acetic acid, CH_3CO_2H

1. For each solution follow steps 2, 3, 4, and 5.

2. Place about 20 mL of the solution to be tested into a 50-mL beaker that has been rinsed with distilled water. A convenient way to rinse the beaker is with a squeezable plastic wash bottle. Direct a stream of water from the wash bottle into the beaker, swirl the water about, and discard the water into the sink.

3. Lower the terminals into the beaker so that the solution covers the terminals. For each test solution, try to keep the same distance between the terminals and the terminals submerged to the same depth.

4. Plug the apparatus into the wall socket. Observe the effect on the light bulb. A solution containing an electrolyte conducts electricity—the circuit is completed and the bulb will light. Strong electrolytes will give a bright light; weak electrolytes will give a dim light; nonelectrolytes will give no light. Note the effect of concentration. Record your observations on the Report Sheet.

5. Between each test, disconnect the conductivity apparatus from the wall socket, raise the terminals from the solution, and rinse the terminals with distilled water from the wash bottle.

Solubility: Solute and Solvent Characteristics

1. Clean and dry 16 test tubes (100 × 13 mm).

2. Place approximately 0.1 g of the following solids into test tubes numbered as indicated (your instructor will weigh exactly 0.1 g of solid; use your spatula to estimate the 0.1 g sample):

   a. No. 1:   table salt, NaCl

   b. No. 2:   table sugar, sucrose, $C_{12}H_{22}O_{11}$

   c. No. 3:   naphthalene, $C_{10}H_8$

   d. No. 4:   iodine, I_2

3. Add 3 mL of distilled water to each test tube and shake the mixture (simply tapping the test tube with your fingers will agitate the contents enough).

4. Record on the Report Sheet whether the solid dissolved completely (soluble), partially (slightly soluble), or not at all (insoluble).

5. With new sets of labeled test tubes containing the solids listed above, repeat the solubility tests using the solvents ethanol (ethyl alcohol), C_2H_5OH, acetone, $(CH_3)_2CO$, and petroleum ether in place of the water. Record your observations.

6. Discard your solutions in waste containers provided. *Do not discard into the sink.*

Chemicals and Equipment

1. Sucrose (solid and solutions)
2. NaCl (solid and solutions)
3. Naphthalene
4. Iodine
5. HCl solutions
6. Acetic acid (glacial and solutions)
7. Ethanol (ethyl alcohol)
8. Acetone
9. Petroleum ether
10. Conductivity apparatus
11. Hot plate
12. Wash bottle

Experiment 15

PRE-LAB QUESTIONS

1. Explain why table salt, NaCl, will dissolve in water but oil will not.

2. Which solvent, methanol, CH_3OH, or hexane, $CH_3CH_2CH_2CH_2CH_2CH_3$, could be used with water, HOH, as an antifreeze in an automobile radiator? Explain.

3. Refer to Fig. 15.1. Look at the solubility curves for potassium nitrate, KNO_3, and sodium nitrate, $NaNO_3$. Which salt is more soluble: (1) at 60°C; (2) at 80°C?

4. What characteristic makes an electrolyte different from a nonelectrolyte?

Experiment 15

REPORT SHEET

Saturated solution

1. Temperature of distilled water　　_____ °C

2. Mass of sucrose　　_____ g/10mL

3. Mass of additional sucrose　　_____ g

4. Total mass of sucrose: (2) + (3)　　_____ g/10mL at 50°C

5. Observations and explanation

Electrical conductivity

Rate the brightness of the light bulb on a scale from 0 to 5: 0 for no light to 5 for very bright light.

Substance	Observation
Distilled water	_____
Tap water	_____
1 M NaCl	_____
0.1 M NaCl	_____
1 M Sucrose	_____
0.1 M Sucrose	_____
1 M HCl	_____
0.1 M HCl	_____
Glacial acetic acid	_____
0.1 M Acetic acid	_____

Solubility: solute and solvent characteristics

Record the solubility as soluble (s), slightly soluble (ss), or insoluble (i).

Solute	Solvent			
	Water	Ethanol	Acetone	Petroleum Ether
Table salt, NaCl				
Table sugar, sucrose				
Naphthalene				
Iodine				

POST-LAB QUESTIONS

1. On the basis of your observations, are acetic acid solutions electrolytes or non-electrolytes?

2. What are the most likely particles present in NaCl solutions and in sucrose solutions? From the brightness of the light bulb in the electrical conductivity experiment, would this account for the observed results?

3. Pure potassium nitrate, KNO_3, is a white, crystalline solid. How would you get white crystals from a brown sample of this salt?

4. Table sugar, sucrose, is soluble in water, but petroleum ether (an organic solvent similar to gasoline) is insoluble in water. Explain this observation.

Water of hydration

Background

Some compounds do not melt when heated but undergo decomposition. In decomposing, the compound can break down irreversibly or reversibly into two or more substances. If it is reversible, recombination leads to reformation of the original material. Hydrates are examples of compounds which do not melt but which decompose upon heating. The decomposition products are an anhydrous salt and water. The original hydrates can be regenerated by addition of water to the anhydrous salt.

The hydrate contains water as an integral part of the crystalline structure of the compound. When salt crystallizes from an aqueous solution, the number of water molecules bound to the metal ion are characteristic of the metal and are in a definite proportion. Thus when copper sulfate crystallizes from water, the blue salt copper(II) sulfate pentahydrate, $CuSO_4 \cdot 5 H_2O$, forms. As indicated by the formula, 5 waters of hydration are bound to the copper(II) ion in copper sulfate. Notice how the formula is written—the waters of hydration are separated from the formula of the salt by a *dot*.

Heat can transform a hydrate into an anhydrous salt. The water can often be seen escaping as steam. For example, the blue crystals of copper(II) sulfate pentahydrate can be changed into a white powder, the anhydrous salt, by heating to approximately 250°C.

$$CuSO_4 \cdot 5 H_2O(s) \xrightarrow[250°C]{} CuSO_4(s) + 5 H_2O(g)$$
$$\text{Blue} \qquad\qquad \text{White}$$

This process is reversible; adding water to the white anhydrous copper sulfate salt will rehydrate the salt and regenerate the blue pentahydrate.

Some anhydrous salts are capable of becoming hydrated upon exposure to the moisture in their surroundings. These salts are called *hygroscopic salts* and can be used as chemical drying agents or *desiccants*. Some salts are such excellent desiccants and are able to absorb so much moisture from their surroundings that they can eventually dissolve themselves! Calcium chloride is such a salt and is said to be *deliquescent*.

Since many hydrates contain water in a stoichiometric quantity, it is possible to determine the molar ratio of water to salt. First, you would determine the weight of the water lost from the hydrate by heating a weighed sample. From the weight of the water lost, you then can calculate the percent of water in the hydrate. From the weight of the water lost you can also determine the number of water molecules in the hydrate salt and thus the molar ratio.

A sample of the hydrate of copper sulfate, 5.320 g, lost water on heating; the anhydrous salt, which remained, weighed 3.401 g.

a. The weight of the water lost:

Weight of hydrate sample (g)	5.320 g
− Weight of the anhydrous salt (g)	− 3.401 g
Weight of the water lost (g)	1.919 g

b. The percent by mass of water:

$$\frac{\text{Weight of water lost (g)}}{\text{Weight of hydrate sample (g)}} \times 100 = \frac{1.919 \text{ g}}{5.320 \text{ g}} \times 100 = 36.08\%$$

c. The number of moles of water lost:

$$\frac{\text{Weight of water lost (g)}}{\text{MW of water (g/mole)}} = \frac{1.919 \text{ g}}{18.00 \text{ g/mole}} = 0.1066 \text{ mole}$$

d. The number of moles of $CuSO_4$:

$$\frac{\text{Weight of } CuSO_4 \text{ anhydrous (g)}}{\text{MW of } CuSO_4 \text{ (g/mole)}} = \frac{3.401 \text{ g}}{159.6 \text{ g/mole}} = 0.02131 \text{ mole}$$

e. The mole ratio of H_2O to anhydrous $CuSO_4$:

$$\frac{\text{Moles of } H_2O}{\text{Moles of } CuSO_4} = \frac{0.1066}{0.02131} = 5$$

Therefore, the formula of the hydrate of copper sulfate is $CuSO_4 \cdot 5\ H_2O$.

Objectives

1. To learn some properties and characteristics of hydrates.
2. To verify the percent of water in the hydrate of copper sulfate.
3. To verify the mole ratio of water to salt in the hydrate of copper sulfate is fixed.

Properties of Anhydrous CaCl₂

1. Take a small spatula full of anhydrous $CaCl_2$ and place it on a watch glass.

2. Set the watch glass to the side, out of the way, and continue the rest of the experiment. From time to time during the period, examine the solid and record your observations.

3. What happened to the solid $CaCl_2$ by the end of the period?

Composition of a Hydrate

1. Obtain from your instructor a porcelain crucible and cover. Clean with soap and water and dry thoroughly with paper towels.

2. Place the crucible and cover in a clay triangle supported by a metal ring on a ringstand (Fig. 16.1). Heat the crucible with a Bunsen burner to red heat for 5 min. Using crucible tongs, place the crucible and cover on a wire gauze and allow it to cool to room temperature.

Figure 16.1

a) Heating the crucible;
b) moving the crucible
with crucible tongs.

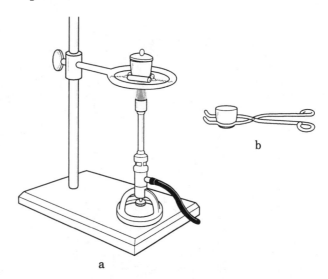

3. Weigh the crucible and cover to the nearest 0.001 g (1).

4. Repeat this procedure (heating, cooling, weighing) until two successive weights of the covered crucible agree to within 0.005 g or less (2).

CAUTION!

Handle the crucible and cover with the crucible tongs from this point on. This will avoid possible burns and will avoid transfer of moisture and oils from your fingers to the porcelain.

Experiment 16

PRE-LAB QUESTIONS

1. Give an example of a salt that is hygroscopic.

2. Give an example of a salt that can be used as a desiccant.

3. Give an example of a salt that is deliquescent.

4. The hydrate of calcium bromide is $CaBr_2 \cdot 3\ H_2O$.

   a. How many atoms of each kind are present?

   b. On heating the hydrate, how many moles of water should be driven off per mole of hydrate?

   c. Calculate the percent of water in the hydrate.

   d. If you heat 15.00 g of the hydrate and drive off the water, what is the weight of the anhydrous salt remaining?

Experiment 16

REPORT SHEET

Observations on the properties of anhydrous $CaCl_2$

Composition of a hydrate	Trial 1	Trial 2
1. Weight of crucible and cover, 1st heating	_____ g	_____ g
2. Weight of crucible and cover, 2nd heating	_____ g	_____ g
3. Weight of covered crucible plus sample	_____ g	_____ g
4. Weight of sample (hydrate): (3) − (2)	_____ g	_____ g
5. Weight of covered crucible plus sample, after 1st heating	_____ g	_____ g
6. Weight of covered crucible plus sample, after 2nd heating	_____ g	_____ g
7. Weight of anhydrous salt: (6) − (2)	_____ g	_____ g
8. Weight of water lost: (4) −(7)	_____ g	_____ g
9. Percent of water in hydrate: [(8)/(4)] × 100	_____ %	_____ %
10. Moles of water lost: (8)/18.00 g/mole	_____ mole	_____ mole
11. Moles of anhydrous $CuSO_4$: (7)/159.6 g/mole	_____ mole	_____ mole

12. Moles of water per mole of $CuSO_4$:
(10)/(11)

_____   _____

13. The formula for the hydrated
copper(II) sulfate

_____   _____

14. Observation: water added to the anhydrous copper(II) sulfate:

POST-LAB QUESTIONS

1. What effect would "spattering" of the solid have on the experimentally determined percent of water in the hydrate?

2. Calculate the percent of water in the following hydrates:

a. $NiCl_2 \cdot 6\ H_2O$

b. $MgSO_4 \cdot 5\ H_2O$

3. Given a 10.00 g sample of $Na_2SO_4 \cdot 10\ H_2O$, what weight of anhydrous sodium sulfate can be obtained after driving off the water?

4. A student found the percent of water in $ZnSO_4 \cdot 7\ H_2O$ to be 45.5%. Determine the experimental error.

Experiment  17

Colligative properties: freezing point depression and osmotic pressure

Background

Certain properties of solutions depend only on the number of solute particles dissolved in a given amount of solvent and not on the nature of these particles. Such properties are called *colligative properties*. For example, one such property is the *freezing point depression*. One mole of any solute dissolved in 1000 g of water lowers the freezing point of the water by 1.86°C. We call this value, 1.86 degree/mole/1000 g water, the *freezing point depression constant* of water, K_f. Each solvent has a characteristic freezing point depression constant that is related to its heat of fusion. The nature of the solute does not matter.

This principle can be used in a number of practical ways. One application is the use of antifreeze in car radiators. Since water expands on freezing, the water in a car's cooling system can crack the engine block of a parked car when the outside temperature falls below 0°C. The addition of a common antifreeze, ethylene glycol, prevents this because the freezing point is depressed and the water-ethylene glycol mixture freezes at a much lower temperature.

The freezing point depression, ΔT, is proportional to the number of particles of the solute (moles) in 1000 g of solvent and the proportionality constant is the freezing point depression constant, K_f.

$$\Delta T = K_f \times \text{mole solute}/1000 \text{ g solvent} \quad (1)$$

For example, if we add 275 g of ethylene glycol (molecular weight 62.0) per 1000 g of water in a car radiator, what will the freezing point of the solution be?

$$\Delta T = \frac{1.86°C}{\text{mole}/1000 \text{ g}} \times \frac{275 \text{ g}}{62.0 \text{ g}} \times \frac{1 \text{ mole}}{1000 \text{ g}} = 8.26°C$$

The freezing point of water will be lowered from 0°C to −8.26°C.

If a solute is ionic, then each mole of solute dissociates. For NaCl we get two moles of ions, and for Na_2SO_4 three moles of ions for each mole of the solute. For convenience in calculation, we define a new term, *osmole,* as moles multiplied by the number of particles produced by one molecule of solute in solution.

In the present experiment, we will obtain the freezing point depression constant, K_f, for lauric acid, $CH_3(CH_2)_{10}COOH$, which will serve as a solvent. We will use benzoic acid, C_6H_5COOH, as a solute. In order to obtain the K_f, you will measure the freezing points of lauric acid and a mixture of the lauric acid-benzoic acid system. In actuality, you will measure the melting points of the solids. Freezing point or melting point is the temperature of transition between solid and liquid. Melting points (going from solid to liquid) can be measured more accurately than freezing points (going from liquid to solid). This is so because in freezing point measurements supercooling may occur which would yield a lower than correct freezing (melting) point.

In addition to freezing point depression, there are several other colligative properties, among which osmotic pressure is the most important biologically. Osmotic pressure develops whenever a semipermeable membrane separates a solution from a solvent. A *semipermeable membrane* is a material that contains tiny holes that are big enough to allow small solvent molecules to pass through but not big enough to allow large solute molecules to pass (Fig. 17.1). The passage of solvent molecules from the solvent side (right compartment) to the solution side (left compartment) of the semipermeable membrane generates the *osmotic pressure* that can be measured by the difference in the heights of the two columns. Living cells, among them the red blood cells, are surrounded by semipermeable membranes. The osmolarity of most cells is 0.30 osmol. For example, a 0.89% w/v NaCl solution, normally referred to as physiological saline solution, has an osmolarity of 0.30. Thus when a cell is put in physiological saline solution, the osmolarity on both sides of the membrane is the same and therefore, no osmotic pressure is generated across the membrane. Such a solution is called *isotonic*. On the other hand, if a cell is put in water (pure solvent) or in a solution which has lower osmolarity than the cell, there will be a net flow of water into the cell driven by the osmotic pressure. Such a solution is called *hypotonic*. A cell placed in a hypotonic solution will swell and eventually may burst. If that happens to a red blood cell, the process is called *hemolysis*. In contrast, a solution with higher osmolarity than the cell is called a *hypertonic* solution. A cell suspended in a hypertonic solution will shrivel; there is a net flow of water from the cell into the surroundings. When that happens to a red blood cell, the process is called *crenation*.

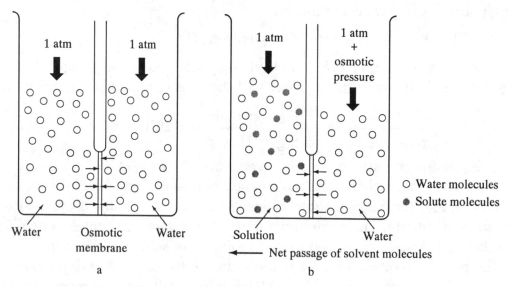

Figure 17.1 • Osmotic pressure. (a) Two compartments separated by an osmotic semipermeable membrane both contain only solvent molecules that can pass through the membrane. (b) The compartment on the right contains only solvent, the one on the left both solute and solvent. Solute molecules cannot pass through the membrane. The solvent molecules move to the left compartment in an effort to dilute the solution, raising the liquid level on that side.

Objectives

1. To demonstrate freezing point depression and obtain the freezing point depression constant.
2. To show the effect of tonicity on cells.

Procedure

Effect of the Tonicity of Solutions on Cells

1. Take five clean test tubes. Label them: a, b, c, d, e.

2. Add 2 mL of the following solutions to the labelled test tubes:
   a. Distilled water
   b. 0.1 M glucose
   c. 0.5 M glucose
   d. 0.89% NaCl
   e. 3% NaCl

3. To each test tube add thin (about 0.5 mm thick) slices of freshly cut carrot, scallion, and celery sections.

4. Put the test tubes in a test tube rack and wait until you have finished all the other experiments.

5. Observe the appearance of the sections with the naked eye and also under a microscope.

6. Repeat steps 1 and 2 using a new set of five clean test tubes.

7. Using an eyedropper, add five drops of fresh whole bovine blood to each test tube. Tap the bottom of the test tubes to ensure proper mixing.

8. Observe the color and the appearance of the solutions after 20 min. both by the naked eye and also under a microscope. *For the proper handling and disposal of blood samples, read the instructions in Appendix, Exp. 17.*

Freezing Point Depression Measurements

1. Assemble a simple freezing point (melting point) apparatus. A beaker will serve as a water bath. A hot plate or Bunsen burner will provide the source of heat. A test tube will serve as a secondary water bath in which a thermometer is suspended (Fig. 17.2).

Figure 17.2

Melting point apparatus.

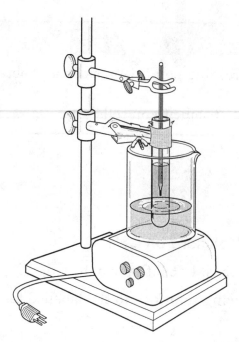

2. Benzoic acid–lauric acid mixtures can be prepared in front of the class as follows (or as an alternative the instructor can prepare this in advance):

Weigh out 3 g of lauric acid and place it in a 25-mL beaker. Weigh out 0.6 g of benzoic acid. Heat the lauric acid gently on a hot plate until it melts (50°C). Add the benzoic acid to the beaker. Mix it thoroughly until a uniform solution is obtained. Cool the beaker in cold water to obtain a solid sample. Grind the sample to a fine powder in a mortar with a pestle.

3. Each student will pack four capillary melting tubes with samples: (a) lauric acid (b) three tubes with the 17% benzoic acid solution.

4. Pack the melting tubes as follows:

a. Scoop up a very small amount of sample into the melting point capillary tube by pressing the open end of the tube vertically into the sample (Fig. 17.3).

Figure 17.3

Sampling

b. Invert the capillary tube. Stroking the capillary with a file, allow the solid to pack at the bottom of the capillary. You only need a 1–5 mm long packed sample in the capillary tube.

5. Attach the capillary tube to the thermometer, using a narrow rubber band near the top of the tube. Be certain to align the tip of the thermometer with the tip of the capillary tube (Fig. 17.4).

Figure 17.4
Positioning the capillary.

Aligned

6. Measure the melting point of each sample as follows:

Clamp the thermometer with the capillary tube attached and immerse it in the secondary thermostat filled with water. Lower the secondary thermostat into the beaker filled with water and start the heating process. Observe the melting point of each sample and record it. Melting occurs when you observe the first shrinkage in your sample or the appearance of tiny bubbles. (**Do not wait until the whole sample in the capillary becomes translucent!**) After taking the melting point of the first sample, allow the thermostat to cool to room temperature by adding some cold water. You should start the heating process to observe the melting point of the second sample only after the water in both the primary and secondary thermostat is at room temperature.

Chemicals and Equipment

1. Capillary tubes
2. Test tubes
3. Thermometer
4. Rubber band
5. Beakers
6. Clamp
7. Hot plate (or Bunsen burner)
8. Microscope
9. Razor blade or dissecting knife
10. Lauric acid
11. Benzoic acid
12. Fresh whole bovine blood
13. 0.1 M and 0.5 M glucose solutions
14. 0.89% and 3% NaCl solutions
15. Fresh carrot, scallion, and celery (cut to 0.5 mm sections)

Experiment 17

PRE-LAB QUESTIONS

1. Write the structure and calculate the molecular weight of (a) lauric acid, $C_{12}H_{24}O_2$, and (b) benzoic acid, $C_7H_6O_2$.

2. What is the expected melting point of lauric acid? (Obtain this information from your textbook: Table 17.1)

3. A 0.89% w/v NaCl solution is isotonic with red blood cells. What would be the tonicity of a 0.65 M KCl solution? Isotonic, hypotonic, or hypertonic?

4. What is the osmolarity of (a) 5% glucose (mol. wt. 180.16) and (b) 5% sodium benzoate, $C_7H_5O_2{}^-Na^+$?

5. What kind of information can we obtain if we know the freezing point depression constant, K_f, of a solvent and the freezing point depression of a 10% solution of an unknown substance in the solvent referred above?

NAME _____   SECTION _____   DATE _____

PARTNER _____   GRADE _____

Experiment 17

REPORT SHEET

Tonicity of solutions

Solutions	Observations			
	Appearance of plant cells		Appearance of red blood cells	
	Naked eye	Microscope	Naked eye	Microscope
Distilled water				
0.1 M glucose				
0.5 M glucose				
0.89% NaCl				
3.0% NaCl				

Freezing point depression

• Melting point of lauric acid _____

• Melting point of 17% w/w benzoic acid   (a) _____

(b) _____

(c) _____

• Freezing point depression of 17% benzoic acid   (a) _____

(b) _____

(c) _____

• Average freezing point depression of 17% benzoic acid   _____

• Mole benzoic acid in 1000 g lauric acid in 17% w/w sample   _____

• K_f calculated from equation (1) for 17% benzoic acid solution   _____

POST-LAB QUESTIONS

1. Was the reproducibility of the melting point measurements within ±2%?

2. Using the average freezing point depression constant obtained for lauric acid in your experiment, calculate what would be the freezing point of a 10.0% w/w benzoic acid solution?

3. Assume that your thermometer was not properly calibrated and showed only 95°C difference between the melting point and boiling point of water. How would that effect your K_f value?

4. Which of your test solutions was (a) isotonic with red blood cells (tonicity = 0.30 osmolar)? (b) Which was hypotonic? (c) Which was hypertonic?

5. Did you observe any difference in the behavior of plant cells versus red blood cells in hypotonic and hypertonic solutions? What were those differences? Red blood cells have only semipermeable plasma membranes, while plant cells have an additional cell wall made of polysaccharides. Would that explain your observations? How?

6. Would the same osmolarity of a solute cause greater, smaller, or the same freezing point depression in water as in lauric acid? Explain.

Experiment 18

● Factors affecting rate of reations

Background

Some chemical reactions take place rapidly; others are very slow. For example, antacid neutralizes stomach acid (HCl) rapidly but hydrogen and oxygen react with each other to form water very slowly. A tank containing a mixture of H_2 and O_2 shows no measurable change even after many years. The study of rates of reactions is called *chemical kinetics*. The *rate of reaction* is the change in concentration of a reactant (or product) per unit time. For example, in the reaction

$$2HCl(aq) + CaCO_3(s) \rightleftharpoons CaCl_2(aq) + H_2O(l) + CO_2(g)$$

we monitor the evolution of CO_2, and we find that 4.4 g of carbon dioxide gas was produced in 10 min. Since 4.4 g corresponds to 0.1 moles of CO_2, the rate of the reaction is 0.01 moles CO_2/min. On the other hand, if we monitor the HCl concentration, we may find that at the beginning we had 6 M HCl and after 10 min. the concentration of HCl was 4 M. This means that we used up 2 M HCl in 10 min. Thus the rate of reaction is 0.2 moles HCl/L-min. From the above we can see that when describing the rate of reaction (it is not sufficient to give a number), we have to specify the units and also the reactant (or product) we monitored.

In order that a reaction should take place, molecules or ions must first collide. Not every collision yields a reaction. In many collisions, the molecules simply bounce apart without reacting. A collision that results in a reaction is called an *effective collision*. The minimum energy necessary for the reaction to happen is called the *activation energy* (Fig. 18.1). In this energy diagram, we see that the rate of reaction depends on this activation energy.

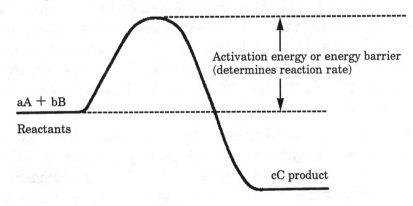

Figure 18.1 • Energy diagram for a typical reaction.

The lower the activation energy the faster the rate of reaction; the higher the activation energy the slower the reaction. This is true for both exothermic and endothermic reactions.

A number of factors affect the rates of reactions. In our experiments we will see how these affect the rates of reactions.

1. **Nature of reactants.** Some compounds are more reactive than others. In general, reactions that take place between ions in aqueous solutions are rapid. Reactions between covalent molecules are much slower.

2. **Concentration.** In most reactions, the rate increases when the concentration of either or both reactants is increased. This is understandable on the basis of the collision theory. If we double the concentration of one reactant, it will collide in each second twice as many times with the second reactant as before. Since the rate of reaction depends on the number of effective collisions per second, the rate is doubled (Fig. 18.2).

Figure 18.2

Concentration affecting the rate of reaction.

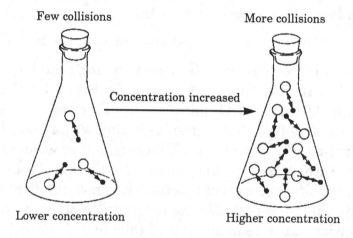

Few collisions

More collisions

Concentration increased

Lower concentration

Higher concentration

3. **Surface area.** If one of the reactants is a solid, the molecules of the second reactant can collide only with the surface of the solid. Thus the surface area of the solid is in effect its concentration. An increase in the surface area of the solid (by grinding to a powder in a mortar) will increase the rate of reaction.

4. **Temperature.** Increasing the temperature makes the reactants more energetic than before. This means that more molecules will have energy equal to or greater than the activation energy. Thus one expects an increase in the rate of reaction with increasing temperature. As a rule of thumb, every time the temperature goes up by 10°C, the rate of reaction doubles. This rule is far from exact, but it applies to many reactions.

5. **Catalyst.** Any substance that increases the rate of reaction without itself being used up in the process is called a *catalyst*. A catalyst increases the rate of reaction by lowering the activation energy (Fig. 18.3). Thus many more molecules can cross the energy barrier (activation energy) in the presence of a catalyst than in its absence. Almost all the chemical reactions in our bodies are catalyzed by specific catalysts called enzymes.

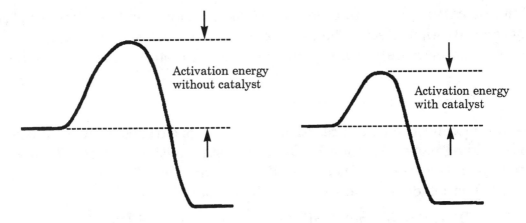

Activation energy
without catalyst

Activation energy
with catalyst

Figure 18.3 • Energy diagrams of a reaction with and without a catalyst.

Objectives

1. To investigate the relationship between the rate and the nature of reactants.

2. To measure the rate of reaction as a function of concentration.

3. To demonstrate the effect of temperature on the rate of reaction.

4. To investigate the effect of surface area and the effect of a catalyst on the rate of reaction.

Procedure

1. **Nature of reactants.** Label five (10 × 75mm) test tubes 1 through 5. Place in each test tube one 1-cm polished strip of magnesium ribbon. Add 1 mL of acid to each test tube as follows: no. 1) 3 M H_2SO_4; no. 2) 6 M HCl; no. 3) 6 M HNO_3; no. 4) 2 M H_3PO_4; and no. 5) 6 M CH_3COOH. The reaction will convert the magnesium ribbon to the corresponding salts with the liberation of hydrogen gas. You can assess the rate of reaction qualitatively, by observing the speed with which the gas is liberated (bubbling) and/or by noticing the time of disappearance of the magnesium ribbon. Do all of the reactions in the five test tubes at the same time; assess the rates of reaction; then list, in decreasing order, the rates of reaction of magnesium with the various acids on your Report Sheet (1).

2. Place 1 mL of 6 M HCl in each of three labeled test tubes. Add a 1-cm polished strip of magnesium to the first, zinc to the second, and copper to the third. Do all of the reactions in the three test tubes at the same time; assess the rates of reaction of the three metals by the speed of evolution of H_2 gas; then list, in decreasing order, the rates of reaction of the metals with the acid on your Report Sheet (2).

3. **Concentration.** The *iodine clock reaction* is a convenient reaction for observing concentration effects. The reaction is between potassium iodate, KIO_3, and sodium bisulfite, $NaHSO_3$; the net ionic reaction is given by the following equation.

$$IO_3^-(aq) + 3HSO_3^-(aq) \rightleftharpoons I^-(aq) + 3SO_4^{2-}(aq) + 3H^+(aq)$$

We can monitor the rate of reaction by the disappearance of the bisulfite. We do so by adding more IO_3^- than HSO_3^- at the start of the reaction. When we have used up all the bisulfite, there is still some iodate left. This will then react with the product iodide, I^-, and results in the formation of I_2.

$$IO_3^-(aq) + 5I^-(aq) + 6H^+(aq) \rightleftharpoons 3I_2(aq) + 3H_2O(l)$$

We can detect the appearance of iodine with the aid of starch indicator; this reagent forms a blue complex with iodine. The time it takes for the blue color to suddenly appear indicates when all the bisulfite was used up in the first reaction. That's why the name: iodine clock. Thus you should measure the time (with a stopwatch, if available) elapsed between mixing the two solutions and the appearance of the blue color. Place the reactants in two separate 150-mL beakers according to the outline in Table 18.1.

Table 18.1	Reactant Concentration and Rate of Reaction				
	BEAKER A			**BEAKER B**	
Trial	**0.1 M KIO_3**	**starch**	**water**	**0.01 M $NaHSO_3$**	**water**
1	2.0 mL	2 mL	46 mL	5 mL	45 mL
2	4.0 mL	2 mL	44 mL	5 mL	45 mL
3	6.0 mL	2 mL	42 mL	5 mL	45 mL

Use a graduated pipet to measure each reactant and a graduated cylinder to measure the water. Simultaneously pour the two reactants into a third beaker and time the appearance of the blue color. Repeat the experiment with the other two trial concentrations. Record your data on the Report Sheet (3).

4. **Surface area.** Using a large mortar and pestle, crush and pulverize about 0.5 g of marble chips. Place the crushed marble chips into one large test tube and 0.5 g of uncrushed marble chips into another. Add 2 mL of 6 M HCl to each test tube and note the speed of bubbling of the CO_2 gas. Record your data on the Report Sheet (4).

5. **Temperature.** Add 5 mL of 6 M HCl to three clean test tubes. Place the first test tube in an ice bath (4°C), the second in a beaker containing warm water (50°C), and the third in a beaker with tap water (20°C). Wait 5 min. To each test tube add a piece of zinc ribbon (1 cm × 0.5 cm × 0.5 mm). Note the time you added the zinc metal. Finally, note the time when the bubbling of gas stops in each test tube and the zinc metal disappears. Record the time of reaction (time of the disappearance of the zinc − the time of the start of the reaction) on your Report Sheet (5).

6. Catalyst. Add 2 mL of 3% H_2O_2 solution to two clean test tubes. The evolution of oxygen bubbles will indicate if hydrogen peroxide decomposed. Note if anything happens. Add a few grains of MnO_2 to one of the test tubes. Note the evolution of oxygen, if any. Record your data on the Report Sheet (6).

Chemicals and Equipment

1. Mortar and pestle
2. 10-mL graduated pipet
3. 5-mL volumetric pipet
4. Magnesium ribbon
5. Zinc ribbon
6. Copper ribbon
7. 3 M H_2SO_4
8. 6 M HCl
9. 6 M HNO_3
10. 2 M H_3PO_4
11. 6 M CH_3COOH
12. 0.1 M KIO_3
13. 0.01 M $NaHSO_3$
14. Starch indicator
15. Marble chips
16. 3% hydrogen peroxide
17. Manganese dioxide

NAME SECTION DATE

PARTNER GRADE

Experiment 18

PRE-LAB QUESTIONS

1. Write the balanced chemical equations for the reactions in Section 6 of the **Procedure** section.

2. Assume that in the above reactions you were able to measure the evolution of oxygen gas. You find that during a 15 min. period, 0.32 g of oxygen was collected in a 1 L vessel. What is the rate of the particular reaction you just monitored?

   (a) in g/L min.

   (b) in moles/L min.

3. If the rate of a certain reaction is 6 M/hr. at room temperature (25°C), what would be the predicted rate at 5°C?

4. Some antacids come in the form of a compressed pill, while others are in the form of loose powder. Which form would give faster relief for heartburn considering that they contained the same ingredients and were taken in equal amounts?

Experiment 18

REPORT SHEET

1. Nature of reactants

Name of the acid

Fastest reaction _____

_____

_____

_____

Slowest reaction _____

2. Nature of reactants

Name of the metal

Fastest reaction _____

_____

Slowest reaction _____

3. Effect of concentration

Trial no.	Time
1	_____
2	_____
3	_____

4. Surface area

Fast reaction _____

Slow reaction _____

5. Effect of temperature

Trial at	4°C	20°C	50°C
Reaction time	_____		

6. Catalyst   **Observation**

No catalyst _____

MnO_2 _____

POST-LAB QUESTIONS

1. If in the reaction between magnesium and HCl we would have used 3 M HCl rather than 6 M HCl, would the rate of reaction increase, decrease, or remain the same?

2. If in the same reaction we would have used globular chunks of magnesium instead of ribbon (both having the same weight), would the rate of reaction increase, decrease, or remain the same? What would be affected (if any) by this change?

3. Assume that the zinc ribbon you added to the test tubes in Section 5 was 0.5 g. Calculate the rate of reaction as moles of Zn per min. for each temperature.

4. In general, we expect that doubling the concentration of a reactant will approximately double the rate of reaction. Was this expectation justified in the iodine clock reaction?

5. Assume we do a reaction of zinc with 6 M HCl at room temperature (20°C). How much faster will these two chemicals react at 40°C (see Section 5)?

Experiment  19

Law of chemical equilibrium and Le Chatelier's principle

Background

Two important questions are asked about every chemical reaction: (a) How much product is produced and (b) How fast is it produced? The first question involves chemical equilibrium and the second question belongs to the domain of chemical kinetics. (We dealt with kinetics in Experiment 18). Some reactions are irreversible and they go to completion (100% yield.) When you ignite methane gas in your gas burner in the presence of air (oxygen), methane burns completely and forms carbon dioxide and water.

$$CH_4(g) + 2O_2(g) \longrightarrow CO_2(g) + 2H_2O(g)$$

Other reactions do not go to completion. They are reversible. In such cases, the reaction can go in either direction: forward or backward. For example, the reaction

$$Fe^{3+}(aq) + SCN^-(aq) \rightleftharpoons FeSCN^{2+}(aq)$$

is often used to illustrate reversible reactions. This is so because it is easy to observe the progress of the reaction visually. The yellow Fe^{3+} ion reacts with thiocyanate ion to form a deep red complex ion, $FeSCN^{2+}$. This is the forward reaction. At the same time, the complex red ion also decomposes and forms the yellow iron(III) ion and thiocyanate ion. This is the backward (reverse) reaction. At the beginning when we mix iron(III) ion and thiocyanate ion, the rate of the forward reaction is at a maximum. As time goes on, this rate decreases because we have less and less iron(III) and thiocyanate to react. On the other hand, the rate of the reverse reaction (which began at zero) gradually increases. Eventually the two rates become equal. When this point is reached, we call the process a *dynamic equilibrium*, or just *equilibrium*. When in equilibrium at a particular temperature, a reaction mixture obeys the *Law of Chemical Equilibrium*. This Law imposes a condition on the concentration of reactants and products expressed in the equilibrium constant (K). For the above reaction between iron(lII) and thiocyanate ions, the equilibrium constant can be written as

$$K = [FeSCN^{2+}]/[Fe^{3+}][SCN^-]$$

or in general

$$K = [products]/[reactants]$$

The brackets, [ ], indicate concentration, in moles/L, at equilibrium. As the name implies, the *equilibrium constant* is a constant at a set temperature for a particular reaction. Its magnitude tells if a reaction goes to completion or if it is far from completion (reversible reaction). A number much smaller than 1 for K indicates that at equilibrium only a few molecules of products are formed, meaning the mixture consists mainly of reactants. We say that the equilibrium lies far to the left. On the other hand, a completion of a reaction (100% yield) would have a very large number (infinite?) for the equilibrium constant. In this case, obviously the equilibrium lies far to the right. The above reaction between iron(III) and thiocyanate has an equilibrium constant of 207, indicating that the equilibrium lies to the right but does not go to completion. Thus at equilibrium, both reactants and product are present, albeit the products far outnumber the reactants.

The Law of Chemical Equilibrium is based on the constancy of the equilibrium constant. This means that if one disturbs the equilibrium, for example by adding more reactant molecules, there will be an increase in the number of product molecules in order to maintain the product/reactant ratio unchanged and thus preserving the numerical value of the equilibrium constant. The Le Chatelier Principle expresses this as follows: *If an external stress is applied to a system in equilibrium, the system reacts in such a way as to partially relieve the stress.* In our present experiment, we demonstrate the Le Chatelier Principle in two manners: (a) disturbing the equilibrium by changing the concentration of a product or reactant; (b) changing the temperature.

(a1) In the first experiment, we add ammonia to a pale blue copper(II) sulfate solution. The ionic reaction is

$$Cu(H_2O)_4^{2+}(aq) + 4NH_3(aq) \rightleftharpoons Cu(NH_3)_4^{2+}(aq) + 4H_2O(l)$$

pale blue · · · · colorless · · · · · · · (color?)

A change in the color indicates the copper-ammonia complex formation. Adding a strong acid, HCl, to this equilibrium causes the ammonia, NH_3, to react with the acid

$$NH_3(aq) + H^+(aq) \rightleftharpoons NH_4^+(aq)$$

Thus we removed some reactant molecules from the equilibrium mixture. As a result we expect the equilibrium to shift to the left, reforming hydrated copper(II) ions with the reappearance of pale blue color.

(a2) In the second reaction, we demonstrate the common ion effect. When we have a mixture of $H_2PO_4^-/HPO_4^{2-}$ solution, the following equilibrium exists:

$$H_2PO_4^-(aq) + H_2O(l) \rightleftharpoons H_3O^+(aq) + HPO_4^{2-}(aq)$$

If we add a few drops of aqueous HCl to the solution, we will have added a common ion, H_3O^+, that already was present in the equilibrium mixture. We expect, on the basis of the Le Chatelier Principle, that the equilibrium will shift to the left. Thus the solution will not become acidic.

(a3) In the iron(III)-thiocyanate reaction

$$Fe^{3+}(aq) + 3Cl^-(aq) + K^+(aq) + SCN^-(aq) \rightleftharpoons Fe(SCN)^{2+}(aq) + 3Cl^-(aq) + K^+(aq)$$

yellow · · · · · · · · · · · colorless · · · · · · · · · · · · · red · · · · · · · · · · colorless the chloride and potassium ions are spectator ions. Nevertheless, their concentration may also influence the equilibrium. For example, when the chloride ions are in excess the yellow color of the Fe^{3+} will disappear with the formation of a colorless $FeCl_4^-$ complex

$$Fe^{3+}(aq) + 4Cl^-(aq) \rightleftharpoons FeCl_4^-(aq)$$
$$\text{yellow} \qquad\qquad\qquad \text{colorless}$$

(b1) Most reactions are accompanied by some energy changes. Frequently, the energy is in the form of heat. We talk of endothermic reactions if heat is consumed during the reaction. In endothermic reactions, we can consider heat as one of the reactants. Conversely, heat is evolved in an exothermic reaction, and we can consider heat as one of the products. Therefore, if we heat an equilibrium mixture of an endothermic reaction, it will behave as if we added one of its reactants (heat) and the equilibrium will shift to the right. Heating the equilibrium mixture of an exothermic reaction, the equilibrium will shift to the left. We will demonstrate the effect of temperature on the reaction:

$$Co(H_2O)_6^{2+}(aq) + 4Cl^-(aq) \rightleftharpoons CoCl_4^{2-}(aq) + 6H_2O(l)$$

You will observe a change in the color depending on whether the equilibrium was established at room temperature or at 100°C (in boiling water). From the color change, you should be able to tell whether the reaction was endothermic or exothermic.

Objectives

1. To study chemical equilibria.
2. To investigate the effects of (a) changing concentrations and (b) changing temperature in equilibrium reactions.

Procedure

a1. Place 20 drops (about 1 mL) of 0.1 M $CuSO_4$ solution into a clean and dry test tube. Add (dropwise) 1 M NH_3 solution, mixing the contents after each drop. Continue to add until the color changes. Note the new color and the number of drops of 1 M ammonia added and record it on your Report Sheet (1). To the equilibrium mixture thus obtained, add (dropwise, counting the number of drops added) 1 M HCl solution until the color changes back to pale blue. Record your observations on your Report Sheet (2).

a2. Place 2 mL of $H_2PO_4^-/HPO_4^{2-}$ solution into a clean and dry test tube. Using red and blue litmus papers, test if the solution is acidic or basic. Record your findings on the Report Sheet (3). Add a drop of 1 M HCl to the litmus papers. Record your observations on the Report Sheet (4). Add one drop of 1 M HCl solution to the test tube. Mix it and test it with litmus papers. Record your observation on the Report Sheet (5).

a3. Prepare a stock solution by adding 1 mL of 0.1 M iron(III) chloride, $FeCl_3$, and 1 mL of 0.1 M potassium thiocyanate, KSCN, to 50 mL of distilled water in a beaker. Set up four clean and dry test tubes and label them. To each test tube, add about 2 mL of the stock equilibrium mixture you just prepared. Use the first test tube as a standard to which you can compare the color of the other solutions. To the second test tube add 10 drops of 0.1 M iron(III) chloride solution; to the third add 10 drops of 0.1 M KSCN solution. To the fourth add five drops of saturated NaCl solution. Observe the color in each test tube and record your observations on the Report Sheet (6) and (7).

b1. Place 5 drops of 1 M $CoCl_2$ in a dry and clean test tube. Add concentrated HCl dropwise until a color change occurs. Record your observations on the Report Sheet (8). Place 1 mL $CoCl_2$ in a clean and dry test tube. Note the color. Immerse the test tube into a boiling water bath. Record your observations on the Report Sheet (9).

CAUTION!

Do not allow skin contact and do not inhale the HCl vapors.

Chemicals and Equipment

1. 0.1 M $CuSO_4$
2. 1 M NH_3
3. 1 M HCl
4. Saturated NaCl
5. Concentrated HCl
6. 0.1 M KSCN
7. 0.1 M $FeCl_3$
8. 1 M $CoCl_2$
9. $H_2PO_4^-/HPO_4^{2-}$ solution
10. Litmus paper

Experiment 19

PRE-LAB QUESTIONS

1. For the reaction $NH_3(aq) + H^+(aq) = NH_4^+(aq)$, at 20°C the equilibrium concentrations were as follows: $[NH_3] = 2 \times 10^{-4}$ M; $[H^+] = 2 \times 10^{-4}$ M; and $[NH_4^+] = 18$ M. Calculate the equilibrium constant for the reaction.

2. If the reaction above is exothermic and it was run at 40°C, would the equilibrium concentration of NH_3 be greater, the same, or smaller than 2×10^{-4} M?

3. If the reaction between iron(III) and thiocyanate ions yielded an equilibrium concentration of 0.2 M for each of these ions, what would be the equilibrium concentration of the red iron-thiocyanate complex? *Hint:* The equilibrium constant can be found in the **Background** section.

NAME _____ SECTION _____ DATE _____

PARTNER _____ GRADE _____

Experiment 19

REPORT SHEET

1. What is the color of the copper-ammonia complex?   _____

   How many drops of 1 M ammonia did you add
   to cause a change in color?   _____

2. How many drops of 1 M HCl did you add to cause
   a change in color back to pale blue?   _____

3. Testing the phosphate solution, what was the color
   of the red litmus paper?   _____

   What was the color of the blue litmus paper?   _____

4. Testing the 1 M HCl solution, what was the color
   of the red litmus paper?   _____

   What was the color of the blue litmus paper?   _____

5. After adding one drop of 1 M HCl to the phosphate
   solution and testing it with litmus paper, what was
   the color of the red litmus paper?   _____

   What was the color of the blue litmus paper?   _____

   Was your phosphate solution acidic, basic, or neutral?
   (a) before the addition of HCl   _____

   (b) after the addition of HCl   _____

   Was your HCl solution acidic, basic, or neutral?   _____

6. Compare the colors in each of the test tubes containing
   the iron(III) chloride-thiocyanate mixtures:

   no. 1 _____

   no. 2 _____

   no. 3 _____

   no. 4 _____

7. In which direction did the equilibrium shift in test tube

no. 2 _____

no. 3 _____

no. 4 _____

8. What is the color of the $CoCl_2$ solution?

(a) before the addition of HCl   _____

(b) after the addition of HCl   _____

9. What is the color of the $CoCl_2$ solution ?

(a) at room temperature   _____

(b) at boiling water temperature   _____

10. In which direction did the equilibrium shift
upon heating?   _____

11. From the above shift, determine if the reaction
was exothermic or endothermic.   _____

POST-LAB QUESTIONS

1. In the first experiment with the copper-ammonia complex, you added ammonia
to change the color and, later, equal strength HCl to change it back to blue. Did
you require more, less, or an equal number of drops from each to accomplish
the color change? On the basis of stoichiometry, what was your expectation?

2. Adding HCl to the $H_2PO_4^-$ / HPO_4^{2-} mixture is called a common ion effect.
Explain why.

3. Silver ion reacts with chloride ion in solution to form the precipitate AgCl.
What would happen to the color of a dilute solution containing $FeCl_4^-$ if we
added a solution containing silver ion?

4. What are the charges on the central cobalt ion in its hexahydrate form?

pH and buffer solutions

We frequently encounter acids and bases in our daily life. Fruits, such as oranges, apples, etc., contain acids. Household ammonia, a cleaning agent, and Liquid Plumber are bases. *Acids* are compounds that can donate a proton (hydrogen ion). *Bases* are compounds that can accept a proton. This classification system was proposed simultaneously by Johannes Brønsted and Thomas Lowry in 1923, and it is known as the Brønsted-Lowry theory. Thus any proton donor is an acid, and a proton acceptor is a base.

When HCl reacts with water

$$HCl + H_2O \rightleftharpoons H_3O^+ + Cl^-$$

HCl is an **acid** and H_2O is a **base** because HCl **donated a proton** thereby becoming Cl^-, and water **accepted a proton** thereby becoming H_3O^+.

In the reverse reaction (from right to left) the H_3O^+ is an acid and Cl^- is a base. As the arrow indicates, the equilibrium in this reaction lies far to the right. That is, out of every 1000 HCl molecules dissolved in water, 990 are converted to Cl^- and only 10 remain in the form of HCl at equilibrium. But H_3O^+ (hydronium ion) is also an acid and can donate a proton to the base, Cl^-. Why do not hydronium ions give up protons to Cl^- with equal ease and re-form more HCl? This is because different acids and bases have different strengths. HCl is a stronger acid than hydronium ion, and water is a stronger base than Cl^-.

In the Brønsted-Lowry theory, every acid base reaction creates its *conjugate acid-base pair*. In the above reaction HCl is an acid which, after giving up a proton, becomes a conjugate base, Cl^-. Similarly, water is a base which, after accepting a proton, becomes a conjugate acid, the hydronium ion.

conjugate base-acid pair

$$HCl + H_2O \rightleftharpoons H_3O^+ + Cl^-$$

conjugate acid-base pair

Some acids can give up only one proton. These are *monoprotic* acids. Examples are (H)Cl, (H)NO₃, HCOO(H), and CH₃COO(H). The hydrogens circled are the ones donated. Other acids yield two or three protons. These are called *diprotic* or *triprotic* acids. Examples are H_2SO_4, H_2CO_3, and H_3PO_4. However, in the Brønsted-Lowry theory, each acid is considered monoprotic, and a diprotic acid (such as carbonic acid) donates its protons in two distinct steps:

1. $H_2CO_3 + H_2O \rightleftharpoons H_3O^+ + HCO_3^-$

2. $HCO_3^- + H_2O \rightleftharpoons H_3O^+ + CO_3^{2-}$

Thus the compound HCO_3^- is a conjugate base in the first reaction and an acid in the second reaction. A compound that can act either as an acid or a base is called *amphoteric*.

In the self-ionization reaction

$$H_2O + H_2O \rightleftharpoons H_3O^+ + OH^-$$

one water acts as an acid (proton donor) and the other as a base (proton acceptor). In pure water, the equilibrium lies far to the left, that is, only very few hydronium and hydroxyl ions are formed. In fact, only 1×10^{-7} moles of hydronium ion and 1×10^{-7} moles of hydroxide ion are found in one liter of water. The dissociation constant for the self-ionization of water is

$$K_d = \frac{[H_3O^+][OH^-]}{[H_2O]^2}$$

This can be rewritten as

$$K_w = K_d [H_2O]^2 = [H_3O^+][OH^-]$$

K_w, the **ion product of water**, is still a constant because very few water molecules reacted to yield hydronium and hydroxide ions; hence the concentration of water essentially remained constant. At room temperature, the K_w has the value of

$$K_w = 1 \times 10^{-14} = [1 \times 10^{-7}] \times [1 \times 10^{-7}]$$

This value of the ion product of water applies not only to pure water but to any aqueous (water) solution. This is very convenient because if we know the concentration of the hydronium ion, we automatically know the concentration of the hydroxide ion and vice versa. For example, if in a 0.01 M HCl solution HCl dissociates completely, the hydronium ion concentration is $[H_3O^+] = 1 \times 10^{-2}$ M. This means that the $[OH^-]$ is

$$[OH^-] = K_w/[H_3O^+] = 1 \times 10^{-14}/1 \times 10^{-2} = 1 \times 10^{-12} \text{ M}$$

To measure the strength of an aqueous acidic or basic solution, P. L. Sorensen introduced the pH scale.

$$pH = -\log[H_3O^+]$$

In pure water, we have seen that the hydronium ion concentration is 1×10^{-7} M. The logarithm of this is -7 and, thus, the pH of pure water is 7. Since water is an amphoteric compound, pH 7 means a neutral solution. On the other hand, in a 0.01 M HCl solution (dissociating completely), we have $[H_3O^+] = 1 \times 10^{-2}$ M. Thus its pH is 2. The pH scale shows that acidic solutions have a pH less than 7 and basic solutions have a pH greater than 7.

pH 0 1 2 3 4 5 6 7 8 9 10 11 12 13 14
       acidic   neutral   basic

The pH of a solution can be measured conveniently by special instruments called pH meters. All that must be done is to insert the electrodes of the pH meter into the solution to be measured and read the pH from a scale. pH of a solution can also be obtained, although less precisely, by using a pH indicator paper. The paper is impregnated with organic compounds that change their color at different pH values. The color shown by the paper is then compared with a color chart provided by the manufacturer.

There are certain solutions that resist a change in the pH even when we add to them acids or bases. Such systems are called *buffers*. A mixture of a weak acid and its conjugate base usually forms a good buffer system. An example is carbonic acid, which is the most important buffer in our blood and maintains it close to pH 7.4. Buffers resist large changes in pH because of the Le Chatelier principle governing equilibrium conditions. In the carbonic acid/bicarbonate (weak acid/conjugate base) buffer system,

$$H_2CO_3 + H_2O \rightleftharpoons HCO_3^- + H_3O^+$$

any addition of an acid, H_3O^+, will shift the equilibrium to the left. Thus this reduces the hydronium ion concentration, returning it to the initial value so that it stays constant; hence the change in pH is small. If a base, OH^-, is added to such a buffer system, it will react with the H_3O^+ of the buffer. But the equilibrium then shifts to the right, replacing the reacted hydronium ions, hence again, the change in pH is small.

Buffers stabilize a solution at a certain pH. This depends on the nature of the buffer and its concentration. For example, the carbonic acid/bicarbonate system has a pH of 6.37 when the two ingredients are at equimolar concentration. A change in the concentration of the carbonic acid relative to its conjugate base can shift the pH of the buffer. The Henderson-Hasselbalch equation below gives the relationship between pH and concentration.

$$pH = pK_a + \log \frac{[A^-]}{[HA]}$$

In this equation the pK_a is the $-\log K_a$, where K_a is the dissociation constant of carbonic acid

$$K_a = \frac{[HCO_3^-][H_3O^+]}{[H_2CO_3]}$$

[HA] is the concentration of the acid and [A$^-$] is the concentration of the conjugate base. The pK_a of the carbonic acid/bicarbonate system is 6.37. When equimolar conditions exist, then [HA] = [A$^-$]. In this case, the second term in the Henderson-Hasselbalch equation is zero. This is so because [A$^-$]/[HA] = 1, and the log 1 = 0. Thus at equimolar concentration of the acid/conjugate base, the pH of the buffer equals the pK_a; in the carbonic acid/bicarbonate system this is 6.37. If, however, we have ten times more bicarbonate than carbonic acid, [A$^-$]/[HA] = 10, then log 10 = 1 and the pH of the buffer will be

$$pH = pK_a + \log [A^-]/[HA] = 6.37 + 1.0 = 7.37$$

This is what happens in our blood—the bicarbonate concentration is ten times that of the carbonic acid and this keeps our blood at a pH of 7.4. Any large change in the pH of our blood may be fatal (acidosis or alkalosis). Other buffer systems work the same way. For example, the second buffer system in our blood is

$$H_2PO_4^- + H_2O \rightleftharpoons HPO_4^{2-} + H_3O^+$$

The pK_a of this buffer system is 7.21. It requires a 1.6/1.0 molar ratio of $HPO_4^{2-}/H_2PO_4^-$ to maintain our blood at pH 7.4.

Objectives

1. To learn how to measure pH of a solution.
2. To understand the operation of buffer systems.

Procedure

Measurement of pH

1. Add one drop of 0.1 M HCl to the first depression of a spot plate. Dip a 2-cm long universal pH paper into the solution. Remove the excess liquid from the paper by touching the plate. Compare the color of the paper to the color chart provided (Fig. 20.1). Record the pH on your Report Sheet (1).

2. Repeat the same procedure with 0.1 M acetic acid, 0.1 M sodium acetate, 0.1 M carbonic acid (or club soda or seltzer), 0.1 M sodium bicarbonate, 0.1 M ammonia, and 0.1 M NaOH. For each solution, use a different depression of the spot plate. Record your results on the Report Sheet (1).

Figure 20.1

pH paper dispenser.

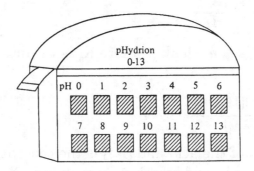

3. Depending on the availability of the number of pH meters this may be a class exercise (demonstration), or 6–8 students may use one pH meter. Add 5 mL of 0.1 M acetic acid to a dry and clean 50-mL beaker. Wash the electrodes over a 200-mL beaker with distilled water contained in a wash bottle. The 200-mL beaker serves to collect the wash water. Gently wipe the electrodes with Kimwipes (or other soft tissues) to dryness. Insert the dry electrodes into the acetic acid solution. Your pH meter has been calibrated by your instructor.

210   Experiment 20

Switch "on" the pH meter and read the pH from the position of the needle on your scale. Alternatively, if you have a digital pH meter, a number corresponding to the pH will appear (Fig. 20.2).

<table>
<tr><td>

CAUTION!

Make sure the electrodes are immersed into the solution but do not touch the walls or the bottom of the beaker. Electrodes are made of thin glass, and they break easily if you don't handle them gently.
</td></tr>
</table>

Figure 20.2

pH meter.

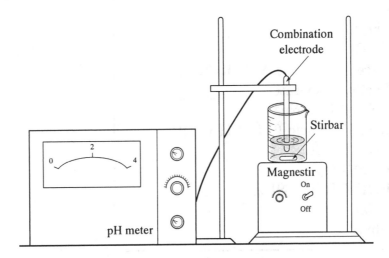

4. Repeat the same procedures with 0.1 M sodium acetate, 0.1 M carbonic acid, 0.1 M sodium bicarbonate, and 0.1 M ammonia. Make certain that for each solution you use a dry and clean beaker, and before each measurement wash the electrodes with distilled water and dry them with Kimwipes. Record your data on the Report Sheet (2).

5. Prepare four buffer systems in four separate, labeled, dry and clean 50-mL beakers, as follows:

   (a) 5 mL 0.1 M acetic acid + 5 mL 0.1 M sodium acetate

   (b) 1 mL 0.1 M acetic acid + 10 mL 0.1 M sodium acetate

   (c) 5 mL 0.1 M carbonic acid + 5 mL 0.1 M sodium bicarbonate

   (d) 1 mL 0.1 M carbonic acid + 10 mL 0.1 M sodium bicarbonate

   Measure the pH of each buffer system with the aid of a universal pH paper. Record your data on the Report Sheet (3), (6), (9), and (12).

6. Divide each of your buffers (a–d) into two halves (5 mL each) and place them into clean and dry 10-mL beakers. To the first sample of buffer (a), add 0.5 mL 0.1 M HCl. Mix and measure the pH with the aid of universal pH paper. Record your data on the Report Sheet (4). To the second sample of buffer (a), add

0.5 mL 0.1 M NaOH. Mix and measure the pH with pH paper. Record your data on the Report Sheet (5).

7. Repeat the same measurements with buffers (b), (c), and (d). Record your data on the Report Sheet for the appropriate buffer system under (7), (8), (10), (11), (13), and (14).

Chemicals and Equipment

1. pH meter
2. pH paper
3. Kimwipes
4. Wash bottle
5. 0.1 M HCl
6. 0.1 M acetic acid (0.1 M CH_3COOH)
7. 0.1 M sodium acetate
   (0.1 M $CH_3COO^-Na^+$)
8. 0.1 M carbonic acid (0.1 M H_2CO_3);
   club soda or seltzer
9. 0.1 M $NaHCO_3$
10. 0.1 M NH_4OH [0.1 NH_3(aq)]
11. 0.1 M NaOH
12. Spot plate

Experiment 20

PRE-LAB QUESTIONS

1. Succinic acid has the formula of $HOOC-CH_2-CH_2-COOH$. It is a diprotic acid. Show the formula of the conjugate base for each proton donated.

2. The pK_a of formic acid is 3.75. What is the pH of a buffer in which formic acid and sodium formate have equimolar concentration? What is the pH of a solution in which the sodium formate is 10 M and the formic acid is 1 M?

3. The pH of blood is 7.4 and that of saliva is 6.4. How much more hydronium ion (H_3O^+) is in the saliva than in the blood?

Experiment 20

REPORT SHEET

pH of solutions	**1. by pH paper**	**2. by pH meter**
0.1 M HCl	_____	— _____
0.1 M acetic acid	_____	_____
0.1 M sodium acetate	_____	_____
0.1 M carbonic acid	_____	_____
0.1 M sodium bicarbonate	_____	_____
0.1 M ammonia	_____	_____
0.1 M NaOH	_____	— _____

Buffer systems	**pH**
3. 5 mL 0.1 M CH_3COOH + 5 mL 0.1 M $CH_3COO^-Na^+$ (**a**)	_____
4. after addition 0.5 mL 0.1 M HCl	_____
5. after addition 0.5 mL 0.1 M NaOH	_____
6. 1 mL 0.1 M CH_3COOH + 10 mL 0.1 M $CH_3COO^-Na^+$ (**b**)	_____
7. after addition 0.5 mL 0.1 M HCl	_____
8. after addition 0.5 mL 0.1 M NaOH	_____
9. 5 mL 0.1 M H_2CO_3 + 5 mL 0.1 M $NaHCO_3$ (**c**)	_____
10. after addition 0.5 mL 0.1 M HCl	_____
11. after addition 0.5 mL 0.1 M NaOH	_____
12. 1 mL 0.1 M H_2CO_3 + 10 mL 0.1 M $NaHCO_3$ (**d**)	_____
13. after addition 0.5 mL 0.1 M HCl	_____
14. after addition 0.5 mL 0.1 M NaOH	_____

POST-LAB QUESTIONS

1. Calculate the expected pH values of the buffer systems from the experiment
(a−d), using the Henderson-Hasselbalch equation and the pK_a values given in
the **Background** section.

(a)

(b)

(c)

(d)

Are they in agreement with your measured pH values?

2. How many units of pH did you observe change upon the addition of 0.5 mL
0.1 M HCl to the four buffer systems? What can you conclude from these
results?

(a)

(b)

(c)

(d)

3. Which of the four buffers you prepared (a−d) is the best buffer?

Analysis of vinegar by titration

In order to measure how much acid or base is present in a solution we often use a method called *titration*. If a solution is acidic, titration consists of adding base to it until all the acid is neutralized. To do this, we need two things: (1) a means of measuring how much base is added and (2) a means of telling just when the acid is completely neutralized.

How much base is added requires the knowledge of the number of equivalents of the base. The number of equivalents is the product of the volume of the base added and the normality of the base.

$$\text{Equivalents} = \mathbf{V} \times \mathbf{N}$$

The titration is completed when the number of equivalents of acid equals the number of equivalents of base.

$$\text{Equivalents}_{acid} = \text{Equivalents}_{base}$$

or

$$\mathbf{V}_{acid}\,\mathbf{N}_{acid} = \mathbf{V}_{base}\,\mathbf{N}_{base}$$

This is called the *titration equation*.

We use an *indicator* to tell us when the titration is completed. Indicators are organic compounds that change their color when there is a change in the pH of the solution. The *end point* of the titration is when a sudden change in the pH of the solution occurs. Therefore, we can tell the completion of the titration when we observe a change in the color of our solution to which a few drops of indicator have been added.

Commercial vinegar contains 5–6% acetic acid. Acetic acid, CH_3COOH, is a monoprotic acid. Therefore, its concentration expressed in molarity or normality is the same. It is a weak acid and when titrated with a strong base such as NaOH, upon completion of the titration, there is a sudden change in the pH in the range from 6.0 to 9.0. The best way to monitor such a change is to use the indicator phenolphthalein, which changes from colorless to a pink hue at pH 8.0–9.0.

With the aid of the titration equation, we can calculate the concentration of acetic acid in the vinegar. To do so, we must know the volume of the acid (5 mL), the normality of the base (0.2 N), and the volume of the base used to reach the end point of the titration. This will be read from the *buret*, which is filled with the 0.2 N NaOH solution at the beginning of the titration to its maximum capacity. The base is then slowly added (dropwise) from the buret to the vinegar in an

Erlenmeyer flask. Continuous swirling ensures proper mixing. The titration is stopped when the indicator shows a permanent pink coloration. The buret is read again. The volume of the base added is the difference between the initial volume (25 mL) and the volume left in the buret at the end of titration.

Objectives

1. To learn the techniques of titration.
2. To determine the concentration of acetic acid in vinegar.

Procedure

1. Rinse a 25-mL buret with about 5 mL of 0.2 N NaOH solution. (Be sure to record the exact concentration of the base.) After rinsing, fill the buret with 0.2 N NaOH solution about 2 mL above the 0.0 mL mark. Use a clean and dry funnel for filling. Tilting the buret at a 45° angle, slowly turn the stopcock to allow the solution to fill the tip. The air bubbles must be completely removed from the tip. If you do not succeed the first time, repeat it until the liquid in the buret forms one continuous column from top to bottom. Clamp the buret onto a ring stand (Fig. 21.1). By slowly opening the stopcock, allow the bottom of the meniscus to drop to the 0.0 mL mark.

Figure 21.1

Titration setup.

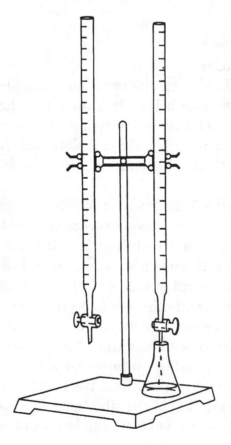

2. With the aid of a 5-mL volumetric pipet, add 5 mL vinegar to a 100-mL Erlenmeyer flask. Allow the vinegar to drain completely from the pipet by holding the pipet in such a manner that its tip touches the wall of the flask. Record the volume of the vinegar for trial 1 on your Report Sheet (1). Record also the normality of the base (2) and the initial reading of the base in the buret on your Report Sheet (3). Add a few drops of phenolphthalein indicator to the flask and about 10 mL of distilled water. The distilled water is added to dilute the natural color that some commercial vinegars have. In this way, the natural color will not interfere with the color change of the indicator.

3. While holding the neck of the Erlenmeyer flask in your left hand and swirling it, open the stopcock of the buret slightly with your right hand and allow the <u>dropwise</u> addition of the base to the flask. At the point where the base hits the vinegar solution the color may turn <u>temporarily</u> pink, but this color will disappear upon mixing the solution by swirling. Continue the titration until a faint <u>permanent</u> pink coloration appears. Stop the titration. Record the readings of the base in your buret on your Report Sheet (4).

CAUTION!

Be careful not to add too much base, an error called "overtitration." If the indicator in your flask turns deep pink or purple, you have overtitrated and will need to repeat the entire titration with a new sample of vinegar.

---

4. Repeat the procedures in nos. 1–3 with a new 5 mL vinegar sample for trial 2. Record these results on your Report Sheet.

5. With the aid of the titration equation, calculate the normality of the vinegar (5) for trial 1 and 2.

6. Average the two normalities. Using the molecular weight of 60 g/mole for acetic acid, calculate the percent concentration of acetic acid in vinegar.

Chemicals and Equipment

1. 25-mL buret
2. Buret clamp
3. 5-mL pipet
4. Small funnel
5. 0.2 N standardized NaOH
6. Phenolphthalein indicator

NAME SECTION DATE

PARTNER GRADE

Experiment 21

PRE-LAB QUESTIONS

1. In a monoprotic acid, such as acetic acid, CH_3COOH, only one hydrogen is donated. In the above structure there are four hydrogens. Which one is donated?

2. What is the molarity of (a) 0.5 N H_3PO_4 (a triprotic acid) and (b) 0.5 N H_2SO_3 (a diprotic acid)?

3. How many equivalents of sulfuric acid are in 20 mL of 0.35 N H_2SO_4 solution? How many grams?

4. What is the normality of an unkown base if 25 mL of the base can be titrated to an end point by 10 mL of 0.2 N HCl?

Experiment 21

REPORT SHEET

Titration	**Trial 1**	**Trial 2**
1. Volume of vinegar sample	_____	_____
2. Normality of NaOH solution	_____	_____
3. Initial reading of NaOH in buret	_____ mL	_____ mL
4. Final reading of NaOH in buret	_____ mL	_____ mL
5. Volume of NaOH used in titration: (4) − (3)	_____ mL	_____ mL
6. Normality of acetic acid in vinegar: (2) × (5)/(1)	_____	_____
7. Average normality of acetic acid		_____
8. Percent (w/v) of acetic acid in vinegar: (7) × 60 × 0.1		_____ %

[% w/v = g/100 mL = (7) equiv./1000 ~~mL~~ × 60 g/equiv. × 0.1 × 1000 ~~mL~~/100 mL]

POST-LAB QUESTIONS

1. Assume your vinegar contained a small amount of citric acid (a triprotic acid). Using the same experimental data, would you expect the normality of this sample to be the same or different than a sample which contained only pure acetic acid?

2. Assume that you overtitrated your vinegar sample, i.e., you added too much NaOH and got a deep purple color at the end point. Would the calculated normality of the acid be smaller, larger, or the same as the solution without overtitration?

3. We added about 10 mL of distilled water to the vinegar, but we did not use this volume in the calculation of normality. Why do you think we can ignore this volume in the calculation?

4. If you would have used an indicator which changes its color at pH 4.0 rather than phenolphthalein which changes at pH 8.0, would your result show a vinegar concentration higher, the same, or lower than the one you obtained?

● Analysis of antacid tablets

The natural environment of our stomach is quite acidic. Gastric juice, which is mostly hydrochloric acid, has a pH of 1.0. Such a strong acidic environment denatures proteins and helps their digestion by enzymes such as pepsin. Not only is the denatured protein more easily digested by enzymes than the native protein, but the acidic environment helps to activate pepsin. The inactive form of pepsin, pepsinogen, is converted to the active form, pepsin, by removing a chunk of its chain, 42 amino acid units. This can only occur in an acidic environment, and pepsin molecules catalyze this reaction (autocatalysis). But too much acid in the stomach is not good either. In the absence of food, the strong acid, HCl, denatures the proteins in the stomach wall itself. If this goes on unchecked it may cause stomach or duodenal ulcers.

We feel the excess acidity in our stomach. Such sensations are called "heartburn" or "sour stomach." To relieve "heartburn," we take antacids in tablet or liquid form. Antacid is a medical term. It implies a substance that neutralizes acid. Drugstore antacids contain a number of different active ingredients. Almost all of them are weak bases (hydroxides and/or carbonates). Table 22.1 lists the active ingredients of some commercial antacids.

Table 22.1 Active Ingredients of Some Drugstore Antacids
Alka-Seltzer: sodium bicarbonate and citrate
Bromo-Seltzer: sodium bicarbonate and citrate
Chooz, Tums: calcium carbonate
Di-gel, Gelusil, Maalox: aluminum hydroxide and magnesium hydroxide
Gaviscon, Remegel: aluminum hydroxide and magnesium carbonate
Rolaids: aluminum sodium dihydroxy carbonate

HCl in the gastric juice is neutralized by these active ingredients in the following reactions:

$$NaHCO_3 + HCl \longrightarrow NaCl + H_2O + CO_2$$
$$CaCO_3 + 2HCl \longrightarrow CaCl_2 + H_2O + CO_2$$
$$Al(OH)_3 + 3HCl \longrightarrow AlCl_3 + 3H_2O$$
$$Mg(OH)_2 + 2HCl \longrightarrow MgCl_2 + 2H_2O$$
$$AlNa(OH)_2CO_3 + 4HCl \longrightarrow AlCl_3 + NaCl + 3H_2O + CO_2$$

Besides the active ingredients, antacid tablets also contain inactive ingredients, such as starch, which act as a binder or filler. The efficacy of an antacid tablet is its ability to neutralize HCl. The more HCl that is neutralized, the more effective the antacid pill. (You must have heard the competing advertisement claims of different commercial antacids: "Tums neutralizes one-third more stomach acid than Rolaids.")

Antacids are not completely harmless. The HCl production in the stomach is regulated by the stomach pH. If too much antacid is taken, the pH becomes too high; the result will be the so-called "acid rebound." This means that ultimately, more HCl will be produced than was present before taking the antacid pill.

In the present experiment, you will determine the amount of HCl neutralized by two different commercial antacid tablets. To do so we use a technique called *back-titration*. We add an **excess** amount of 0.2 N HCl to the antacid tablet. The excess acid (more than is needed for neutralization) helps to dissolve the tablet. Then the active ingredients in the antacid tablet will neutralize part of the added acid. The remaining HCl is determined by titration with NaOH. A standardized NaOH solution of known concentration (0.2 N) is used and added slowly until all the HCl is neutralized. We observe this end point of the titration when the added indicator, thymol blue, changes its color from red to yellow. The volume of the excess 0.2 N HCl (the volume not neutralized by the antacid) is obtained from the titration equation:

$$V_{acid} \times N_{acid} = V_{base} \times N_{base}$$
$$V_{acid} = (V_{base} \times N_{base})/N_{acid}$$

Once this is known, the amount of HCl neutralized by the antacid pill is obtained as the difference between the initially added volume and the back-titrated volume:

$$V_{HCl\ neutralized\ by\ the\ pill} = V_{HCl\ initially\ added} - V_{HCl\ backtitrated}$$

In this way we can compare the effectiveness of different drugstore antacids.

Objectives

1. To learn the technique of back-titration.
2. To compare the efficacies of drugstore antacid tablets.

Procedure

1. Rinse a 25-mL buret with about 5 mL 0.2 N NaOH. After rinsing, fill the buret with 0.2 N NaOH solution about 2 mL above the top mark. Use a clean and dry funnel for filling. Tilting the filled buret at a 45° angle, turn the stopcock open to allow the solution to fill the tip of the buret. The air bubbles should be completely removed from the tip by this maneuver. If you do not succeed the first time, repeat it until the liquid in the buret forms one continuous column from

top to bottom. Clamp the buret onto a ring stand (Fig. 21.1). By slowly opening the stopcock, allow the bottom of the meniscus to drop to the 0.0 mL mark.

2. Repeat the above procedure with a 100-mL buret and fill it to the 0.0 mL mark with 0.2 N HCl. Clamp this, too, onto a ring stand.

3. Obtain two different antacid tablets from your instructor. Note the name of the tablets on your Report Sheet (1). Weigh each tablet on a balance to the nearest 0.001 g. Report the weight on your Report Sheet (2). Place each tablet in separate 250-mL Erlenmeyer flasks. Label the flasks. Add about 10 mL water to each flask. With the help of stirring rods (one for each flask), break up the tablets.

4. Add exactly 50 mL 0.2 N HCl to each Erlenmeyer flask from the buret. Also, add a few drops of thymol blue indicator. Gently stir with the stirring rods to disperse the tablets. (Some of the inactive ingredients may not go into solution and will settle as a fine powder on the bottom of the flask). At this point the solution should be red (the color of thymol blue at acidic pH). If either of your solutions does not have red coloration, add 10 mL 0.2 N HCl from the refilled buret and make certain that the red color will persist for more than 30 sec. Record the total volume of 0.2 N HCl added to each flask on your Report Sheet (3).

5. Place the Erlenmeyer flask under the buret containing the 0.2 N NaOH. Record the level of the meniscus of the NaOH solution in the buret before you start the titration (4). While holding and swirling the neck of the Erlenmeyer flask with your left hand, titrate the contents of your solution by adding (dropwise) 0.2 N NaOH by opening the stopcock of the buret with your right hand. Continue to add NaOH until the color changes to yellow and stays yellow for 30 sec. after the last drop. Record the level of the NaOH solution in the buret by reading the meniscus at the end of titration (5).

6. Refill the buret with 0.2 N NaOH as before and repeat the titration for the second antacid.

7. Calculate the volume of the acid obtained in the back-titration and record it on your Report Sheet (6). Calculate the volume of the 0.2 N HCl neutralized by the antacid tablets (7). Calculate the g HCl neutralized by 1 g antacid tablet. Record it on your Report Sheet (8).

Chemicals and Equipment

1. 25-mL buret
2. 100-mL buret
3. Buret clamp
4. Ring stand
5. Balance
6. Antacid tablets
7. 0.2 N NaOH
8. 0.2 N HCl
9. Thymol blue indicator

NAME _____ SECTION _____ DATE _____

PARTNER _____ GRADE _____

Experiment 22

PRE-LAB QUESTIONS

1. In addition to relieving "heartburn," some antacids advertise that they provide an essential mineral ingredient—calcium. Which antacids can make such a claim?

2. What does the term back-titration mean?

3. Looking at Table 8.4 in your textbook (Bettelheim & March), which beverage is most likely to cause "heartburn"?
(a) water   (b) milk   (c) tomato juice   (d) black coffee

4. Some patients with elevated blood pressure must restrict the sodium intake in their diet. Which antacid would you recommend for such a patient?

Experiment 22

REPORT SHEET

	(a)	(b)
1. Name of the antacid tablet	_____	_____
2. Weight of the antacid tablet	_____ g	_____ g
3. Total volume of 0.2 N HCl added to the antacid before titration	_____ mL	_____ mL
4. Reading of 0.2 N NaOH in the buret before titration	_____ mL	_____ mL
5. Reading of 0.2 N NaOH in the buret after titration	_____ mL	_____ mL
6. Volume of 0.2 N HCl obtained in back-titration: (5)−(4)	_____ mL	_____ mL
7. Volume of 0.2 N HCl neutralized by one antacid tablet: (3)−(6)	_____ mL	_____ mL
8. Gram HCl neutralized by 1 g antacid tablet: [(7)/(2)] × 0.2 × 36.5	_____	_____

POST-LAB QUESTIONS

1. Which antacid tablet neutralized more stomach acid (a) per tablet and (b) per gram tablet?

2. Certain TV commercials claim that their antacid product neutralizes 47 times its own weight in stomach acid. On the basis of your results, is such a claim justified? (In your calculation, you should keep in mind that stomach acid is 0.1 M HCl. In other words, there are 3.65 g HCl in 1 L of stomach acid, or 1 g HCl equals 274 g stomach acid.)

3. How many grams of $Mg(OH)_2$ do you need in an antacid tablet to neutralize 25 mL of stomach acid (0.1 M HCl)?

4. You used an indicator, thymol blue, which changed its color from red (acidic) to yellow (neutral) at a pH of 2.0. If you would have selected another indicator, for example litmus, which changes color from red (acidic) to blue (neutral) at pH 6.0, would you need the same, more, or less volume of NaOH in the back titration? Explain.

Structure in organic compounds: use of molecular models (I)

Background

The study of organic chemistry usually involves those molecules which contain carbon. Thus a convenient definition of *organic chemistry* is the chemistry of carbon compounds.

There are several characteristics of organic compounds that make their study interesting:

a. Carbon forms strong bonds to itself as well as to other elements; the most common elements found in organic compounds, other than carbon, are hydrogen, oxygen, and nitrogen.

b. Carbon atoms are generally tetravalent. This means that carbon atoms in most organic compounds are bound by four covalent bonds to adjacent atoms.

c. Organic molecules are three-dimensional and occupy space. The covalent bonds which carbon makes to adjacent atoms are at discrete angles to each other. Depending on the type of organic compound, the angle may be 180°, 120°, or 109.5°. These angles correspond to compounds which have triple bonds (1), double bonds (2), and single bonds (3), respectively.

$$-C\equiv C- \qquad\qquad {>}C{=}C{<} \qquad\qquad -\overset{|}{\underset{|}{C}}-\overset{|}{\underset{|}{C}}-$$

$$(1) \qquad\qquad\qquad (2) \qquad\qquad\qquad (3)$$

d. Organic compounds can have a limitless variety in composition, shape, and structure.

Thus while a molecular formula tells the number and type of atoms present in a compound, it tells nothing about the structure. The structural formula is a two-dimensional representation of a molecule and shows the sequence in which the atoms are connected and the bond type. For example, the molecular formula C_4H_{10} can be represented by two different structures: butane (4) and 2-methylpropane (isobutane) (5).

H H H H

H−C−C−C−C−H

H H H H

Butane (4)

H H H

H−C−C−C−H

H H H

H−C−H

H

2-Methylpropane (5)

Isobutane

Consider also the molecular formula C_2H_6O. There are two structures which correspond to this formula: dimethyl ether (6) and ethanol (ethyl alcohol) (7).

H H

H−C−O−C−H

H H

Dimethyl ether (6)

H H

H−C−C−O−H

H H

Ethanol (7)

Ethyl alcohol

In the pairs above, each structural formula represents a different compound. Each compound has its own unique set of physical and chemical properties. Compounds with the same molecular formula but with different structural formulas are called *isomers*.

The three-dimensional character of molecules is expressed by its stereochemistry. By looking at the *stereochemistry* of a molecule, the spatial relationships between atoms on one carbon and the atoms on an adjacent carbon can be examined. Since rotation can occur around carbon-carbon single bonds in open chain molecules, the atoms on adjacent carbons can assume different spatial relationships with respect to each other. The different arrangements that atoms can assume as a result of a rotation about a single bond are called *conformations*. A specific conformation is called a *conformer*. While individual isomers can be isolated, conformers cannot since interconversion, by rotation, is too rapid.

Conformers may be represented by the use of two conventions as shown in Fig. 23.1.

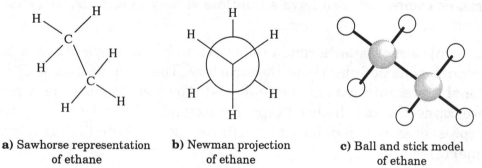

a) Sawhorse representation of ethane

b) Newman projection of ethane

c) Ball and stick model of ethane

Figure 23.1 • Molecular representations.

The *sawhorse* views the carbon-carbon bond at an angle and, by showing all the bonds and atoms, shows their spatial arrangements. The *Newman projection* provides a view along a carbon-carbon bond by sighting directly along the carbon-carbon bond. The near carbon is represented by a circle, and bonds attached to it are represented by lines going to the center of the circle. The carbon behind is not visible (since it is blocked by the near carbon), but the bonds attached to it are partially visible and are represented by lines going to the edge of the circle. With Newman projections, rotations show the spatial relationships of atoms on adjacent carbons easily. Two conformers that represent extremes are shown in Fig. 23.2.

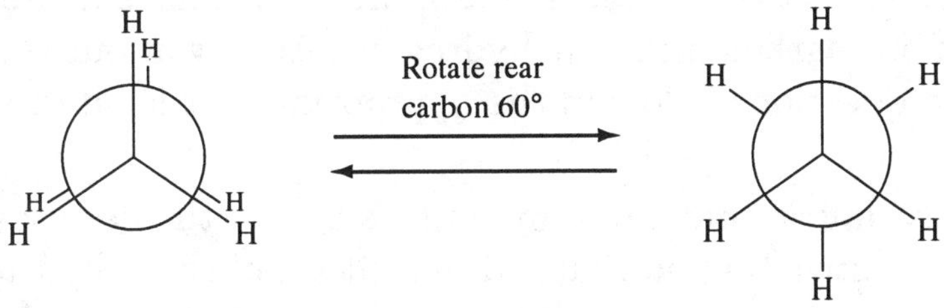

a) Eclipsed conformation of ethane

b) Staggered conformation of ethane

Figure 23.2 • Two conformers of ethane.

The *eclipsed* conformation has the bonds (and the atoms) on the adjacent carbons as close as possible. The *staggered* conformation has the bonds (and the atoms) on adjacent carbons as far as possible. One conformation can interconvert into the other by rotation around the carbon-carbon bond axis.

The three-dimensional character of molecular structure is shown through molecular model building. With molecular models, the number and types of bonds between atoms and the spatial arrangements of the atoms can be visualized for the molecules. This allows comparison of isomers and of conformers for a given set of compounds.

Objectives

1. To use models to visualize structure in organic molecules.
2. To build and compare isomers having a given molecular formula.
3. To explore the three-dimensional character of organic molecules.

Procedure

Obtain a set of ball-and-stick molecular models from the laboratory instructor. The set contains the following parts (other colored spheres may be substituted as available):

- 2 Black spheres representing *Carbon*; this tetracovalent element has four holes;

- 6 Yellow spheres representing *Hydrogen*; this monovalent element has one hole;
- 2 Colored spheres representing the *halogen Chlorine*; this monovalent element has one hole;
- 1 Blue sphere representing *Oxygen*; this divalent element has two holes;
- 8 Sticks to represent bonds.

1. With your models, construct the molecule methane. Methane is a simple hydrocarbon consisting of one carbon and four hydrogens. After you put the model together, answer the questions below in the appropriate space on the Report Sheet.

   a. With the model resting so that three hydrogens are on the desk, examine the structure. Move the structure so that a different set of three hydrogens are on the desk each time. Is there any difference between the way that the two structures look (1a)?

   b. Does the term "equivalent" adequately describe the four hydrogens of methane (1b)?

   c. Tilt the model so that only two hydrogens are in contact with the desk and imagine pressing the model flat onto the desk top. Draw the way in which the methane molecule would look in two dimensional space (1c). This is the usual way that three-dimensional structures are written.

   d. Using a protractor, measure the angle H−C−H on the model (1d).

2. Replace one of the hydrogens of the methane model with a colored sphere, which represents the halogen chlorine. The new model is chloromethane (methyl chloride), CH_3Cl. Position the model so that the three hydrogens are on the desk.

   a. Grasp the atom representing chlorine and tilt it to the right, keeping two hydrogens on the desk. Write the structure of the projection on the Report Sheet (2a).

   b. Return the model to its original position and then tilt as before, but this time to the left. Write this projection on the Report Sheet (2b).

   c. While the projection of the molecule changes, does the structure of chloromethane change (2c)?

3. Now replace a second hydrogen with another chlorine sphere. The new molecule is dichloromethane, CH_2Cl_2.

   a. Examine the model as you twist and turn it in space. Are the projections given below isomers of the molecule CH_2Cl_2 or representations of the same structure only seen differently in three dimensions (3a)?

4. Construct the molecule ethane, C_2H_6. Note that you can make ethane from the methane model by removing a hydrogen and replacing the hydrogen with a methyl group, $-CH_3$.

  a. Write the structural formula for ethane (4a).

  b. Are all the hydrogens attached to the carbon atoms equivalent (4b)?

  c. Draw a sawhorse representation of ethane. Draw a staggered and an eclipsed Newman projection of ethane (4c).

  d. Replace any hydrogen in your model with chlorine. Write the structure of the molecule chloroethane (ethyl chloride), C_2H_5Cl (4d).

  e. Twist and turn your model. Draw two Newman projections of the chloroethane molecule (4e).

  f. Do the projections that you drew represent different isomers or conformers of the same compound (4f)?

5. Dichloroethane, $C_2H_4Cl_2$

  a. In your molecule of chloroethane, if you choose to remove another hydrogen note that you now have a choice among the hydrogens. You can either remove a hydrogen from the carbon to which the chlorine is attached, or you can remove a hydrogen from the carbon that has only hydrogens attached. First, remove the hydrogen from the carbon that has the chlorine attached and replace it with a second chlorine. Write its structure on the Report Sheet (5a).

  b. Compare this structure to the model which would result from removal of a hydrogen from the other carbon and its replacement by chlorine. Write its structure (5b) and compare it to the previous example. One isomer is 1,1-dichloroethane; the other is 1,2-dichloroethane. Label the structures drawn on the Report Sheet with the correct name.

  c. Are all the hydrogens of chloroethane equivalent? Are some of the hydrogens equivalent? Label those hydrogens which are equivalent to each other (5c).

6. Butane

  a. Butane has the formula C_4H_{10}. With help from a partner, construct a model of butane by connecting the four carbons in a series (C−C−C−C) and then adding the needed hydrogens. First, orient the model in such a way that the carbons appear as a straight line. Next, tilt the model so that the carbons appear as a zig-zag line. Then, twist around any of the C−C bonds so that a part of the chain is at an angle to the remainder. Draw each of these structures in the space on the Report Sheet (6a). Note that the structures you draw are for the same molecule but represent only a different orientation and projection.

  b. Sight along the carbon-carbon bond of C_2 and C_3 on the butane chain: $\overset{1}{C}H_3-\overset{2}{C}H_2-\overset{3}{C}H_2-\overset{4}{C}H_3$. Draw a staggered Newman projection. Rotate the C_2 carbon clockwise by 60°; draw the eclipsed Newman projection. Again, rotate the C_2 carbon clockwise by 60°; draw the Newman projection. Is the last projection staggered or eclipsed (6b)? Continue rotation of the C_2 carbon clockwise by 60° increments and observe the changes that take place.

c. Examine the structure of butane for equivalent hydrogens. In the space on the Report Sheet (6c), redraw the structure of butane and label those hydrogens which are equivalent to each other. On the basis of this examination, predict how many monochlorobutane isomers (C_4H_9Cl) that could be obtained (6d). Test your prediction by replacement of hydrogen by chlorine on the models. Draw the structures of these isomers (6e).

d. Reconstruct the butane system. First, form a three-carbon chain, then connect the fourth carbon to the center carbon of the three-carbon chain. Add the necessary hydrogens. Draw the structure of 2-methylpropane (isobutane) (6f). Can any manipulation of the model, by twisting or turning of the model or by rotation of any of the bonds, give you the butane system? If these two, butane and 2-methylpropane (isobutane), are *isomers*, then how may we recognize that any two structures are isomers (6g)?

e. Examine the structure of 2-methylbutane for equivalent hydrogens. In the space on the Report Sheet (6h), redraw the structure of 2-methylbutane and label the equivalent hydrogens. Predict how many monochloroisomers of 2-methylbutane could be formed (6i) and test your prediction by replacement of hydrogen by chlorine on the model. Draw the structures of these isomers (6j).

7. C_2H_6O

a. There are two isomers with the molecular formula C_2H_6O, ethanol (ethyl alcohol) and dimethyl ether. With your partner, construct both of these isomers. Draw these isomers on the Report Sheet (7a) and name each one.

b. Manipulate each model. Can either be turned into the other by a simple twist or turn (7b)?

c. For each compound, label those hydrogens which are equivalent. How many sets of equivalent hydrogens are there in ethanol (ethyl alcohol) and dimethyl ether (7c)?

8. Optional: Butenes

a. If springs are available for the construction of double bonds, construct 2-butene, $CH_3-CH=CH-CH_3$. There are two isomers for compounds of this formulation: the isomer with the 2 $-CH_3$ groups on the same side of the double bond, *cis*-2-butene; and the isomer with the 2 $-CH_3$ groups on opposite sides of the double bond, *trans*-2-butene. Draw these two structures on the Report Sheet (8a).

b. Can you twist, turn, or rotate one model into the other? Explain (8b).

c. How many bonds are connected to any single carbon of these structures (8c)?

d. With the protractor, measure the C−C=C angle (8d).

e. Construct methylpropene, $CH_3-C=CH_2$. Can you have a *cis*- or
$$\underset{CH_3}{|}$$
a *trans*- isomer in this system (8e)?

9. Optional: Butynes

    a. If springs are available for the construction of triple bonds, construct 2-butyne, $CH_3-C\equiv C-CH_3$. Can you have a *cis*- or a *trans*- isomer in this system (9a)?

    b. With the protractor, measure the $C-C\equiv C$ angle (9b).

    c. Construct a second butyne with your molecular models and springs. How does this isomer differ from the one in (a) above (9c)?

Experiment 23

PRE-LAB QUESTIONS

1. How many bonds can each of the elements below form with neighboring atoms in a compound?

        C   H   O   N   Br   S   Cl

2. What is a convenient definition for organic chemistry?

3. Write structural formulas for the four (4) compounds with the molecular formula C_4H_9Br.

4. How are the compounds in no. 3 (above) related?

NAME

SECTION

DATE

PARTNER

GRADE

Experiment 23

REPORT SHEET

1. Methane

a.

b.

c.

d.

2. Chloromethane (methyl chloride)

a.

b.

c.

3. Dichloromethane

a.

4. Ethane and chloroethane (ethyl chloride)

    a.

    b.

    c.

    d.

    e.

    f.

5. Dichloroethane

    a.

    b.

    c.

6. Butane

a.

b.

c.

d.

e.

f.

g.

h.

i.

j.

7. C_2H_6O

   a.

   b.

   c. Ethanol (ethyl alcohol) has _____ set(s) of equivalent hydrogens.

      Dimethyl ether has _____ set(s) of equivalent hydrogens.

8. Butenes

   a.

   b.

   c.

   d. C−C=C angle

   e.

9. Butynes

   a.

   b.

   c.

POST-LAB QUESTIONS

1. There are three (3) isomers of formula, C_3H_8O. Write structural formulas for these compounds.

2. Draw the structure of propane and identify equivalent hydrogens. Identify equivalent sets by letters, e.g., H_a, H_b, etc.

3. Draw the structural formulas for the four (4) isomers of the butenes, C_4H_8, which are alkenes. Label *cis-* and *trans-* isomers.

4. Draw a staggered and an eclipsed conformer for propane, $\overset{1}{C}H_3 - \overset{2}{C}H_2 - \overset{3}{C}H_3$, sighting along the $C_1 - C_2$ bond.

Experiment | 24

Stereochemistry: use of molecular models (II)

Background

In Experiment 23, we looked at some molecular variations that acyclic organic molecules can take:

1. *Constitutional isomerism*: molecules can have the same molecular formula but different arrangements of atoms.
   a. *skeletal isomerism*: structural isomers where differences are in the order in which atoms that make up the skeleton are connected; e.g., C_4H_{10}

$$CH_3CH_2CH_2CH_3$$

Butane

$$\overset{\displaystyle CH_3}{\underset{\displaystyle }{CH_3-CH-CH_3}}$$

2-Methylpropane

   b. *positional isomerism*: structural isomers where differences are in the location of a functional group; e.g., C_3H_7Cl

$$CH_3CH_2CH_2-Cl$$

1-Chloropropane

$$\overset{\displaystyle Cl}{\underset{\displaystyle }{CH_3-CH-CH_3}}$$

2-Chloropropane

2. *Stereoisomerism*: molecules which have the same order of attachment of atoms but differ in the arrangement of the atoms in 3-dimensional space.
   a. *cis/trans isomerism*: molecules that differ due to the geometry of substitution around a double bond; e.g., C_4H_8

cis-2-Butene   *trans*-2-Butene

   b. *conformational isomerism*: variation in acyclic molecules as a result of a rotation about a single bond; e.g., ethane, CH_3-CH_3

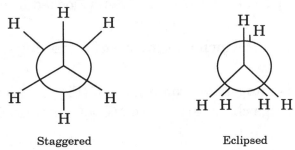

Staggered   Eclipsed

In this experiment, we will further investigate stereoisomerism by examining a cyclic system, cyclohexane, and several acyclic tetrahedral carbon systems. The latter possess more subtle characteristics as a result of the spatial arrangement of the component atoms. We will do this by building models of representative organic molecules, then studying their properties.

Objectives

1. To use models to study the conformations of cyclohexane.
2. To use models to distinguish between chiral and achiral systems.
3. To define and illustrate enantiomers, diastereomers, and meso forms.
4. To learn how to represent these systems in 2-dimensional space.

Procedure

You will build models and then you will be asked questions about the models. You will provide answers to these questions in the appropriate places on the Report Sheet. In doing this laboratory, it will be convenient if you tear out the Report Sheet and keep it by the Procedure as you work through the exercises. In this way, you can answer the questions without unnecessarily turning pages back and forth.

Cyclohexane

Obtain a model set of "atoms" that contain the following:

- 8 Carbon components—model atoms with 4 holes at the tetrahedral angle (e.g., black);
- 2 Substituent components—model atoms with 1 hole (e.g., red);
- 18 Hydrogen components—model atoms with 1 hole (optional) (e.g., white);
- 24 Connecting links—bonds.

1. Construct a model of cyclohexane by connecting 6 carbon atoms in a ring; then into each remaining hole insert a connecting link (bond) and, if available, add a hydrogen to each.

   a. Is the ring rigid or flexible, that is, can the ring of atoms move and take various arrangements in space, or is the ring of atoms locked into only one configuration (1a)?

   b. Of the many configurations, which appears best for the ring—a planar or a puckered arrangement (1b)?

   c. Arrange the ring atoms into a *chair* conformation (Fig. 24.1a) and compare it to the picture of the lounge chair (Fig. 24.1b). (Does the term fit the picture?)

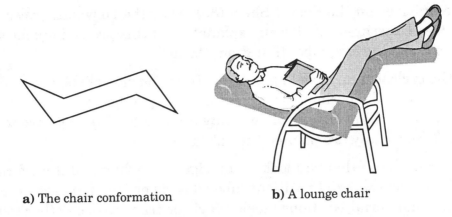

a) The chair conformation

b) A lounge chair

Figure 24.1 • The chair conformation for a 6-carbon ring.

2. With the model in the chair conformation, rest it on the tabletop.

   a. How many hydrogens are in contact with the tabletop (2a)?

   b. How many hydrogens point in a direction 180° opposite to these (2b)?

   c. Take your pencil and place it into the center of the ring perpendicular to the table. Now, rotate the ring around the pencil; we'll call this an *axis of rotation*. How many hydrogens are on bonds parallel to this axis (2c)? These hydrogens are called the *axial* hydrogens, and the bonds are called the *axial* bonds.

   d. If you look at the perimeter of the cyclohexane system, the remaining hydrogens lie roughly in a ring perpendicular to the axis through the center of the molecule. How many hydrogens are on bonds lying in this ring (2d)? These hydrogens are called *equatorial* hydrogens, and the bonds are called the *equatorial* bonds.

   e. Compare your model to the diagrams in Fig. 24.2 and be sure you are able to recognize and distinguish between axial and equatorial positions.

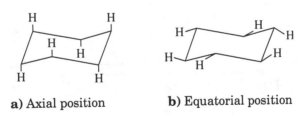

a) Axial position

b) Equatorial position

Figure 24.2 • Axial and equatorial hydrogens in the chair conformation.

In the space provided on the Report Sheet (2e), draw the structure of cyclohexane in the chair conformation with all 12 hydrogens attached. Label all the axial hydrogens H_a and all the equatorial hydrogens H_e. How many hydrogens are labeled H_a (2f)? How many hydrogens are labeled H_e (2g)?

3. Look along any bond connecting any two carbon atoms in the ring. (Rotate the ring and look along a new pair of carbon atoms.) How are the bonds connected to these two carbons arranged? Are they staggered or are they eclipsed (3a)? In

the space provided on the Report Sheet (3b), draw the Newman projection for the view (see Experiment 23 for an explanation of this projection); for the bond connecting a ring carbon, label that group "ring."

4. Pick up the cyclohexane model and view it from the side of the chair. Visualize the "ring" around the perimeter of the system perpendicular to the axis through the center. Of the 12 hydrogens, how many are pointed "up" relative to the plane (4a)? How many are pointed "down" (4b)?

5. Orient your model so that you look at an edge of the ring and it conforms to Fig. 24.3. Are the two axial positions labeled A cis or trans to each other (5a)? Are the two equatorial positions labeled B cis or trans to each other (5b)? Are the axial and equatorial positions A and B cis or trans to each other (5c)? Rotate the ring and view new pairs of carbons in the same way. See whether the relationships of positions vary from the above. Position your eye as in Fig. 24.3 and view along the carbon-carbon bond. In the space provided on the Report Sheet (5d), draw the Newman projection. Using this projection, review your answers to 5a, 5b, and 5c.

Figure 24.3

Cyclohexane ring viewed on edge.

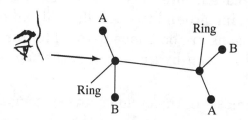

6. Replace one of the axial hydrogens with a colored component atom. Do a "ring flip" by moving one of the carbons *up* and moving the carbon farthest away from it *down* (Fig. 24.4). In what position is the colored component after the ring flip (6a)—axial or equitorial? Do another ring flip. In what position is the colored component now (6b)? Observe all the axial positions and follow them through a ring flip.

Figure 24.4

A "ring flip."

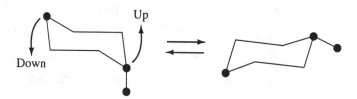

7. Refer to Fig. 24.3 and replace both positions labeled A by colored component atoms. Are they cis or trans (7a)? Do a ring flip. Are the two colored components cis or trans (7b)? Does the geometry change for the two components as the ring undergoes a ring flip (7c)? Repeat the exercise, replacing atoms in positions labeled A and B and answer the same three questions for this model.

8. Replace one of the colored components with a methyl, -CH$_3$, group. Manipulate the model so that the -CH$_3$ group is in an axial position; examine the model. Do

a ring flip placing the -CH$_3$ in an equatorial position; examine the model. Which of the chair conformations, -CH$_3$ axial or -CH$_3$ equatorial, is more crowded (8a)? What would account for one of the conformations being more crowded than the other (8b)? Which would be of higher energy and thus less stable (8c)? In the space provided on the Report Sheet (8d), draw the two conformations and connect with equilibrium arrows. Given your answers to 8a, 8b, and 8c, toward which conformation will the equilibrium lay (indicate by drawing one arrow bigger and thicker than the other)?

9. *A substituent group in the equatorial position of a chair conformation is more stable than the same substituent group in the axial position.* Do you agree or disagree? Explain your answer (9).

For the exercises in 10–15, although we will *not* be asking you to draw each and every conformation, we encourage you to practice drawing them in order to gain experience and facility in creating drawings on paper. Your instructor may make these exercises optional.

10. Construct *trans*-1,2-dimethylcyclohexane. By means of ring flips, examine the model with the two -CH$_3$ groups axial and the two -CH$_3$ groups equatorial. Which is the more stable conformation? Explain your answer (10).

11. Construct *cis*-1,2-dimethylcyclohexane by placing one -CH$_3$ group axial and the other equatorial. Do ring flips and examine the two chair conformations. Which is the more stable conformation? Explain your answer (11a). Given the two isomers, *trans*-1,2-dimethylcyclohexane and *cis*-1,2-dimethylcyclohexane, which is the more stable isomer? Explain your answer (11b).

12. Construct *cis*-1,3-dimethylcyclohexane by placing both -CH$_3$ groups in the axial positions. Do ring flips and examine the two chair conformations. Which is the more stable conformation? Explain your answer (12).

13. Construct *trans*-1,3-dimethylcyclohexane by placing one -CH$_3$ group axial and the other equatorial. Do ring flips and examine the two chair conformations. Which is the more stable conformation? Explain your answer (13a). Given the two isomers, *trans*-1,3-dimethylcyclohexane and *cis*-1,3-dimethylcyclohexane, which is the more stable isomer? Explain your answer (13b).

14. Construct *trans*-1,4-dimethylcyclohexane by placing both -CH$_3$ groups axial. Do ring flips and examine the two chair conformations. Which is the more stable conformation? Explain your answer (14).

15. Construct *cis*-1,4-dimethylcyclohexane by placing one -CH$_3$ group axial and the other equatorial. Do ring flips and examine the two chair conformations. Which is the more stable conformation? Explain your answer (15a). Given the two isomers, *trans*-1,4-dimethylcyclohexane and *cis*-1,4-dimethylcyclohexane, which is the more stable isomer? Explain your answer (15b).

16. Before we leave the cyclohexane ring system, there are some additional ring conformations we can examine. As we move from one cyclohexane chair conformation to another, the *boat* is a transitional conformation between them

(Fig. 24.5). Examine a model of the boat conformation by viewing along a carbon-carbon bond, as shown by Fig. 24.5. In the space provided on the Report Sheet (16a), draw the Newman projection for this view and compare with the Newman projection of 5d. By examining the models and comparing the Newman projections, explain which conformation, the chair or the boat, is more stable (16b). Replace the "flagpole" hydrogens by -CH_3 groups. What happens when this is done (16c)? The steric strain can be relieved by twisting the ring and separating the two bulky groups. What results is a *twist boat*.

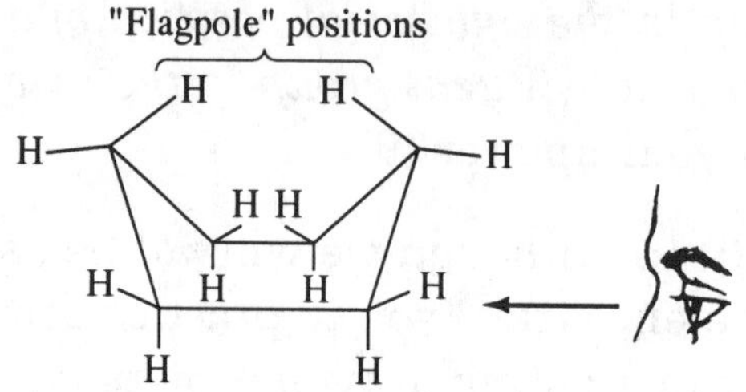

Figure 24.5

The boat conformation.

17. Review the conformations the cyclohexane ring can assume as it moves from one chair conformation to another:

$$\text{chair} \rightleftharpoons \text{twist boat} \rightleftharpoons \text{boat} \rightleftharpoons \text{twist boat} \rightleftharpoons \text{chair}$$

Chiral Molecules

For this exercise, obtain a small hand mirror and a model set of "atoms" which contain the following:

- 8 Carbon components—model atoms with four holes at the tetrahedral angle (e.g., black);
- 32 Substituent components—model atoms with one hole in four colors (e.g., 8 red; 8 white; 8 blue; 8 green; or any other colors which your set may have);
- 28 Connecting links—bonds.

Enantiomers

1. Construct a model consisting of a tetrahedral carbon center with four different component atoms attached: red, white, blue, green; each color represents a *different* group or atom attached to carbon. Does this model have a *plane of symmetry* (1a)? A plane of symmetry can be described as a cutting plane—a plane that when passed through a model or object *divides it into two equivalent halves*; the elements on one side of the plane are the exact reflection of the elements on the other side. If you are using a pencil to answer these questions, examine the pencil. Does it have a plane of symmetry (1b)?

2. Molecules without a plane of symmetry are *chiral*. In the model you constructed in no. 1, the tetrahedral carbon is the chiral center; the molecule is chiral. A simple test for a chiral center in a molecule is to look for a carbon center with four different atoms or groups attached to it; this molecule will have no plane of symmetry. On the Report Sheet (2) are three structures; label the chiral center in each structure with an asterisk (*).

3. Now take the model you constructed in no. 1 and place it in front of a mirror. Construct the model of the image projected in the mirror. You now have two models. If one is the object, what is the other (3a)? Do either have a plane of symmetry (3b)? Are both chiral (3c)? Now try to superimpose one model onto the other, that is, to place one model on top of the other in such a way that all five elements (i.e., the colored atoms) fall exactly one-on-top-of-the-other. Can you superimpose one model onto the other (3d)? *Enantiomers* are two molecules that are related to each other such that they are *nonsuperimposable mirror images of each other*. Are the two models you have a pair of enantiomers (3e)?

4. Molecules with a plane of symmetry are *achiral*. Replace the blue substituent with a second green one. The model should now have three different substituents attached to the carbon. Does the model now have a plane of symmetry (4a)? Passing the cutting plane through the model, what colored elements does it cut in half (4b)? What is on the left and right half of the cutting plane (4c)? Place this model in front of the mirror. Construct the model of the image projected in the mirror. You now have two models—an object and its mirror image. Are these two models superimposable on each other (4d)? Are the two models representative of different molecules or identical molecules (4e)?

Each stereoisomer in a pair of enantiomers has the property of being able to rotate monochromatic plane-polarized light. The instrument chemists use to demonstrate this property is called a *polarimeter* (see your text for a further description of the instrument). A pure solution of a single one of the enantiomers (referred to as an *optical isomer*) can rotate the light in either a clockwise (dextro-rotatory, +) or a counterclockwise (levorotatory, −) direction. Thus those molecules that are optically active possess a "handedness" or chirality. Achiral molecules are optically inactive and do not rotate the light.

Meso Forms and Diastereomers

5. With your models, construct a pair of enantiomers. From each of the models, remove the same common element (e.g., the white component) and the connecting links (bonds). Reconnect the two central carbons by a bond. What you have constructed is the *meso form* of a molecule, such as *meso*-tartaric acid. How many chiral carbons are there in this compound (5a)?

$$HOOC - C_1H - C_2H - COOH$$
$$\quad\quad\quad\ \ | \quad\quad\ |$$
$$\quad\quad\quad OH \quad\ OH$$

Tartaric acid

Is there a plane of symmetry (5b)? Is the molecule chiral or achiral (5c)?

6. In the space provided on the Report Sheet (6), use circles to indicate the four different groups for carbon C_1 and squares to indicate the four different groups for carbon C_2.

7. Project the model into a mirror and construct a model of the mirror image. Are these two models superimposable or nonsuperimposable (7a)? Are the models identical or different (7b)?

8. Now take one of the models you constructed in no. 7, and on one of the carbon centers exchange any two colored component groups. Does the new model have a plane of symmetry (8a)? Is it chiral or achiral (8b)? How many chiral centers are present (8c)? Take this model and one of the models you constructed in no. 7 and see whether they are superimposable. Are the two models superimposable (8d)? Are the two models identical or different (8e)? Are the two models mirror images of each other (8f)? Here we have a pair of molecular models, each with two chiral centers, that are not mirror images of each other. These two examples represent *diastereomers*, stereoisomers that are not related as mirror images.

9. Take the new model you constructed in no. 8 and project it into a mirror. Construct a model of the image in the mirror. Are the two models superimposable (9a)? What term describes the relationship of the two models (9b)?

Thus if we let these three models represent different isomers of tartaric acid, we find that there are three stereoisomers for tartaric acid—a meso form and a pair of enantiomers. A meso form with any one of the enantiomers of tartaric acid represents a pair of diastereomers. Although it may not be true for this compound because of the meso form, in general, if you have **n** chiral centers, there are 2^n stereoisomers possible (see Post-Lab question no. 3).

Drawing Stereoisomers

This section will deal with conventions for representing these 3-dimensional systems in 2-dimensional space.

10. Construct models of a pair of enantiomers; use tetrahedral carbon and four differently colored components for the four different groups: red, green, blue, white. Hold one of the models in the following way:

a. Grasp the blue group with your fingers and rotate the model until the green and red groups are pointing toward you (Fig. 24.6a). (Use the model which has the green group on the left and the red group on the right.)

b. Holding the model in this way, the blue and white groups point away from you.

c. If we use a drawing that describes a bond pointing toward you as a wedge and a bond pointing away from you as a dashed-line, the model can be drawn as shown in Fig. 24.6b.

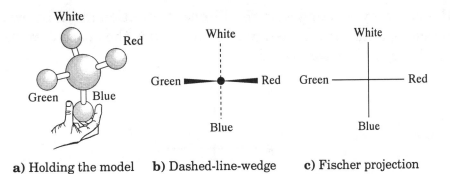

a) Holding the model   **b)** Dashed-line-wedge   **c)** Fischer projection

Figure 24.6 • Projections in 2-dimensional space.

If this model were compressed into 2-dimensional space, we would get the projection shown in Fig. 24.6c. This is termed a *Fischer projection* and is named after a pioneer in stereochemistry, Emile Fischer. The Fischer projection has the following requirements:

(1) the center of the cross represents the chiral carbon and is in the plane of the paper;

(2) the horizontal line of the cross represents those bonds projecting *out front* from the plane of the paper;

(3) the vertical line of the cross represents bonds projecting *behind* the plane of the paper.

d. In the space provided on the Report Sheet (10), use the enantiomer of the model in Fig. 24.6a and draw both the dashed-line-wedge and Fischer projection.

11. Take the model shown in Fig. 24.6a and rotate by 180° (turn upside down). Draw the Fischer projection (11a). Does this keep the requirements of the Fischer projection (11b)? Is the projection representative of the same system or of a different system (i.e., the enantiomer) (11c)?

In general, if you have a Fischer projection and rotate it in the plane of the paper by 180°, the resulting projection is of the *same* system. Test this assumption by taking the Fischer projection in Fig. 24.6c, rotating it in the plane of the paper by 180°, and comparing it to the drawing you did for no. 11a.

12. Again, take the model shown in Fig. 24.6a. Exchange the red and the green components. Does this exchange give you the enantiomer (12a)? Now exchange the blue and the white components. Does this exchange return you to the original model (12b)?

In general, for a given chiral center, whether we use the dashed-line-wedge or the Fischer projection, an odd-numbered exchange of groups leads to the mirror image of that center; an even-numbered exchange of groups leads back to the original system.

13. Test the above by starting with the Fischer projection given below and carrying out the operations directed in a, b, and c; use the space provided on the Report Sheet (13) for the answers.

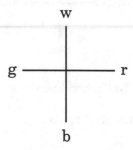

a. Exchange *r* and *g*; draw the Fischer projection you obtain; label this new projection as either the same as the starting model or the enantiomer.

b. Using the new Fischer projection from above, exchange *b* and *w*; draw the Fischer projection you now have.

c. Now rotate the last Fischer projection you obtained by 180°; draw the Fischer projection you now have; label this as either the same as the starting model or the enantiomer.

14. Let us examine models with two chiral centers by using tartaric acid as the example, HOOC−CH(OH)−CH(OH)−COOH; use your colored components to represent the various groups. Hold your models so that each stereoisomer is oriented as in Fig. 24.7. In the space provided on the Report Sheet (14), draw each of the corresponding Fischer projections.

COOH	COOH	COOH
H▶C◀OH	H▶C◀OH HO▶C◀H	HO▶C◀H
H▶C◀OH	HO▶C◀H H▶C◀OH	H▶C◀OH
COOH	COOH	COOH
a) Meso	**b)** Enantiomers	

Figure 24.7 • The stereoisomers of tartaric acid.

Circle the Fischer projection that shows a plane of symmetry. Underline all the Fischer projections that would be optically active.

15. Use the Fischer projection of *meso*-tartaric acid and carry out even and odd exchanges of the groups; follow these exchanges with a model. Does an odd exchange lead to an enantiomer, a diastereomer, or to a system identical to the meso form (15a)? Does an even exchange lead to an enantiomer, a diastereomer, or to a system identical to the meso form (15b)?

Chemicals and Equipment

Model kits vary in size and color of components. Use what is available; other colors may be substituted.

1. Cyclohexane model kit: 8 carbons (black, 4 holes); 18 hydrogens (white, 1 hole); 2 substituents (red, 1 hole); 24 bonds.
2. Chiral model kit: 8 carbons (black, 4 holes); 32 substituents (8 red, 1 hole; 8 white, 1 hole; 8 blue, 1 hole; 8 green, 1 hole); 28 bonds.
3. Hand mirror

Experiment 24

PRE-LAB QUESTIONS

1. Draw a chair conformation of cyclohexane which contains only axial hydrogens.

2. Look at your hands. Which term best explains the relationship of the two hands: identical, constitutional, conformational, enantiomers?

3. Define stereoisomers.

4. Label the chiral carbons in the molecules below with an asterisk (*).

$$CH_3 - \overset{\overset{\displaystyle Cl}{|}}{C}H - \overset{\overset{\displaystyle Cl}{|}}{C}H - CH_3$$

5. Define optical isomer.

NAME

SECTION

DATE

PARTNER

GRADE

Experiment 24

REPORT SHEET

Cyclohexane

1. a.

b.

2. a.

b.

c.

d.

e.

f.

g.

3. a.

b.

4. a.

b.

5. a.

b.

   c.

   d.

6. a.

  b.

7.   *Trial 1*   *Trial 2*

  a.

  b.

  c.

8. a.

  b.

  c.

  d.

9.

10.   e,e or a,a

11. a. a,e or e,a

  b.

12.   a,a or e,e

13. a. a,e or e,a

  b.

14. a,a or e,e

15. a. a,e or e,a

b.

16. a.

b.

c.

Enantiomers

1. a.

b.

2.

$$\begin{array}{c} OH \\ | \\ CH_3\text{-}CH\text{-}CH_2CH_3 \end{array} \qquad \begin{array}{c} OH \\ | \\ CH_3\text{-}CH\text{-}COOH \end{array} \qquad \begin{array}{c} Br \\ | \\ ClCH_2\text{-}CH\text{-}CH_3 \end{array}$$

3. a.

b.

c.

d.

e.

4. a.

b.

c.

d.

e.

Meso forms and diastereomers

5. a.

   b.

   c.

6.

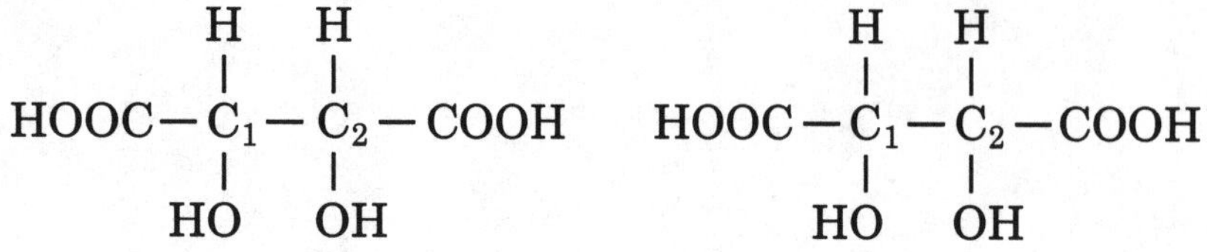

$$HOOC-\underset{\underset{HO}{|}}{\overset{\overset{H}{|}}{C_1}}-\underset{\underset{OH}{|}}{\overset{\overset{H}{|}}{C_2}}-COOH \qquad HOOC-\underset{\underset{HO}{|}}{\overset{\overset{H}{|}}{C_1}}-\underset{\underset{OH}{|}}{\overset{\overset{H}{|}}{C_2}}-COOH$$

7. a.

   b.

8. a.

   b.

   c.

   d.

   e.

   f.

9. a.

   b.

Drawing stereoisomers

10.

11. a.

    b.

    c.

12. a.

    b.

13. a.

    b.

    c.

14.

15. a.

    b.

POST-LAB QUESTIONS

1. Draw the most stable conformation for methylcyclohexane. Explain why you drew the molecule the way you did.

2. Draw the Fischer projections for the pair of enantiomers of lactic acid, CH₃-CH(OH)-COOH.

3. For 2,3-dibromopentane:

   a. How many stereoisomers are possible for this compound?

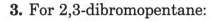

$$\underset{\underset{\displaystyle CH_3-CH-CH-CH_2CH_3}{\overset{\displaystyle |}{}}}{\overset{\displaystyle Br \quad\; Br}{\overset{\displaystyle |\qquad |}{}}}$$

   b. Draw Fischer projections for each stereoisomer; label enantiomers. Label any meso isomers (if there are any).

4. Determine the relationship between the following pairs of structures: identical, enantiomers, diastereomers.

a. 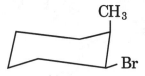   and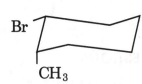

b.

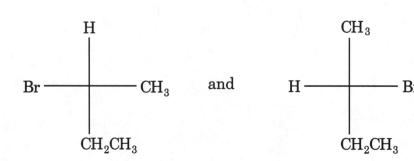

Experiment 25

Identification of hydrocarbons

The number of known organic compounds totals into the millions. Of these compounds, the simplest types are those which contain only hydrogen and carbon atoms. These are known as *hydrocarbons*. Because of the number and variety of hydrocarbons that can exist, some means of classification is necessary.

One means of classification depends on the way in which carbon atoms are connected. *Aliphatic* hydrocarbons are compounds consisting of carbons linked either in a single chain or in a branched chain. *Cyclic* hydrocarbons are compounds having the carbon atoms linked in a closed polygon. For example, hexane and 2-methylpentane are aliphatic molecules, while cyclohexane is a cyclic system.

$$CH_3 CH_2 CH_2 CH_2 CH_2 CH_3$$

Hexane

$$CH_3 CHCH_2 CH_2 CH_3$$
$$\;\;\;\;\;\;|$$
$$CH_3$$

2-Methylpentane

Cyclohexane

Another means of classification depends on the type of bonding that exists between carbons. Hydrocarbons which contain only carbon-to-carbon single bonds are called *alkanes*. These are also referred to as *saturated* molecules. Hydrocarbons containing at least one carbon-to-carbon double bond are called *alkenes*, and those compounds with at least one carbon-to-carbon triple bond are *alkynes*. These are compounds that are referred to as *unsaturated* molecules. Finally, a class of cyclic hydrocarbons that contain a closed loop (sextet) of electrons are called *aromatic* (see your text for further details). Table 25.1 distinguishes between the families of hydrocarbons.

With so many compounds possible, identification of the bond type is an important step in establishing the molecular structure. Quick, simple tests on small samples can establish the physical and chemical properties of the compounds by class.

Some of the observed physical properties of hydrocarbons result from the non-polar character of the compounds. In general, hydrocarbons do not mix with polar solvents such as water or ethyl alcohol. On the other hand, hydrocarbons mix with relatively non-polar solvents such as ligroin (a mixture of alkanes), carbon tetrachloride, or dichloromethane. Since the density of most hydrocarbons is less than that of water, they will float. Crude oil and crude oil products (home heating oil and gasoline) are mixtures of hydrocarbons; these substances, when spilled on water, spread quickly along the surface because they are insoluble in water.

Table 25.1	Types of Hydrocarbons		
Class	**Characteristic Bond Type**		**Example**
I. Aliphatic			
1. Alkane*	$-\overset{\mid}{\underset{\mid}{C}}-\overset{\mid}{\underset{\mid}{C}}-$ single		$CH_3CH_2CH_2CH_2CH_2CH_3$ hexane
2. Alkene†	$>C=C<$ double		$CH_3CH_2CH_2CH_2CH=CH_2$ 1-hexene
3. Alkyne†	$-C\equiv C-$ triple		$CH_3CH_2CH_2CH_2C\equiv CH$ 1-hexyne
II. Cyclic			
1. Cycloalkane*	$-\overset{\mid}{\underset{\mid}{C}}-\overset{\mid}{\underset{\mid}{C}}-$ single		cyclohexane
2. Cycloalkene†	$>C=C<$ double		cyclohexene
3. Aromatic			benzene
			CH_3 toluene

*Saturated   †Unsaturated

The chemical reactivity of hydrocarbons is determined by the type of bond in the compound. Although saturated hydrocarbons (alkanes) will burn (undergo *combustion*), they are generally unreactive to most reagents. (Alkanes do undergo a substitution reaction with halogens but require ultraviolet light.) Unsaturated hydrocarbons, alkenes and alkynes, not only burn, but also react by *addition* of reagents to the double or triple bonds. The addition products become saturated, with fragments of the reagent becoming attached to the carbons of the multiple bond. Aromatic compounds, with a higher carbon-to-hydrogen ratio than non-aromatic compounds, burn with a sooty flame as a result of unburned carbon particles being present. These compounds undergo *substitution* in the presence of catalysts rather than an addition reaction.

1. *Combustion*. The major component in "natural gas" is the hydrocarbon methane. Other hydrocarbons used for heating or cooking purposes are propane and butane. The products from combustion are carbon dioxide and water (heat is evolved, also).

$$CH_4 + 2O_2 \longrightarrow CO_2 + 2H_2O$$

$$(CH_3)_2CHCH_2CH_3 + 8O_2 \longrightarrow 5CO_2 + 6H_2O$$

2. *Reaction with bromine.* Unsaturated hydrocarbons react rapidly with bromine in a solution of carbon tetrachloride or cyclohexane. The reaction is the addition of the elements of bromine to the carbons of the multiple bonds.

$$CH_3CH = CHCH_3 + Br_2 \longrightarrow \underset{\text{Colorless}}{CH_3\overset{\overset{\displaystyle Br}{|}}{CH} - \overset{\overset{\displaystyle Br}{|}}{CH}CH_3}$$

Red

$$CH_3C \equiv CCH_3 + 2Br_2 \longrightarrow CH_3\overset{\overset{\displaystyle Br}{|}}{\underset{\underset{\displaystyle Br}{|}}{C}} - \overset{\overset{\displaystyle Br}{|}}{\underset{\underset{\displaystyle Br}{|}}{C}}CH_3$$

Red

Colorless

The bromine solution is red; the product that has the bromine atoms attached to carbon is colorless. Thus a reaction has taken place when there is a loss of color from the bromine solution and a colorless solution remains. Since alkanes have only single C-C bonds present, no reaction with bromine is observed; the red color of the reagent would persist when added. Aromatic compounds resist addition reactions because of their "aromaticity": *the possession of a closed loop (sextet) of electrons.* These compounds react with bromine in the presence of a catalyst such as iron filings or aluminum chloride.

Note that a substitution reaction has taken place and the gas HBr is produced.

3. *Reaction with concentrated sulfuric acid.* Alkenes react with cold concentrated sulfuric acid by addition. Alkyl sulfonic acids form as products and are soluble in H_2SO_4.

$$CH_3 - CH \overset{\cancel{}}{=} CH - CH_3 + \underset{(H_2SO_4)}{HOSO_2OH} \longrightarrow CH_3 - \overset{\overset{\displaystyle H}{|}}{CH} - \overset{\overset{\displaystyle OSO_2OH}{|}}{CH} - CH_3$$

Saturated hydrocarbons are unreactive (additions are not possible); alkynes react slowly and require a catalyst ($HgSO_4$); aromatic compounds also are unreactive since addition reactions are difficult.

4. *Reaction with potassium permanganate.* Dilute or alkaline solutions of $KMnO_4$ oxidize unsaturated compounds. Alkanes and aromatic compounds are generally unreactive. Evidence that a reaction has occurred is by the loss of the purple color of $KMnO_4$ and the formation of the brown precipitate manganese dioxide, MnO_2.

$$3CH_3-CH=CH-CH_3 + 2KMnO_4 + 4H_2O \rightarrow 3CH_3-CH-CH-CH_3 + 2MnO_2 + 2KOH$$
$$\underset{\text{Purple}}{} \qquad \underset{\text{OH OH}}{} \qquad \underset{\text{Brown}}{}$$

Note that the product formed from an alkene is a glycol.

Objectives

1. To investigate the physical properties, solubility and density, of some hydrocarbons.
2. To compare the chemical reactivity of an alkane, an alkene, and an aromatic compound.
3. To use physical and chemical properties to identify an unknown.

Procedure

CAUTION!

Assume the organic compounds are highly flammable. Use only small quantities. Keep away from open flames. Assume the organic compounds are toxic and can be absorbed through the skin. Avoid contact; wash if any chemical spills on your person. Handle concentrated sulfuric acid carefully. Flush with water if any spills on your person. Potassium permanganate and bromine are toxic; bromine solutions are also corrosive. Although the solutions are dilute, they may cause burns to the skin. Wear gloves when working with these chemicals.

General Instructions

1. The hydrocarbons hexane, cyclohexene, and toluene (alkane, alkene, and aromatic) are available in dropper bottles.
2. The reagents 1% Br_2 in cyclohexane, 1% aqueous $KMnO_4$, and concentrated H_2SO_4 are available in dropper bottles.
3. Unknowns are in dropper bottles labeled A, B, and C. They may include an alkane, an alkene, or an aromatic compound.
4. Record all data and observations in the appropriate places on the Report Sheet.
5. Dispose of all organic wastes as directed by the instructor. *Do not pour into the sink!*

Physical properties of hydrocarbons

1. A test tube of 100 x 13 mm will be suitable for this test. When mixing the components, grip the test tube between thumb and forefinger; it should be held firmly enough to keep from slipping but loosely enough so that when the third and fourth fingers tap it, the contents will be agitated enough to mix.

2. *Water solubility of hydrocarbons.* Label six test tubes with the name of the substance to be tested. Place into each test tube 5 drops of the appropriate hydrocarbon: hexane, cyclohexene, toluene, unknown A, unknown B, unknown C. Add about 5 drops of water dropwise into each test tube. Is there any separation of components? Which component is on the bottom; which component is on the top? Mix the contents as described above. What happens when the contents are allowed to settle? What do you conclude about the density of the hydrocarbon? Record your observations. Save these solutions for comparison with the next part.

3. *Solubility of hydrocarbons in ligroin .* Label six test tubes with the name of the substance to be tested. Place into each test tube 5 drops of the appropriate hydrocarbon: hexane, cyclohexene, toluene, unknown A, unknown B, unknown C. Add about 5 drops of ligroin dropwise into each test tube. Is there a separation of components? Is there a bottom layer and top layer? Mix the contents as described above. Is there any change in the appearance of the contents before and after mixing? Compare these test tubes to those from the previous part. Can you make any conclusion about the density of the hydrocarbon? Record your observations.

Chemical properties of hydrocarbons

1. *Combustion.* The instructor will demonstrate this test in the fume hood. Place 5 drops of each hydrocarbon and unknown on separate watch glasses. Carefully ignite each sample with a match. Observe the flame and color of the smoke for each of the samples. Record your observations on the Report Sheet.

2. *Reaction with bromine.* Label six clean, dry test tubes with the name of the substance to be tested. Place into each test tube 5 drops of the appropriate hydrocarbon: hexane, cyclohexene, toluene, unknown A, unknown B, unknown C. Carefully add (dropwise and with shaking) 1% Br_2 in cyclohexane. Keep count of the number of drops needed to have the color persist; do not add more than 10 drops. Record your observations. To the test tube containing toluene, add 5 more drops of 1% Br_2 solution and a small quantity of iron filings; shake the mixture. Place a piece of moistened blue litmus paper on the test tube opening. Record any change in the color of the solution and the litmus paper.

CAUTION!

Use 1% Br_2 solution in the hood; wear gloves when using this chemical.

---

3. *Reaction with KMnO₄.* Label six clean, dry test tubes with the name of the substance to be tested. Place into each test tube 5 drops of the appropriate hydrocarbon: hexane, cyclohexene, toluene, unknown A, unknown B, unknown C. Carefully add (dropwise) 1% aqueous $KMnO_4$ solution; after each drop, shake to mix the solutions. Keep count of the number of drops needed to have the color of the permanganate solution persist; do not add more than 10 drops. Record your observations.

4. *Reaction with concentrated H₂SO₄.* Label six clean, dry test tubes with the name of the substance to be tested. Place into each test tube 5 drops of the appropriate hydrocarbon: hexane, cyclohexene, toluene, unknown A, unknown B, unknown C. Place all of the test tubes in an ice bath. *Wear gloves and carefully* add (with shaking) 3 drops of cold, concentrated sulfuric acid to each test tube. Note whether heat is evolved by feeling the test tube. Note whether the solution has become homogeneous or whether a color is produced. (The evolution of heat or the formation of a homogeneous solution or the appearance of a color is evidence that a reaction has occurred.) Record your observations.

5. *Unknowns.* By comparing the observations you made for your unknowns with that of the known hydrocarbons, you can identify unknowns A, B, and C. Record their identities on your Report Sheet.

Chemicals and Equipment

1. 1% aqueous $KMnO_4$
2. 1% Br_2 in cyclohexane
3. Blue litmus paper
4. Concentrated H_2SO_4
5. Cyclohexene
6. Hexane
7. Iron filings or powder
8. Test tubes
9. Ligroin
10. Toluene
11. Unknowns A, B, and C
12. Watch glasses
13. Ice

NAME SECTION DATE

PARTNER GRADE

Experiment 25

PRE-LAB QUESTIONS

1. Why is hexane known as a *hydrocarbon*?

2. Show the structural feature that distinguishes whether a hydrocarbon is an

alkane

alkene

alkyne

aromatic

3. Hydrocarbons mix with solvents like ligroin and dichloromethane. What accounts for this property?

Experiment 25

REPORT SHEET

Physical properties of hydrocarbons

Hydrocarbon	H₂O		Ligroin	
	Solubility	Density	Solubility	Density
Hexane				
Cyclohexene				
Toluene				
Unknown A				
Unknown B				
Unknown C				

Chemical properties of hydrocarbons

Hydrocarbon	Combustion	Bromine Test	KMnO₄ Test	H₂SO₄
Hexane				
Cyclohexene				
Toluene				
Unknown A				
Unknown B				
Unknown C				

Unknown A is _____ .

Unknown B is _____ .

Unknown C is _____ .

POST-LAB QUESTIONS

1. Write the structure of the major organic product for the following reactions; if no reaction, write NR.

   a. $CH_3-CH=CH_2$ + Br_2 $\longrightarrow$

   b. 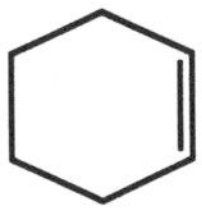 + $KMnO_4$ + H_2O $\longrightarrow$

   c. $CH_3-CH=CH-CH_3$ + H_2SO_4 $\longrightarrow$

   d.  + $KMnO_4$ $\longrightarrow$

2. Of the compounds used in this experiment (hexane, cyclohexene, toluene), which burned with a dirty, i. e. sooty, flame? Explain why this sooty burning took place.

3. On the basis of the chemical tests used in this experiment, is it possible to *clearly* distinguish between an alkane and an alkene? Explain your conclusion.

Column and paper chromatography; separation of plant pigments

Background

Chromatography is a widely used experimental technique by which a mixture of compounds can be separated into its individual components. Two kinds of chromatographic experiments will be explored. In column chromatography, a mixture of components dissolved in a solvent is poured over a column of solid adsorbent and is eluted with the same or a different solvent. This is therefore a solid-liquid system; the stationary phase (the adsorbent) is solid and the mobile phase (the eluent) is liquid. In paper chromatography, the paper adsorbs water from the atmosphere of the developing chromatogram. (The water is present in the air as vapor, and it may be supplied as one component in the eluting solution.) The water is the stationary phase. The (other) component of the eluting solvent is the mobile phase and carries with it the components of the mixture. This is a liquid-liquid system.

Column chromatography is used most conveniently for preparative purposes, when one deals with a relatively large amount of the mixture and the components need to be isolated in mg or g quantities. Paper chromatography, on the other hand, is used mostly for analytical purposes. Microgram or even picogram quantities can be separated by this technique, and they can be characterized by their R_f number. This number is an index of how far a certain spot moved on the paper.

$$R_f = \frac{\text{Distance of the center of the sample spot from the origin}}{\text{Distance of the solvent front from the origin}}$$

For example, in Fig. 26.1 the R_f values are as follows:

R_f (substance 1) = 3.1 cm/11.2 cm = 0.28 and

R_f (substance 2) = 8.5 cm/11.2 cm = 0.76

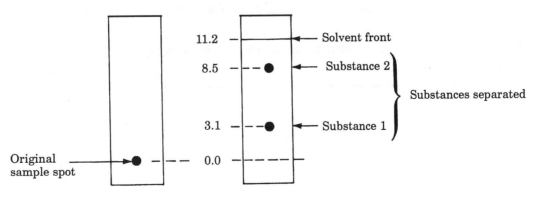

Figure 26.1 • Illustration of chromatograms before and after elution.

Using the R_f values, one is able to identify the components of the mixture with the individual components. The two main pigment components of tomato paste are β-carotene (yellow-orange) and lycopene (red) pigments. Their structures are given below:

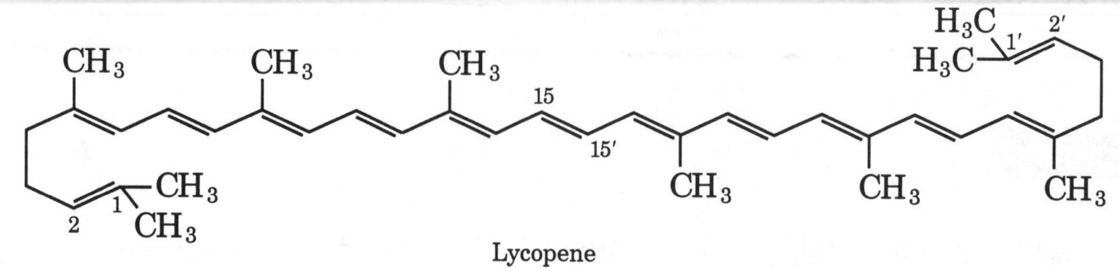

Lycopene

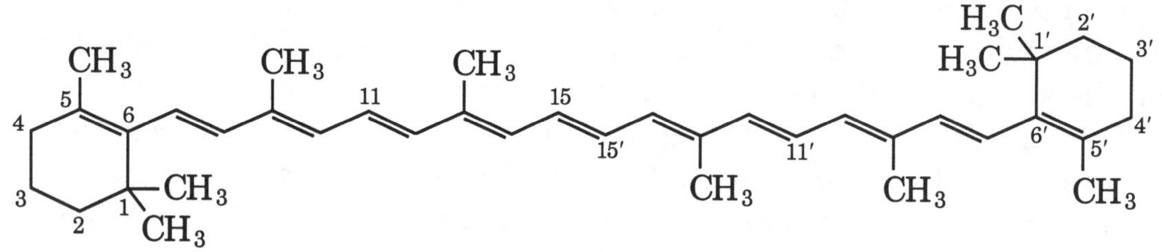

β-Carotene

The colors of these pigments are due to the numerous double bonds in their structure. When bromine is added to double bonds, it saturates them and the color changes accordingly. In the tomato juice "rainbow" experiment, we stir bromine water into the tomato juice. The slow stirring allows the bromine water to penetrate deeper and deeper into the cylinder in which the tomato juice was placed. As the bromine penetrates, more and more double bonds will be saturated. Therefore, you may be able to observe a continuous change, a "rainbow" of colors, starting with the reddish tomato color at the bottom of the cylinder where no reaction occurred (since the bromine did not reach the bottom). Lighter colors will be observed on the top of the cylinder where most of the double bonds have been saturated.

Objectives

1. To compare separation of components of a mixture by two different techniques.
2. To demonstrate the effect of bromination on plant pigments of tomato juice.

Paper Chromatography

1. Obtain a sheet of Whatman no.1 filter paper, cut to size.

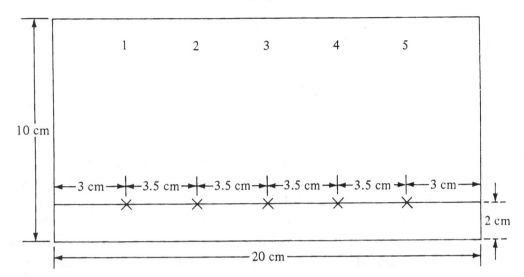

Figure 26.2 • Preparation of chromatographic paper for spotting.

2. Plan the **spotting** of the samples as illustrated on Fig. 26.2. Five spots will be applied. The first and fifth will be β-carotene solutions supplied by your instructor. The second, third, and fourth lanes will have your tomato paste extracts in different concentrations. Use a pencil to mark the spots lightly according to Fig. 26.2.

3. Pigments of tomato paste will be **extracted** in two steps.

(a) Weigh about 10 g of tomato paste in a 50-mL beaker. Add 15 mL of 95% ethanol. Stir the mixture vigorously with a spatula until the paste will not stick to the stirrer. Place a small amount of glass wool (the size of a pea) in a small funnel blocking the funnel exit. Place the funnel into a 50-mL Erlenmeyer flask and pour the tomato paste-ethanol mixture into the funnel. When the filtration is completed, squeeze the glass wool lightly with your spatula. In this step, we removed the water from the tomato paste and the aqueous components are in the filtrate, which we discard. The residue in the glass wool will be used to extract the pigments.

(b) Place the residue from the glass wool in a 50-mL beaker. Add 10 mL petroleum ether and stir the mixture for about two min. to extract the pigments. Filter the extract as before through a new funnel with glass wool blocking the exit into a new and clean 50-mL beaker. Place the beaker under the hood on a hot plate (or water bath) and evaporate the solvent to about 1 mL volume. Use low heat and take care not to evaporate all the solvent. After evaporation, cover the beaker with aluminum foil.

Figure 26.3
Withdrawing samples
by a capillary tube.

Spotting

4. Place your chromatographic paper on a clean area (another filter paper) in order not to contaminate it. Use separate capillaries for your tomato paste extract and for the β-carotene solution. First, apply your capillary to the extracted pigment by dipping it into the solution as illustrated in Fig. 26.3. Apply the capillary lightly to the chromatographic paper by touching sequentially the spots marked 2, 3, and 4. Make sure you apply only small spots, not larger than 2 mm diameter, by quickly withdrawing the capillary from the paper each time you touch it. (See Fig. 26.4.)

Figure 26.4
Spotting.

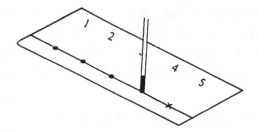

While allowing the spots to dry, use your second capillary to apply spots of β-carotene in lanes 1 and 5. Return to the first capillary and apply another spot of the extract on top of the spots of lanes 3 and 4. Let them dry (Fig. 26.5). Finally, apply one more spot on top of lane 4. Let the spots dry. The unused extract in your beaker should be covered with aluminum foil. Place it in your drawer in the dark to save it for the second part of your experiment.

Figure 26.5
Drying chromatographic spots.

Developing the paper chromatogram

5. Curve the paper into a cylinder and staple the edges above the 2 cm line as it is shown in Fig. 26.6.

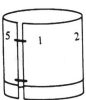

Figure 26.6 • Stapling.

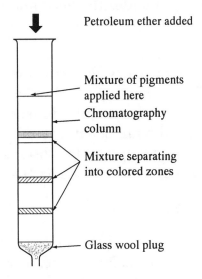

Figure 26.7 • Developing the chromatogram.

6. Pour 20 mL of the eluting solvent (petroleum ether : toluene : acetone in 45:1:5 ratio, supplied by your instructor) into a 600-mL beaker.

7. Place the stapled chromatogram into the 600-mL beaker, the spots being at the bottom near the solvent surface but **not covered by it.** Cover the beaker with aluminum foil (Fig. 26.7). Allow the solvent front to migrate up to 0.5–1 cm below the edge of the paper. This may take from 15 min. to one hr. Make certain by frequent inspection that the **solvent front does not run over the edge of the paper.** Remove the chromatogram from the beaker when the solvent front reaches 0.5–1 cm from the edge; then **proceed to step no. 11.**

Column Chromatography

8. While you are waiting for the chromatogram to develop (step no. 7), you can perform the column chromatography experiment. Take a 25-mL buret. (You may use a chromatographic column, if available, of 1.6 cm diameter and about 13 cm long; see Fig. 26.8. If you use the column instead of the buret, all subsequent quantities below should be doubled.)

Figure 26.8
Chromatographic column.

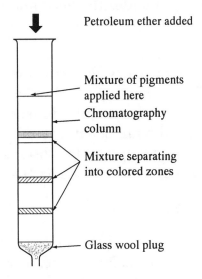

Petroleum ether added

Mixture of pigments applied here

Chromatography column

Mixture separating into colored zones

Glass wool plug

Add a small piece of glass wool and with the aid of a glass rod push it down near the stopcock. Add 15–16 mL of petroleum ether to the buret. Open the stopcock slowly and allow the solvent to fill the tip of the buret. Close the stopcock. You should have 12–13 mL of solvent above the glass wool. Weigh 20 g of aluminum oxide (alumina) in a 100-mL beaker. Place a small funnel on top of your buret. Pour the alumina into the buret. Allow the alumina to settle in order to form a 20-cm column. *Drain* the solvent but *do not allow the column to run dry. Always have at least a 0.5 mL of clear solvent above the alumina in the column.* If alumina adheres to the walls of the buret, wash it down with more solvent.

9. Transfer by pipet 0.5-1 mL of the extract you stored in your drawer onto the column. The pipet containing the extract should be placed near the surface of the solvent on top of the column. Touching the walls of the buret with the tip of the pipet, allow the extract to drain slowly on top of the column. Open the stopcock slightly. Allow the sample to enter the column, *but make sure there is a small amount of solvent above the alumina in the column. (The column should never run dry.)* Add 10 or more mL of petroleum ether and wash the sample into the column by opening the stopcock and collecting the eluted solvent in a beaker.

10. As the solvent elutes the sample, you observe the migration of the pigments and their separation into at least two bands. When the fastest-moving pigment band reaches near the bottom of the column, close the stopcock and observe the color of the pigment bands and how far they migrated from the top of the column. Record your observation on the Report Sheet. This concludes the column chromatographic part of the experiment. Discard your solvent in a bottle supplied by your instructor for a later redistillation.

11. Meanwhile your paper chromatogram has developed. You must remove the filter paper from the 600-mL beaker before the solvent front reaches the edges of the paper. Mark the position of the solvent front with a pencil. Put the paper standing on its edges under the hood and let it dry.

Tomato Juice "Rainbow"

12. While waiting for the paper to dry, you can perform the following short experiment. Weigh about 15 g of tomato paste in a beaker. Add about 30 mL of water and stir. Transfer the tomato juice into a 50-mL graduated cylinder and, with the aid of a pipet, add 5 mL of saturated bromine water (dropwise). With a glass rod, stir the solution very gently. Observe the colors and their positions in the cylinder. Record your observations on the Report Sheet.

Paper Chromatography *(continued)*

13. Remove the staples from the dried chromatogram. Mark the spots of the pigments by circling with a pencil. Note the colors of the spots. Measure the distance of the center of each spot from its origin. Calculate the R_f values.

14. If the spots on the chromatogram are faded, we can visualize them by exposing the chromatogram to iodine vapor. Place your chromatogram into a wide-mouthed jar containing a few iodine crystals. Cap the jar and warm it slightly on a hot plate to enhance the sublimation of iodine. The iodine vapor will interact with the faded pigment spots and make them visible. After a few minutes exposure to iodine vapor, remove the chromatogram and mark the spots **immediately** with pencil. The spots will fade again with exposure to air. Measure the distance of the center of the spots from the origin and calculate the R_f values.

15. Record the results of the Paper Chromatography on the Report Sheet.

Chemicals and Equipment

1. Melting point capillaries open at both ends
2. 25-mL buret or chromatographic column
3. Glass wool
4. Whatman no.1 filter paper, 10×20 cm, cut to size
5. Heat lamp (optional)
6. Stapler
7. Hot plate (with or without water bath)
8. Tomato paste
9. Aluminum oxide (alumina)
10. Petroleum ether (b.p. 30–60°C)
11. 95% ethanol
12. Toluene
13. Acetone
14. 0.5% β-carotene in petroleum ether
15. Saturated bromine water
16. Iodine crystals
17. Ruler
17. Widemouthed jar

Experiment 26

PRE-LAB QUESTIONS

1. (a) What is the "stationary phase" in column chromatography?

(b) What is the "mobile phase" in column chromatography?

2. The structures of the two main pigments, lycopene and β-carotene, are given in the first part (**Background**):

(a) What is the basic difference between the structures of these two pigments?

(b) To what class of hydrocarbons do these pigments belong?

(c) Do the names of these pigments indicate their chemical classification?

3. Write the structure of β-carotene after it completely reacted with Br_2.

NAME _____ SECTION _____ DATE _____

PARTNER _____ GRADE _____

Experiment 26

REPORT SHEET

Paper chromatography

Sample	Distance from origin to solvent front (cm) (a)	Distance from origin to center of spot (cm) (b)	R_f (b)/(a)	Color
β-carotene lane 1 lane 5				
Tomato extract lane 2 **(a)** **(b)** **(c)** **(d)** lane 3 **(a)** **(b)** **(c)** **(d)** lane 4 **(a)** **(b)** **(c)** **(d)**				

Column chromatography

Number of bands	Distance migrated from top of the column (cm)	Color
1		
2		
3		

"Rainbow"

Describe the colors observed in the tomato juice "rainbow" experiment, starting from the bottom of the cylinder:

1. red   **2.**   **3.**

4.   **5.**   **6.**

POST-LAB QUESTIONS

1. Did your tomato paste contain lycopene? What support is there for your answer?

2. Did your "rainbow" experiment indicate that the bromine penetrated to the bottom of your cylinder?

3. What is the effect of the amount of sample applied to the paper on the separation of the tomato pigments? Compare the results on lanes 2, 3, and 4 of the paper chromatogram.

4. Based on the R_f value you calculated for lycopene, how far would this pigment travel if the solvent front moved 20 cm?

5. Which technique used in this experiment would be appropriate to isolate (not just to detect) β-carotene from tomato juice?

Identification of alcohols and phenols

Background

Specific groups of atoms in an organic molecule can determine its physical and chemical properties. These groups are referred to as *functional groups*. Organic compounds which contain the functional group -OH, the hydroxyl group, are called *alcohols*.

Alcohols are important commercially and include use as solvents, drugs, and disinfectants. The most widely used alcohols are methanol or methyl alcohol, CH_3OH, ethanol or ethyl alcohol, CH_3CH_2OH, and 2-propanol or isopropyl alcohol, $(CH_3)_2CHOH$. Methyl alcohol is found in automotive products such as antifreeze and "dry gas." Ethyl alcohol is used as a solvent for drugs and chemicals, but is more popularly known for its effects as an alcoholic beverage. Isopropyl alcohol, also known as "rubbing alcohol," is an antiseptic.

Alcohols may be classified as either primary, secondary, or tertiary:

$$R-CH_2-OH$$
Primary alcohol

$$R-\underset{\underset{OH}{|}}{CH}-R'$$
Secondary alcohol

$$R-\underset{\underset{R''}{|}}{\overset{\overset{R'}{|}}{C}}-OH$$
Tertiary alcohol

Note that the classification depends on the number of carbon-containing groups, R (alkyl or aromatic), attached to the carbon bearing the hydroxyl group. Examples of each type are as follows:

$$CH_3CH_2-OH$$
Ethanol
(Ethyl alcohol)
a primary alchol

$$CH_3-\underset{\underset{CH_3}{|}}{CH}-OH$$
2-Propanol
(Isopropyl alcohol)
a secondary alcohol

$$CH_3-\underset{\underset{CH_3}{|}}{\overset{\overset{CH_3}{|}}{C}}-OH$$
2-Methyl-2-propanol
(*t*-Butyl alcohol)
a tertiary alcohol

Phenols bear a close resemblance to alcohols structurally since the hydroxyl group is present. However, since the -OH group is bonded directly to a carbon that is part of an aromatic ring, the chemistry is quite different from that of alcohols. Phenols are more acidic than alcohols; concentrated solutions of the compound phenol are quite toxic and can cause severe skin burns. Phenol derivatives are found in medicines; for example, thymol is used to kill fungi and hookworms. (Also see Table 27.1.)

Phenol

Thymol
(2-isopropyl-5-methylphenol)

In this laboratory, you will examine physical and chemical properties of representative alcohols and phenols. You will be able to compare the differences in chemical behavior between these compounds and use this information to identify an unknown.

Table **27.1**	Selected Alcohols and Phenols
Compound	**Name and Use**
CH_3OH	Methanol: solvent for paints, shellacs, and varnishes
CH_3CH_2OH	Ethanol: alcoholic beverages; solvent for medicines, perfumes, and varnishes
$CH_3-CH-CH_3$ $\quad\quad\ \ |$ $\quad\quad\ \ OH$	Isopropyl alcohol (2-propanol): rubbing alcohol; astringent; solvent for cosmetics, perfumes, and skin creams
CH_2-CH_2 $\ |\quad\quad\ |$ $OH\quad\ OH$	Ethylene glycol: antifreeze
$CH_2-CH-CH_2$ $\ |\quad\quad |\quad\quad\ |$ $OH\quad OH\quad OH$	Glycerol (glycerin): sweetening agent; solvent for medicines; lubricant; moistening agent
	Phenol (carbolic acid): cleans surgical and medical instruments; topical antipruritic (relieves itching)
	Vanillin: flavoring agent (vanilla)
	Tetrahydrourushiol: irritant in poison ivy

Physical Properties

Since the hydroxyl group is present in alcohols and phenols, these compounds are polar. The polarity of the hydroxyl group, coupled with its ability to form hydrogen bonds, enables alcohols and phenols to mix with water. Since these compounds also contain non-polar portions, they show additional solubility in many organic solvents, such as dichloromethane and diethyl ether.

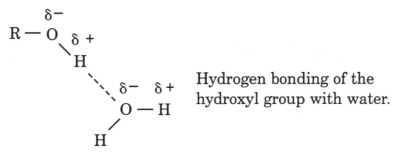

Hydrogen bonding of the hydroxyl group with water.

Chemical Properties

The chemical behavior of the different classes of alcohols and of phenols can be used as a means of identification. Quick, simple tests that can be carried out in test tubes will be performed.

1. *Lucas test*. This test is used to distinguish between primary, secondary, and tertiary alcohols. Lucas reagent is a mixture of zinc chloride, $ZnCl_2$, in concentrated HCl. Upon addition of this reagent, a tertiary alcohol reacts rapidly and immediately gives an insoluble white layer. A secondary alcohol reacts slowly and, after heating slightly, gives the white layer within 10 min. A primary alcohol does not react. Any formation of a heterogeneous phase or appearance of an emulsion is a positive test.

$$CH_3CH_2-OH \ + \ HCl \ + \ ZnCl_2 \ \longrightarrow \ \text{no reaction}$$
primary alcohol

$$(CH_3)_2CH-OH \ + \ HCl \ + \ ZnCl_2 \ \longrightarrow \ (CH_3)_2CH-Cl \downarrow \ + \ H_2O \ \text{(10 min. heat)}$$
secondary alcohol   insoluble

$$(CH_3)_3C-OH \ + \ HCl \ + \ ZnCl_2 \ \longrightarrow \ (CH_3)_3C-Cl \downarrow \ + \ H_2O \ \text{(<5 min.)}$$
tertiary alcohol   insoluble

2. *Chromic acid test*. This test is able to distinguish primary and secondary alcohols from tertiary alcohols. Using acidified dichromate solution, primary alcohols are oxidized to aldehydes; secondary alcohols are oxidized to ketones; tertiary alcohols are not oxidized. (Note that in those alcohols which are oxidized, the carbon that has the hydroxyl group *loses a hydrogen*.) In the oxidation, the brown-red color of the chromic acid changes to a blue-green solution. Phenols

are oxidized to nondescript brown tarry masses. (Aldehydes are also oxidized under these conditions to carboxylic acids, but ketones remain intact; see Experiment 28 for further discussion.)

$$3CH_3CH_2-OH + 4H_2CrO_4 + 6H_2SO_4 \longrightarrow 3CH_3-\overset{\overset{\displaystyle O}{\|}}{C}-OH + 2Cr_2(SO_4)_3 + 13H_2O$$

primary alcohol   brown-red   carboxylic acid   blue-green

$$3\,CH_3-\overset{\overset{\displaystyle OH}{|}}{CH}-CH_3 + 2H_2CrO_4 + 3H_2SO_4 \longrightarrow 3CH_3-\overset{\overset{\displaystyle O}{\|}}{C}-CH_3 + Cr_2(SO_4)_3 + 8H_2O$$

secondary alcohol   brown-red   ketone   blue-green

$$(CH_3)_3C-OH + H_2CrO_4 + H_2SO_4 \longrightarrow \text{no reaction}$$

tertiary alcohol

3. *Iodoform test*. This test is more specific than the previous two tests. Only ethanol (ethyl alcohol) and alcohols with the part structure $CH_3CH(OH)$ react. These alcohols react with iodine in aqueous sodium hydroxide to give the yellow precipitate iodoform.

$$\overset{\overset{\displaystyle OH}{|}}{RCHCH_3} + 4I_2 + 6NaOH \longrightarrow \overset{\overset{\displaystyle O}{\|}}{RC}-O^-\,Na^+ + 5NaI + 5H_2O + HCl_3\,(s)$$

iodoform
yellow

Phenols also react under these conditions. With phenol, the yellow precipitate triiodophenol forms.

triiodophenol
yellow

4. *Acidity of phenol*. Phenol is also called carbolic acid. Phenol is an acid and will react with base; thus phenols readily dissolve in base solutions. In contrast, alcohols are not acidic.

5. *Ferric chloride test.* Addition of aqueous ferric chloride to a phenol gives a colored solution. Depending on the structure of the phenol, the color can vary from green to purple.

$$\langle\bigcirc\rangle - OH \; + \; FeCl_3 \; \longrightarrow \; \langle\bigcirc\rangle - O - Fe - Cl \; + \; HCl$$

light yellow   Cl

violet color

Objectives

1. To learn characteristic chemical reactions of alcohols and phenols.

2. To use these chemical characteristics for identification of an organic compound.

Procedure

CAUTION!

Chromic acid is very corrosive. Any spill should be immediately flushed with water. Phenol is toxic. Also, contact with the solid will cause burns to skin; any contact should be thoroughly washed with large quantities of water. Solid phenol should be handled only with a spatula or forceps. Use gloves with these reagents. Dispose of reaction mixtures and excess reagents in proper containers as directed by your instructor.

Physical Properties of Alcohols and Phenols

1. You will test the alcohols 1-butanol (a primary alcohol), 2-butanol (a secondary alcohol), 2-methyl-2-propanol (a tertiary alcohol), and phenol; you will also have as an unknown one of these compounds (labeled A, B, C, or D). As you run a test on a known, test the unknown at the same time for comparison. Note that the phenol will be provided as an aqueous solution.

2. Into separate test tubes (100 × 13 mm) labeled 1-butanol, 2-butanol, 2-methyl-2-propanol, and unknown, place 10 drops of each sample; dilute by mixing with 3 mL of distilled water. Into a separate test tube, place 2 mL of a prepared water solution of phenol. Are all the solutions homogeneous?

3. Test the pH of each of the aqueous solutions. Do the test by first dipping a clean glass rod into the solutions and then transferring a drop of liquid to pH paper. Use a broad indicator paper (e.g., pH range 1–12) and read the value of the pH by comparing the color to the chart on the dispenser. Record the results on the Report Sheet (1).

Chemical Properties of Alcohols and Phenols

1. *Lucas test.* Place 5 drops of each sample into separate clean, dry test tubes (100 × 13 mm), labeled 1-butanol, 2-butanol, 2-methyl-2-propanol, phenol, and unknown. Add 1 mL of Lucas reagent; mix well by stoppering each test tube with a cork, tapping the test tube sharply with your finger for a few seconds to mix; remove the cork after mixing and allow each test tube to stand for 5 min. Look carefully for any cloudiness that may develop during this time period. If there is no cloudiness after 10 min., warm the test tubes that are clear for 15 min. in a 60°C water bath. Record your observations on the Report Sheet (2).

2. *Chromic acid test.* Place into separate clean, dry test tubes (100 × 13 mm), labeled as before, 5 drops of sample to be tested. To each test tube add 10 drops of reagent grade acetone and 2 drops of chromic acid. Place the test tubes in a 60°C water bath for 5 min. Note the color of each solution. (Remember, the loss of the brown-red and the formation of a blue-green color is a positive test.) Record your observations on the Report Sheet (3).

3. *Iodoform test.* Place into separate clean, dry test tubes (150 × 18 mm), labeled as before, 5 drops of sample to be tested. Add to each test tube (dropwise) 15 drops of 6 M NaOH; tap the test tube with your finger to mix. The mixture is warmed in a 60°C water bath, and the prepared solution of I_2-KI test reagent is added dropwise (with shaking) until the solution becomes brown (approx. 25 drops). Add 6 M NaOH (dropwise) until the solution becomes colorless. Keep the test tubes in the warm water bath for 5 min. Remove the test tubes from the water, let cool, and look for a light yellow precipitate. Record your observations on the Report Sheet (4).

4. *Ferric chloride test.* Place into separate clean, dry test tubes (100 × 13 mm), labeled as before, 5 drops of sample to be tested. Add 2 drops of ferric chloride solution to each. Note any color changes in each solution. (Remember, a purple color indicates the presence of a phenol.) Record your observations on the Report Sheet (5).

5. From your observations identify your unknown.

Chemicals and Equipment

1. Aqueous phenol
2. Acetone (reagent grade)
3. 1-Butanol
4. 2-Butanol
5. 2-Methyl-2-propanol (*t*-butyl alcohol)
6. Chromic acid solution
7. Ferric chloride solution
8. I$_2$-KI solution
9. Lucas reagent
10. Corks
11. Hot plate
12. pH paper
13. Unknown

NAME SECTION DATE

PARTNER GRADE

Experiment 27

PRE-LAB QUESTIONS

1. Give an example for the functional groups given below:

a. a primary alcohol

b. a secondary alcohol

c. a tertiary alcohol

2. The compound below, vanillin, is an example of a compound which contains the phenolic functional group. Circle the phenol part of the molecule.

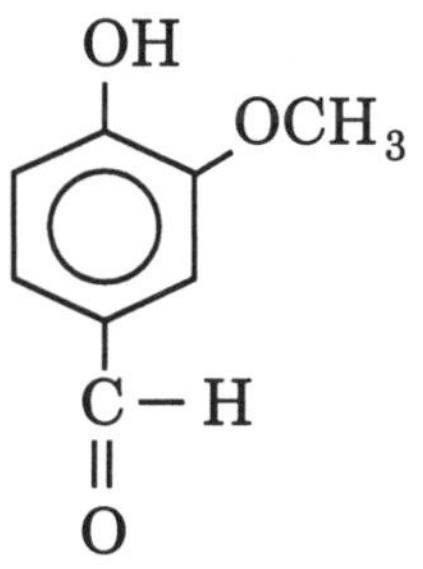

3. In general, why are alcohols more soluble in water than hydrocarbons?

NAME ______________________ SECTION ____________ DATE ____________

PARTNER ______________________ GRADE ______________________

Experiment 27

REPORT SHEET

Test	1-Butanol	2-Butanol	2-Methyl-2-propanol	Phenol	Unknown
1. pH					
2. Lucas					
3. Chromic acid					
4. Iodoform					
5. Ferric chloride					

Identity of unknown:
Unknown no.____________. The unknown compound is____________________.

POST-LAB QUESTIONS

1. Write the structures for the following alcohols. Also, indicate whether the alcohol is either primary, secondary, or tertiary.

a. 1-butanol

b. 2-butanol

c. 2-methyl-2-propanol

2. Write the structure of the major organic product expected from each of the following reactions. If no reaction is expected write "No Reaction."

a.

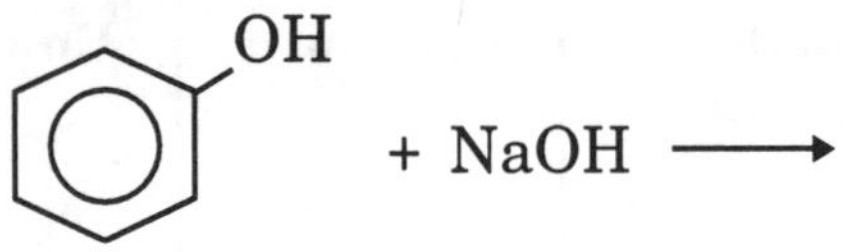

$$+ \; NaOH \longrightarrow$$

b.

$$CH_3 - \overset{\underset{\displaystyle |}{}}{CH} - CH_2CH_3 \; + \; H_2CrO_4 \xrightarrow{\; H_2SO_4 \;}$$
$$\overset{\displaystyle |}{OH}$$

c.

$$CH_3CH_2 - \overset{\overset{\displaystyle CH_3}{\displaystyle |}}{\underset{\underset{\displaystyle CH_3}{\displaystyle |}}{C}} - OH \; + \; H_2CrO_4 \xrightarrow{\; H_2SO_4 \;}$$

d.

$$CH_3 - \overset{\overset{\displaystyle CH_3}{\displaystyle |}}{\underset{\underset{\displaystyle CH_3}{\displaystyle |}}{C}} - OH \; + \; HCl \xrightarrow{\; ZnCl_2 \;}$$

3. Tetrahydrourushiol (see Table 27.1) gives a purple color with ferric chloride solution. What part of the structure is responsible for the reaction that gives this test?

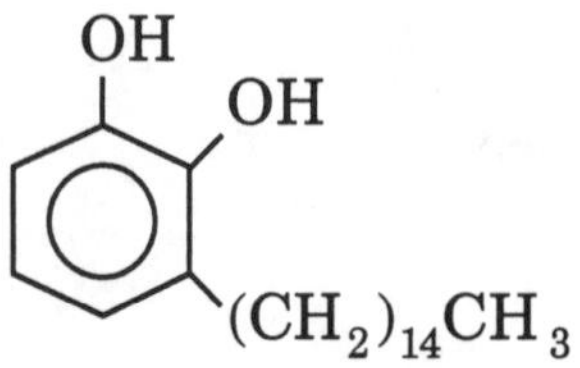

4. Glycerol (see Table 27.1) is a liquid at room temperature and is soluble in water in all proportions. However, hexane, $CH_3CH_2CH_2CH_2CH_2CH_3$, a compound of similar molecular weight and also a liquid at room temperature, is *insoluble* in water. How do you account for this difference?

5. Primary alcohols and secondary alcohols can be oxidized with chromic acid, but tertiary alcohols cannot. How do the structural differences between the alcohols account for the observed reactions?

6. What is the main chemical characteristic of phenol?

Experiment 28

Identification of aldehydes and ketones

Aldehydes and ketones are representative of compounds which possess the carbonyl group:

$$\gtrsim C=O$$
The carbonyl group.

Aldehydes have at least one hydrogen attached to the carbonyl carbon; in ketones, no hydrogens are directly attached to the carbonyl carbon, only carbon containing R-groups:

$$R-\underset{\underset{O}{\|}}{C}-H \qquad\qquad R-\underset{\underset{O}{\|}}{C}-R'$$

      Aldehyde   Ketone

(R and R' can be alkyl or aromatic)

Aldehydes and ketones of low molecular weight have commercial importance. Many others occur naturally. Table 28.1 has some representative examples.

Table 28.1 Representative Aldehydes and Ketones		
Compound		**Source and Use**
$\underset{O}{\overset{O}{\underset{\|}{HCH}}}$	Formaldehyde	Oxidation of methanol; plastics; preservative
$CH_3\overset{\overset{O}{\|}}{C}CH_3$	Acetone	Oxidation of isopropyl alcohol; solvent
	Citral	Lemon grass oil; fragrance
	Jasmone	Oil of jasmine; fragrance

In this experiment you will investigate the chemical properties of representative aldehydes and ketones.

Classification tests

1. *Chromic acid test.* Aldehydes are oxidized to carboxylic acids by chromic acid; ketones are not oxidized. A positive test results in the formation of a blue-green solution from the brown-red color of chromic acid.

$$3R-\overset{\overset{\displaystyle O}{\|}}{C}-H \;+\; 2H_2CrO_4 \;+\; 3H_2SO_4 \longrightarrow \; 3R-\overset{\overset{\displaystyle O}{\|}}{C}-OH \;+\; Cr_4(SO_4)_3 \;+\; 5H_2O$$

aldehyde   brown-red   blue-green

$$\begin{matrix} R \\ R \end{matrix} \!\!\diagdown\!\!\diagup\, C=O \quad \xrightarrow[H_2SO_4]{H_2CrO_4} \quad \text{no reaction}$$

ketone

2. *Tollens' test.* Most aldehydes reduce Tollens' reagent (ammonia and silver nitrate) to give a precipitate of silver metal. The free silver forms a silver mirror on the sides of the test tube. (This test is sometimes referred to as the "silver mirror" test.) The aldehyde is oxidized to a carboxylic acid.

$$R-\overset{\overset{\displaystyle O}{\|}}{C}-H \;+\; 2Ag(NH_3)_2OH \longrightarrow \; 2Ag(s) \;+\; R-\overset{\overset{\displaystyle O}{\|}}{C}-O^-NH_4^+ \;+\; H_2O \;+\; 3NH_3$$

aldehyde   silver
mirror

3. *Iodoform test.* Methyl ketones give the yellow precipitate iodoform when reacted with iodine in aqueous sodium hydroxide.

$$R-\overset{\overset{\displaystyle O}{\|}}{C}-CH_3 \;+\; 3I_2 \;+\; 4NaOH \longrightarrow \; 3NaI \;+\; 3H_2O \;+\; R-\overset{\overset{\displaystyle O}{\|}}{C}-O^-Na^+ \;+\; HCI_3(s)$$

methyl
ketone   iodoform
yellow

4. *2,4-Dinitrophenylhydrazine test.* All aldehydes and ketones give an immediate precipitate with 2,4-dinitrophenylhydrazine reagent. This reaction is general for both these functional groups. The color of the precipitate varies from yellow to red. (Note that alcohols do not give this test.)

Identification by forming a derivative

The classification tests (summarized in Table 28.2), when properly done, can distinguish between various types of aldehydes and ketones. However, these tests alone may not allow for the identification of a specific unknown aldehyde or ketone. A way to correctly identify an unknown compound is by using a known chemical reaction to convert it into another compound that is known. The new compound is referred to as a *derivative*. Then, by comparing the physical properties of the unknown and the derivative to the physical properties of known compounds listed in a table, an identification can be made.

Table **28.2** Summary of Classification Tests	
Compound	**Reagent for Positive Test**
Aldehydes and ketones	2,4-Dinitrophenylhydrazine
Aldehydes	Chromic acid
	Tollens' reagent
Methyl ketones	Iodoform

The ideal derivative is a solid. A solid can be easily purified by crystallization and easily characterized by its melting point. Thus two similar aldehydes or two similar ketones usually have derivatives that have *different melting points*. The most frequently formed derivatives for aldehydes and ketones are the

2,4-dinitrophenylhydrazone (2,4-DNP), oxime, and semicarbazone. Table 28.3 (p. 315) lists some aldehydes and ketones along with melting points of their derivatives. If, for example, we look at the properties of valeraldehyde and crotonaldehyde, though the boiling points are virtually the same, the melting points of the 2,4-DNP, oxime, and semicarbazone are different and provide a basis for identification.

1. *2,4-Dinitrophenylhydrazone.* 2,4-Dinitrophenylhydrazine reacts with aldehydes and ketones to form 2,4-dinitrophenylhydrazones (2,4-DNP).

2,4-dinitrophenylhydrazine   2,4-dinitrophenylhydrazone (2,4-DNP)

(R' = H or alkyl or aromatic)

The 2,4-DNP product is usually a colored solid (yellow to red) and is easily purified by recrystallization.

2. *Oxime.* Hydroxylamine reacts with aldehydes and ketones to form oximes.

hydroxylamine   oxime

(R' = H or alkyl or aromatic)

These are usually lower-melting derivatives.

3. *Semicarbazone.* Semicarbazide, as its hydrochloride salt, reacts with aldehydes and ketones to form semicarbazones.

semicarbazide   semicarbazone

(R' = H or alkyl or aromatic)

A pyridine base is used to neutralize the hydrochloride in order to free the semicarbazide so it may react with the carbonyl substrate.

Objectives

1. To learn the chemical characteristics of aldehydes and ketones.
2. To use these chemical characteristics in simple tests to distinguish between examples of aldehydes and ketones.
3. To identify aldehydes and ketones by formation of derivatives.

Procedure

Classification Tests

1. Classification tests are to be carried out on four known compounds and one unknown. Any one test should be carried out on all five samples at the same time for comparison. Label test tubes as shown in Table 28.4.

Table 28.4 Labeling Test Tubes	
Test Tube No.	**Compound**
1	Isovaleraldehyde (an aliphatic aldehyde)
2	Benzaldehyde (an aromatic aldehyde)
3	Cyclohexanone (a ketone)
4	Acetone (a methyl ketone)
5	Unknown

> ## CAUTION!
>
> Chromic acid is toxic and corrosive. Handle with care and **promptly wash any spill**. Use gloves with this reagent.

2. *Chromic acid test.* Place 5 drops of each substance into separate, labeled test tubes (100 × 13 mm). Dissolve each compound in 20 drops of reagent-grade acetone (to serve as solvent); then add to each test tube 4 drops of chromic acid reagent, one drop at a time; after each drop, mix by sharply tapping the test tube with your finger. Let stand for 10 min. Aliphatic aldehydes should show a change within a minute; aromatic aldehydes take longer. Note the approximate time for any change in color or formation of a precipitate on the Report Sheet.

[handwritten: → Use ~2 mls.]

3. *Tollens' test.*

> ## CAUTION!
>
> The reagent must be freshly prepared before it is to be used and any excess disposed of immediately after use. Organic residues should be discarded in appropriate waste containers; unused Tollens' reagent should be flushed away in the sink with large quantities of water. **Do not store Tollens' reagent since it is explosive when dry.**

Enough reagent for your use can be prepared in a 25-mL Erlenmeyer flask by mixing 5 mL of Tollens' solution A with 5 mL of Tollens' solution B. To the silver oxide precipitate which forms, add (dropwise, with shaking) 10% ammonia solution until the brown precipitate just dissolves. *Avoid an excess of ammonia.*

Place 5 drops of each substance into separately labeled clean, dry test tubes (100 × 13 mm). Dissolve the compound in bis(2-ethoxyethyl)ether by adding this solvent dropwise until a homogeneous solution is obtained. Then, add 2 mL (40 drops) of the prepared Tollens' reagent and mix by sharply tapping the test tube with your finger. Place the test tube in a 60°C water bath for 5 min. Remove the test tubes from the water and look for a silver mirror. Record your results on the Report Sheet.

→ diethyl ether is okay

4. *Iodoform test.* Place 10 drops of each substance into separately labeled clean, dry test tubes (150 × 18 mm). Add to each test tube (dropwise, with shaking) 25 drops of 6 M NaOH. The mixture is warmed in a water bath (60°C), and the prepared solution of I_2-KI test reagent is added (dropwise, with shaking) until the solution becomes brown (approx. 35 drops). Add 6 M NaOH until the solution becomes colorless. Keep the test tubes in the warm water bath for 5 min. Remove the test tubes from the water, let cool, and look for a light yellow precipitate. Record your observations on the Report Sheet.

5. *2,4-Dinitrophenylhydrazine test.* Place 5 drops of each substance into separately labeled clean, dry test tubes (100 × 13 mm) and add 20 drops of the 2,4-dinitrophenylhydrazine reagent to each. If no precipitate forms immediately, heat for 5 min. in a warm water bath (60°C); cool. Record your observations on the Report Sheet.

Formation of derivatives

CAUTION!

The chemicals used to prepare derivatives and some of the derivatives are potential carcinogens. Exercise care in using the reagents and in handling the derivatives. Avoid skin contact by wearing gloves.

---

1. This section is *optional*. Consult your instructor to determine whether this section is to be completed. Your instructor will indicate how many derivatives and which derivatives you should make.

2. *General procedure for recrystallization.* Heat a small volume (10–20 mL) of solvent to boiling on a steam bath (or carefully on a hot plate). Place crystals into a test tube (100 × 13 mm) and add the hot solvent (dropwise) until the crystals just dissolve (keep the solution hot, also). Allow the solution to cool to room temperature; then cool further in an ice bath. Collect the crystals on a Hirsch funnel by vacuum filtration (use a trap between the Hirsch funnel set-up and

the aspirator (Fig. 28.1); wash the crystals with 10 drops of ice cold solvent. Allow the crystals to dry by drawing air through the Hirsch funnel. Take a melting point (see Experiment 13 for a review of the technique).

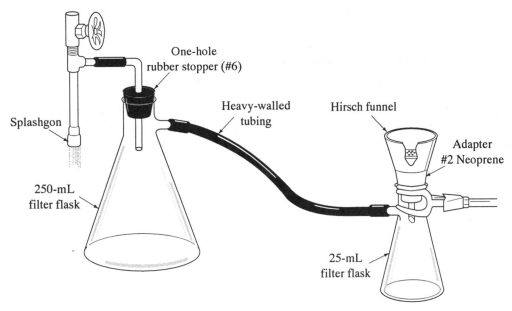

Figure 28.1 • Vacuum filtration with a Hirsch funnel.

3. *2,4-Dinitrophenylhydrazone.* Place 5 mL of the 2,4-dinitrophenylhydrazine reagent in a test tube (150 × 18 mm). Add 10 drops of the unknown compound; sharply tap the test tube with your finger to mix. If crystals do not form immediately, gently heat in a water bath (60°C) for 5 min. Cool in an ice bath until crystals form. Collect the crystals by vacuum filtration using a Hirsch funnel (Fig. 28.1). Allow the crystals to dry on the Hirsch funnel by drawing air through the crystals. Take a melting point and record on the Report Sheet. (The crystals are usually pure enough to give a good melting point. However, if the melting point range is too large, recrystallize from a minimum volume of ethanol.)

4. *Oxime.* Prepare fresh reagent by dissolving 1.0 g of hydroxylamine hydrochloride and 1.5 g of sodium acetate in 4 mL of distilled water in a test tube (150 × 18 mm). Add 20 drops of unknown and sharply tap the test tube with your finger to mix. Warm in a hot water bath (60°C) for 5 min. Cool in an ice bath until crystals form. (If no crystals form, scratch the inside of the test tube with a glass rod.) Collect the crystals on a Hirsch funnel by vacuum filtration (Fig. 28.1). Allow the crystals to air dry on the Hirsch funnel by drawing air through the crystals. Take a melting point and record on the Report Sheet. (Recrystallize, if necessary, from a minimum volume of ethanol.)

5. *Semicarbazone.* Place 2.0 mL of the semicarbazide reagent in a test tube (150 × 18 mm); add 10 drops of unknown. If the solution is not clear, add methanol (dropwise) until a clear solution results. Add 2.0 mL of pyridine and gently warm in a hot bath (60°C) for 5 min. Crystals should begin to form. (If there

→ Run only this one.

are no crystals, place the test tube in an ice bath and scratch the inside of the test tube with a glass rod.) Collect the crystals on a Hirsch funnel by vacuum filtration (Fig. 28.1). Allow the crystals to air dry on the Hirsch funnel by drawing air through the crystals. Take a melting point and record on the Report Sheet. (Recrystallize, if necessary, from a minimum volume of ethanol.)

6. *Waste.* Place all the waste solutions from these preparations in designated waste containers for disposal by your instructor.

7. Based on the observations you recorded on the Report Sheet, and by comparing the melting points of the derivatives for your unknown to the knowns listed in Table 28.3, identify your unknown.

Table 28.3 Selection of Aldehydes and Ketones with Derivatives

Compound	Formula	b.p.°C	2,4-DNP m.p.°C	Oxime m.p.°C	Semi-carbazone m.p.°C
Aldehydes Isovaleraldehyde (3-methylbutanal)	CH₃ H │ │ CH₃–CH–CH₂–C=O	93	123	49	107
Valeraldehyde (pentanal)	H │ CH₃CH₂CH₂CH₂–C=O	103	106	52	——
Crotonaldehyde (2-butenal)	H │ CH₃–CH=CH–C=O	104	190	119	199
Caprylaldehyde (octanal)	H │ CH₃CH₂CH₂CH₂CH₂CH₂CH₂C=O	171	106	60	101
Benzaldehyde	(phenyl)–C=O with H	178	237	35	222
Ketones Acetone (2-propanone)	O ║ CH₃–C–CH₃	56	126	59	187
2-Pentanone	O ║ CH₃–C–CH₂CH₂CH₃	102	144	58 (b.p. 167°C)	112
3-Pentanone	O ║ CH₃CH₂–C–CH₂CH₃	102	156	69 (b.p. 165°C)	139
Cyclopentanone	(cyclopentane)=O	131	146	56	210
Cyclohexanone	(cyclohexane)=O	156	162	90	166
Acetophenone	O ║ (phenyl)–C–CH₃	202	238	60	198

Source: Compiled by Zvi Rappoport, *CRC Handbook of Tables for Organic Compound Identification, 3rd ed.*, The Chemical Rubber Co.: Cleveland (1967).

Chemicals and Equipment

1. Acetone (reagent grade)
2. 10% ammonia solution
3. Benzaldehyde
4. Bis(2-ethoxyethyl) ether
5. Chromic acid reagent
6. Cyclohexane
7. 2,4-dinitrophenylhydrazine reagent
8. Ethanol
9. Hydroxylamine hydrochloride
10. I_2-KI test solution
11. Isovaleraldehyde
12. Methanol
13. 6 M NaOH, sodium hydroxide
14. Pyridine
15. Semicarbazide reagent
16. Sodium acetate
17. Tollens' reagent (solution A and solution B)
18. Hirsch funnel
19. Hot plate
20. Neoprene adapter (no. 2)
21. Rubber stopper (no. 6, one-hole), with glass tubing
22. 50-mL side-arm filter flask
23. 250-mL side-arm filter flask
24. Vacuum tubing (heavy walled)

NAME | SECTION | DATE

PARTNER | GRADE

Experiment 28

PRE-LAB QUESTIONS

1. Write a general structure for the following functional groups:

   a. an aldehyde

   b. an aromatic aldehyde

   c. a ketone

   d. a methyl ketone

2. What reagent gives a characteristic test for both aldehydes and ketones?

3. Name a reagent that can distinguish between an aldehyde and a ketone.

4. What do you expect to see for a positive Tollens' test?

Experiment 28

REPORT SHEET

Test	Isovaler-aldehyde	Benz-aldehyde	Cyclo-hexanone	Acetone	Unknown
Chromic acid					
Tollens'					
Iodoform					
2,4-Dinitro-phenyl-hydrazine					

Optional

Derivative	Observed m.p.	Literature m.p.
2,4-DNP		
Oxime		
Semicarbazone		

Unknown no. _____. The unknown compound is _____.

1. A compound of molecular formula $C_5H_{10}O$ forms a yellow precipitate with 2,4-dinitrophenylhydrazine reagent and a yellow precipitate with reagents for the iodoform test. Draw a structural formula for a compound that fits these tests.

2. What kind of results do you see when the following compounds are mixed together with the given test solution?

   a. ⬠=O   with 2,4-dinitrophenylhydrazine

   b. CH_3CH_2-C-H with chromic acid
      $\quad\quad\quad\quad\;\;\overset{\|}{O}$

   c. ⬡$-\overset{\overset{\displaystyle O}{\|}}{C}-CH_3$ with I_2-KI reagent

   d. ⬡$-\underset{\underset{\displaystyle O}{\|}}{C}-H$ Tollens' reagent

3. Write the structure of the compound that gives a silver mirror with Tollens' reagent and a yellow precipitate with sodium hydroxide and iodine.

4. Using the laboratory tests of this experiment, show how you could distinguish between the following compounds (Hint: use tests that would give a clear, positive test result for a listed compound that is unique to it and thus different from the others):

Test	Cyclohexane	Cyclohexanone	Hexanal

Experiment 29

Properties of carboxylic acids and esters

Carboxylic acids are structurally like aldehydes and ketones in that they contain the carbonyl group. However, an important difference is that carboxylic acids contain a hydroxyl group attached to the carbonyl carbon.

$$-\overset{\underset{\|}{O}}{C}-OH \quad \text{The carboxylic acid group.}$$

This combination gives the group its most important characteristic; it behaves as an acid.

As a family, carboxylic acids are weak acids that ionize only slightly in water. As aqueous solutions, typical carboxylic acids ionize to the extent of only one percent or less.

$$R-\overset{\underset{\|}{O}}{C}-OH + H_2O \rightleftharpoons R-\overset{\underset{\|}{O}}{C}-O^- + H_3O^+$$

At equilibrium, most of the acid is present as un-ionized molecules. Dissociation constants, K_a, of carboxylic acids, where R is an alkyl group, are 10^{-5} or less. Water solubility depends to a large extent on the size of the R-group. Only a few low molecular weight acids (up to four carbons) are very soluble in water.

Although carboxylic acids are weak, they are capable of reacting with bases stronger than water. Thus while benzoic acid shows limited water solubility, it reacts with sodium hydroxide to form the soluble salt sodium benzoate. (Sodium benzoate is a preservative in soft drinks.)

Benzoic acid
Insoluble

Sodium benzoate
Soluble

Sodium carbonate, Na_2CO_3, and sodium bicarbonate, $NaHCO_3$, solutions can neutralize carboxylic acids, also.

The combination of a carboxylic acid and an alcohol gives an ester; water is eliminated. Ester formation is an equilibrium process, catalyzed by an acid catalyst.

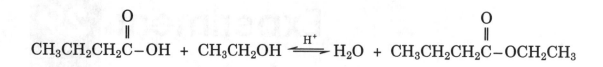

Butyric acid   Ethyl alcohol   Ethyl butyrate (Ester)

Esterification ⟶
⟵ Hydrolysis

The reaction typically gives 60% to 70% of the maximum yield. The reaction is a reversible process. An ester reacting with water, giving the carboxylic acid and alcohol, is called *hydrolysis*; it is acid catalyzed. The base-promoted decomposition of esters yields an alcohol and a salt of the carboxylic acid; this process is called *saponification*. Saponification means "soap making," and the sodium salt of a fatty acid (e.g., sodium stearate) is a soap.

$$CH_3CH_2CH_2C-OCH_2CH_3 + NaOH \longrightarrow CH_3CH_2CH_2C-O^-Na^+ + CH_3CH_2OH$$

Saponification

A distinctive difference between carboxylic acids and esters is in their characteristic odors. Carboxylic acids are noted for their sour, disagreeable odors. On the other hand, esters have sweet and pleasant odors often associated with fruits, and fruits smell the way they do because they contain esters. These compounds are used in the food industry as fragrances and flavoring agents. For example, the putrid odor of rancid butter is due to the presence of butyric acid, while the odor of pineapple is due to the presence of the ester, ethyl butyrate. Only those carboxylic acids of low molecular weight have odor at room temperature. Higher molecular weight carboxylic acids form strong hydrogen bonds, are solid, and have a low vapor pressure. Thus few molecules reach our nose. Esters, however, do not form hydrogen bonds among themselves; they are liquid at room temperature, even when the molecular weight is high. Thus they have high vapor pressure and many molecules can reach our nose, providing odor.

Objectives

1. To study the physical properties of carboxylic acids: solubility, acidity, aroma.
2. To prepare a variety of esters and note their odors.
3. To demonstrate saponification.

Carboxylic Acids and Their Salts

Characteristics of acetic acid

1. Place into a clean, dry test tube (100 × 13 mm) 2 mL of water and 10 drops of glacial acetic acid. Note its odor. Of what does it remind you?

2. Take a glass rod and dip it into the solution. Using wide range indicator paper (pH 1–12), test the pH of the solution by touching the paper with the wet glass rod. Determine the value of the pH by comparing the color of the paper with the chart on the dispenser.

3. Now, add 2 mL of 2 M NaOH to the solution. Cork the test tube and sharply tap it with your finger. Remove the cork and determine the pH of the solution as before; if not basic, continue to add more base (dropwise) until the solution is basic. Note the odor and compare to the odor of the solution before the addition of base.

4. By dropwise addition of 3 M HCl, carefully reacidify the solution from step no. 3 (above); test the solution as before with pH paper until the solution tests acid. Does the original odor return?

Characteristics of benzoic acid

1. Your instructor will weigh out 0.1 g of benzoic acid for sample size comparison. With your microspatula, take some sample equivalent to the preweighed sample (an exact quantity is not important here). Add the solid to a test tube (100 × 13 mm) along with 2 mL of water. Is there any odor? Mix the solution by sharply tapping the test tube with your finger. How soluble is the benzoic acid?

2. Now add 1 mL of 2 M NaOH to the solution from step no. 1 (above), cork and mix by sharply tapping the test tube with your finger. What happens to the solid benzoic acid? Is there any odor?

3. By dropwise addition of 3 M HCl, carefully reacidify the solution from step no. 2 (above); test as before with pH paper until acidic. As the solution becomes acidic, what do you observe?

Esterification

1. Into five clean, dry test tubes (100 × 13 mm), add 10 drops of liquid carboxylic acid or 0.1 g of solid carboxylic acid and 10 drops of alcohol according to the scheme in Table 29.1. Note the odor of each reactant.

Table 29.1	Acids and Alcohols	
Test Tube No.	**Carboxylic Acid**	**Alcohol**
1	Formic	Isobutyl
2	Acetic	Benzyl
3	Acetic	Isopentyl
4	Acetic	Ethyl
5	Salicylic	Methyl

2. Add 5 drops of concentrated sulfuric acid to each test tube and mix the contents thoroughly by sharply tapping the test tube with your finger.

CAUTION!

Sulfuric acid causes severe burns. Flush any spill with lots of water. Use gloves with this reagent.

---

3. Place the test tubes in a warm water bath at 60°C for 15 min. Remove the test tubes from the water bath, cool, and add 2 mL of water to each. Note that there is a layer on top of the water in each test tube. With a Pasteur pipet, take a few drops from this top layer and place on a watch glass. Note the odor. Match the ester from each test tube with one of the following odors: banana, peach, raspberry, nail polish remover, wintergreen.

Saponification

This part of the experiment can be done while the esterification reactions are being heated.

1. Place into a test tube (150 × 18 mm) 10 drops of methyl salicylate and 5 mL of 6 M NaOH. Heat the contents in a boiling water bath for 30 min. Record on the Report Sheet what has happened to the ester layer (1).

2. Cool the test tube to room temperature by placing it in an ice-water bath. Determine the odor of the solution and record your observation on the Report Sheet (2).

3. Carefully add 6 M HCl to the solution, 1 mL at a time, until the solution is acidic. After each addition, mix the contents and test the solution with litmus. When the solution is acidic, what do you observe? What is the name of the compound formed? Answer these questions on the Report Sheet (3).

Chemicals and Equipment

1. Glacial acetic acid
2. Benzoic acid
3. Formic acid
4. Salicylic acid
5. Benzyl alcohol
6. Ethanol (ethyl alcohol)
7. 2-Methyl-1-propanol (isobutyl alcohol)
8. 3-Methyl-1-butanol (isoamyl alcohol)
9. Methanol (methyl alcohol)
10. Methyl salicylate
11. 3 M HCl
12. 6 M HCl
13. 2 M NaOH
14. 6 M NaOH
15. Concentrated H_2SO_4
16. pH paper (broad range pH 1–12)
17. Litmus paper
18. Pasteur pipet
19. Hot plate

NAME SECTION DATE

PARTNER GRADE

Experiment 29

PRE-LAB QUESTIONS

1. Write the structures of the following carboxylic acids:

a. formic acid

b. benzoic acid

c. acetic acid

2. Write the products from the reaction of acetic acid and sodium hydroxide.

3. Ethyl formate has the flavor of rum. What alcohol and carboxylic acid would you use to synthesize this ester?

4. Esters can be decomposed by either hydrolysis or saponification. Explain the difference between the methods in terms of conditions and end products.

Experiment 29

REPORT SHEET

Carboxylic acids and their salts

Characteristics of Acetic Acid			
Property	Water Solution	NaOH Solution	HCl Solution
Odor			
Solubility			
pH			

Characteristics of Benzoic Acid			
Property	Water Solution	NaOH Solution	HCl Solution
Odor			
Solubility			
pH			

Esterification

Test Tube	Acid	Odor	Alcohol	Odor	Ester	Odor
1	Formic		Isobutyl			
2	Acetic		Benzyl			
3	Acetic		Isopentyl			
4	Acetic		Ethyl			
5	Salicylic		Methyl			

Saponification

1. What has happened to the ester layer?

2. What has happened to the odor of the ester?

3. What forms on reacidification of the solution? Name the compound.

4. Write the chemical equation for the saponification of methyl salicylate.

POST-LAB QUESTIONS

1. Explain why acetic acid has an odor but benzoic acid does not.

2. Write equations for each of the five esterification reactions.

   a.

   b.

   c.

   d.

   e.

3. Butyric acid has a putrid odor (like rancid butter). Suppose you got some on your hands. How could you rid your hands of the odor? (Remember, butyric acid has marginal solubility in water.)

Properties of amines and amides

Amines and amides are two classes of organic compounds which contain nitrogen. Amines behave as organic bases and may be considered as derivatives of ammonia. Amides are compounds which have a carbonyl group connected to a nitrogen atom and are neutral. In this experiment, you will learn about the physical and chemical properties of some members of the amine and amide families.

If the hydrogens of ammonia are replaced by alkyl or aryl groups, amines result. Depending on the number of organic groups attached to nitrogen, amines are classified as either primary (one group), secondary (two groups), or tertiary (three groups) (Table 30.1).

Table 30.1 Types of Amines		
Primary Amines	**Secondary Amines**	**Tertiary Amines**
NH_3 Ammonia		
CH_3NH_2 Methylamine	$(CH_3)_2N$ Dimethylamine	$(CH_3)_3N$ Trimethylamine
Aniline (NH_2)	N-Methylaniline ($NH-CH_3$)	N,N-Dimethylaniline ($N-CH_3$ with CH_3)

There are a number of similarities between ammonia and amines that carry beyond the structure. Consider odor. The smell of amines resembles that of ammonia but is not as sharp. However, amines can be quite pungent. Anyone handling or working with raw fish knows how strong the amine odor can be, since raw fish contains low-molecular weight amines such as dimethylamine and trimethylamine. Other amines associated with decaying flesh have names suggestive of their odors: putrescine and cadaverine.

$$NH_2CH_2CH_2CH_2CH_2NH_2$$

Putrescine
(1,4-Diaminobutane)

$$NH_2CH_2CH_2CH_2CH_2CH_2NH_2$$

Cadaverine
(1,5-Diaminopentane)

The solubility of low molecular weight amines in water is high. In general, if the total number of carbons attached to nitrogen is six or less, the amine is water soluble; amines with a carbon content greater than six are water insoluble. However, all amines are soluble in organic solvents such as diethyl ether or methylene chloride.

Since amines are organic bases, water solutions show weakly basic properties. If the basicity of aliphatic amines and aromatic amines are compared to ammonia, aliphatic amines are stronger than ammonia, while aromatic amines are weaker. Amines characteristically react with acids to form ammonium salts; the non-bonded electron pair on nitrogen bonds the hydrogen ion.

$$RNH_2 \ + \ HCl \longrightarrow RNH_3^+Cl^-$$

Amine   Ammonium Salt

If an amine is insoluble, reaction with an acid produces a water-soluble salt. Since ammonium salts are water soluble, many drugs containing amines are prepared as ammonium salts. After working with fish in the kitchen, a convenient way to rid one's hands of fish odor is to rub a freshly cut lemon over the hands. The citric acid found in the lemon reacts with the amines found on the fish; a salt forms which can be easily rinsed away with water.

Amides are carboxylic acid derivatives. The amide group is recognized by the nitrogen connected to the carbonyl group. Amides are neutral compounds; the electrons are delocalized into the carbonyl (resonance) and thus, are not available to bond to a hydrogen ion.

Amide group   Acetamide   Benzamide

Under suitable conditions, amide formation can take place between an amine and a carboxylic acid, an acyl halide, or an acid anhydride. Along with ammonia, primary and secondary amines yield amides with carboxylic acids or derivatives. Table 30.2 relates the nitrogen base with the amide class (based on the number of alkyl or aryl groups on the nitrogen of the amide).

$$CH_3NH_2 \ + \ CH_3COOH \longrightarrow CH_3COO^-(CH_3NH_3^+) \xrightarrow{\Delta} CH_3CONHCH_3 \ + \ H_2O$$

$$CH_3NH_2 \ + \ CH_3COCl \longrightarrow CH_3CONHCH_3 \ + \ HCl$$

Table 30.2	Classes of Amides	
Nitrogen Base	**(reacts to form)**	**Amide (-C-N-)**
Ammonia		Primary amide (no R groups)
Primary amine		Secondary amide (one R group)
Secondary amine		Tertiary amide (two R groups)

Hydrolysis of amides can take place in either acid or base. Primary amides hydrolyze in acid to ammonium salts and carboxylic acids. Neutralization of the acid and ammonium salts releases ammonia which can be detected by odor or by litmus.

$$\underset{\underset{\displaystyle\|}{\overset{\displaystyle O}{}}}{R-C}-NH_2 + HCl + H_2O \longrightarrow \underset{\underset{\displaystyle\|}{\overset{\displaystyle O}{}}}{R-C}-OH + NH_4Cl$$

$$NH_4Cl + NaOH \longrightarrow NH_3 + NaCl + H_2O$$

Secondary and tertiary amides would release the corresponding alkyl ammonium salts which, when neutralized, would yield the amine.

In base, primary amides hydrolyze to carboxylic acid salts and ammonia. The presence of ammonia (or amine from corresponding amides) can be detected similarly by odor or litmus. The carboxylic acid would be generated by neutralization with acid.

$$\underset{\underset{\displaystyle\|}{\overset{\displaystyle O}{}}}{R-C}-NH_2 + NaOH \longrightarrow \underset{\underset{\displaystyle\|}{\overset{\displaystyle O}{}}}{R-C}-O^-Na^+ + NH_3$$

$$\underset{\underset{\displaystyle\|}{\overset{\displaystyle O}{}}}{R-C}-O^-Na^+ + HCl \longrightarrow \underset{\underset{\displaystyle\|}{\overset{\displaystyle O}{}}}{R-C}-OH + NaCl$$

Objectives

1. To show some physical and chemical properties of amines and amides.
2. To demonstrate the hydrolysis of amides.

Procedure

CAUTION!

Amines are toxic chemicals. Avoid excessive inhaling of the vapors and use gloves to avoid direct skin contact. Anilines are more toxic and are readily absorbed through the skin. Wash any amine or aniline spill with large quantities of water.

Properties of amines

1. Place 5 drops of liquid or 0.1 g of solid from the compounds listed in the following table into labeled clean, dry test tubes (100 × 13 mm).

Test Tube No.	Nitrogen Compound
1	6 M NH$_3$
2	Triethylamine
3	Aniline
4	N,N-Dimethylaniline
5	Acetamide

2. Carefully note the odors of each compound. Do not inhale deeply. Merely wave your hand across the mouth of the test tube toward your nose in order to note the odor. Record your observations on the Report Sheet.

3. Add 2 mL of distilled water to each of the labeled test tubes. Mix thoroughly by sharply tapping the test tube with your finger. Note on the Report Sheet whether the amines are soluble or insoluble.

4. Take a glass rod, and test each solution for its pH. Carefully dip one end of the glass rod into a solution and touch a piece of pH paper. Between each test, be sure to clean and dry the glass rod. Record the pH by comparing the color of the paper with the chart on the dispenser.

5. Carefully add 2 mL of 6 M HCl to each test tube. Mix thoroughly by sharply tapping the test tube with your finger. Compare the odor and solubility of this solution to previous observations.

6. Place 5 drops of liquid or 0.1 g of solid from the compounds listed in the table into labeled clean, dry test tubes (100 × 13 mm). Add 2 mL of diethyl ether (ether) to each test tube. Stopper with a cork and mix thoroughly by shaking. Record the observed solubilities.

7. Carefully place on a watch glass, side-by-side, without touching, a drop of triethylamine and a drop of concentrated HCl. Record your observations.

Hydrolysis of acetamide

1. Dissolve 0.5 g of acetamide in 5 mL of 6 M H$_2$SO$_4$ in a large test tube (150 × 18 mm). Heat the solution in a boiling water bath for 5 min.

2. Hold a small strip of moist pH paper over the mouth of the test tube; note any changes in color; record the pH reading. Remove the test tube from the water bath, holding it in a test tube holder. Carefully note any odor.

3. Place the test tube in an ice-water bath until cool to the touch. Now *carefully add, dropwise with shaking*, 6 M NaOH to the cool solution until basic. (You will need more than 7 mL of base.) Hold a piece of moist pH paper over the mouth. Record the pH reading. Carefully note any odor.

Chemicals and Equipment

1. Acetamide
2. 6 M NH_3, ammonia water
3. Aniline
4. N,N-Dimethylaniline
5. Triethylamine
6. Diethyl ether (ether)
7. 6 M NaOH
8. Concentrated HCl
9. 6 M HCl
10. 6 M H_2SO_4
11. pH papers
12. Hot plate

Experiment 30

PRE-LAB QUESTIONS

1. Draw the structure of the functional group that is found in an amine and an amide.

2. Dimethyl amine, $(CH_3)_2NH$, is soluble in water but hexyl amine, $CH_3CH_2CH_2CH_2CH_2CH_2NH_2$, is not. Explain this observation.

3. What happens when methyl amine, CH_3NH_2, is mixed with hydrochloric acid?

4. Compare the basicity of the following nitrogen compounds: ethylamine, $CH_3CH_2NH_2$, versus acetamide, $CH_3-\overset{\overset{\displaystyle O}{\|}}{C}-NH_2$.

NAME _______________ SECTION _______ DATE _______

PARTNER _______________ GRADE _______

Experiment 30

REPORT SHEET

Properties of amines

	Odor		Solubility			pH
	Original Sol.	with HCl	H_2O	Ether	HCl	H_2O
6 M NH_3						
Triethylamine						
Aniline						
N,N-Dimethylaniline						
Acetamide						

Triethylamine and concentrated hydrochloric acid observation:

Hydrolysis of acetamide

1. Acid solution

a. pH reading:

b. Odor noted:

2. Base solution

a. pH reading:

b. Odor noted:

POST-LAB QUESTIONS

1. The active ingredient in the commercial insect repellent "OFF" is N,N-diethyl-*m*-toluamide. Does this compound contain an amine or an amide functional group?

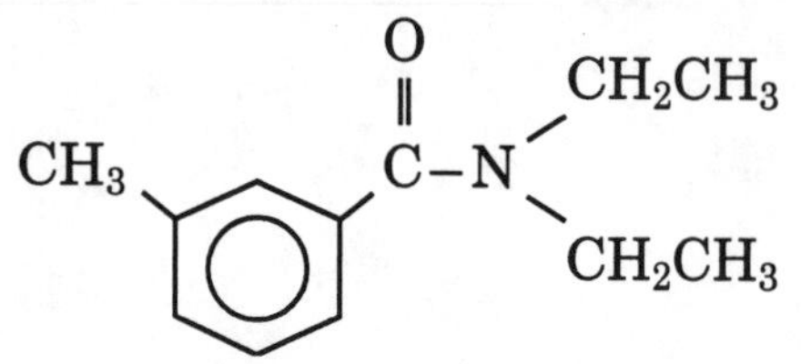

2. Of the amines tested, which was the least soluble in water? Why?

3. Why would an amine like putrescine lose its odor when mixed with hydrochloric acid?

4. Write the equations that account for what happens in the hydrolysis of the acetamide solution in (a) acid and in (b) base.

a.

b.

Experiment 31

Polymerization reactions

Background

Polymers are giant molecules made of many (poly-) small units. The starting material, which is a single unit, is called the monomer. Many of the most important biological compounds are polymers. Cellulose and starch are polymers of glucose units, proteins are made of amino acids, and nucleic acids are polymers of nucleotides. Since the 1930s, a large number of man-made polymers have been manufactured. They contribute to our comfort and gave rise to the previous slogan of DuPont Co.: "Better living through chemistry." Man-made fibers such as nylon and polyesters, plastics such as the packaging materials made of polyethylene and polypropylene films, polystyrene, and polyvinyl chloride, just to name a few, all became household words. Man-made polymers are parts of buildings, automobiles, machinery, toys, appliances, etc.; we encounter them daily in our life.

We focus our attention in this experiment on man-made polymers and the basic mechanism by which some of them are formed. The two most important types of reactions that are employed in polymer manufacturing are the addition and condensation polymerization reactions. The first is represented by the polymerization of styrene and the second by the formation of nylon.

Styrene is a simple organic monomer which, by its virtue of containing a double bond, can undergo addition polymerization.

$$H_2C = CH + H_2C = CH \longrightarrow H_3C - CH - CH = CH$$

The reaction is called an *addition reaction* because two monomers are added to each other with the elimination of a double bond. However, the reaction as such does not go without the help of an unstable molecule, called an *initiator*, that starts the reaction. Benzoyl peroxide or *t*-butyl benzoyl peroxide are such initiators. Benzoyl peroxide splits into two halves under the influence of heat or ultraviolet light and thus produces two free radicals. A *free radical* is a molecular fragment that has one unpaired electron. Thus, when the central bond was broken in the benzoyl peroxide, each of the shared pair of electrons went with one half of the molecule, each containing an unpaired electron.

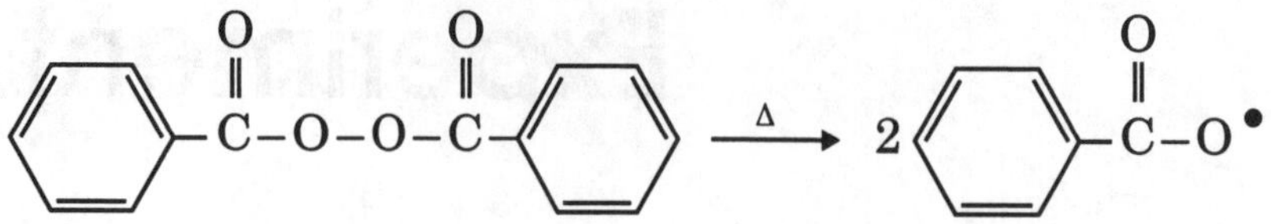

Benzoyl peroxide

Similarly, *t*-butyl benzoyl peroxide also gives two free radicals:

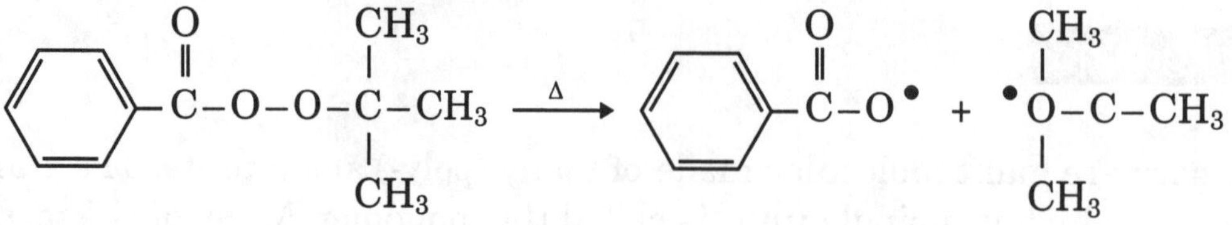

t-butyl benzoyl peroxide

The dot indicates the unpaired electron. The free radical reacts with styrene and initiates the reaction:

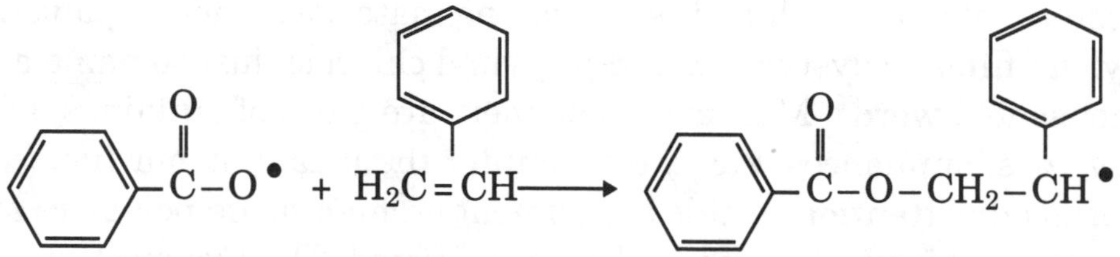

styrene

After this, the styrene monomers are added to the growing chain one by one until giant molecules containing hundreds and thousands of styrene-repeating units are formed. Please note the distinction between the monomer and the repeating unit. The monomer is the starting material, the repeating unit is part of the polymer chain. Chemically they are not identical. In the case of styrene, the monomer contains a double bond, while the repeating unit (in the brackets in the following structure) does not.

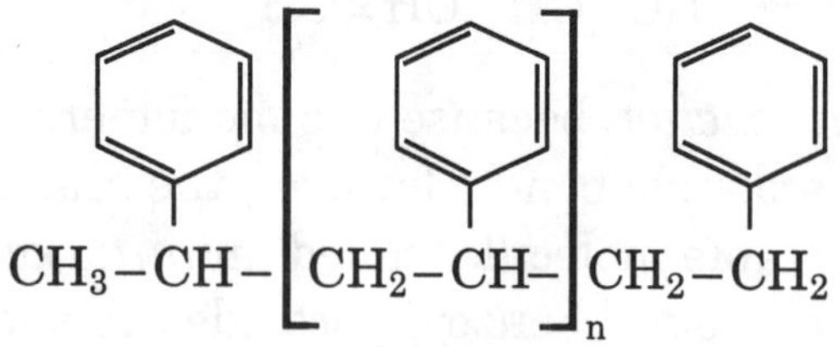

polystyrene

Since the initiators are unstable compounds, care should be taken not to keep them near flames or heat them directly. If a bottle containing a peroxide initiator is dropped, a minor explosion can even occur.

The second type of reaction is called a *condensation reaction* because we condense two monomers into a longer unit, and at the same time we eliminate—expel—a small molecule. Nylon 6-6 is made of adipoyl chloride and hexamethylene diamine:

$$n \; Cl-\overset{\overset{\displaystyle O}{\|}}{C}-CH_2-CH_2-CH_2-CH_2-\overset{\overset{\displaystyle O}{\|}}{C}-Cl + n \; H_2N-CH_2-CH_2-CH_2-CH_2-CH_2-CH_2-NH_2 \rightarrow$$

adipoyl chloride   hexamethylene diamine

$$Cl-\overset{\overset{\displaystyle O}{\|}}{C}(CH_2)_4-\overset{\overset{\displaystyle O}{\|}}{C}-\left[NH-(CH_2)_6-NH-\overset{\overset{\displaystyle O}{\|}}{C}-(CH_2)_4-\overset{\overset{\displaystyle O}{\|}}{C}-\right]_n NH-(CH_2)_6-NH_2 \; + \; n \; HCl$$

repeating unit

Nylon 6-6

We form an amide linkage between the adipoyl chloride and the amine with the elimination of HCl. The polymer is called nylon 6-6 because there are six carbon atoms in the acyl chloride and six carbon atoms in the diamine. Other nylons, such as nylon 10-6, are made of sebacoyl chloride (a ten carbon atom containing acyl chloride) and hexamethylene diamine (a six carbon atom containing diamine). We use an acyl chloride rather than a carboxylic acid to form the amide bond because the former is more reactive. NaOH is added to the polymerization reaction in order to neutralize the HCl that is released every time an amide bond is formed.

The length of the polymer chain formed in both reactions depends on environmental conditions. Usually the chains formed can be made longer by heating the products longer. This process is called *curing*.

Objectives

1. To acquaint students with the conceptual and physical distinction between monomer and polymer.
2. To perform addition and condensation polymerization and solvent casting of films.

Preparation of Polystyrene

1. Set up your hot plate in the hood. Place 25 mL of styrene in a 150-mL beaker. Add 20 drops of *t*-butyl benzoyl peroxide (*t*-butyl peroxide benzoate) initiator. Mix the solution.

2. Heat the mixture on the hot plate to about 140°C. The mixture will turn yellow.

3. When bubbles appear, remove the beaker from the hot plate with beaker tongs. The polymerization reaction is exothermic and thus it generates its own heat. Overheating will create sudden boiling. When the bubbles disappear, put the beaker back on the hot plate. However, every time the mixture starts boiling you must remove the beaker.

4. Continue the heating until the mixture in the beaker has a syrup-like consistency.

5. Pour the contents of the beaker onto a clean watch glass and let it solidify. The beaker can be cleaned from residual polystyrene by adding xylene and warming it on the hot plate under the hood until the polymer is dissolved.

6. Pour a few drops of the warm xylene solution on a microscope slide and let the solvent evaporate. A thin film of polystyrene will be obtained. This is one of the techniques—the so-called solvent-casting technique—used to make films from bulk polymers.

7. Discard the remaining xylene solution into a special jar labeled "Waste." Wash your beaker with soap and water.

8. Investigate the consistency of the solidified polystyrene on your watch glass. You can remove the solid mass by prying it off with a spatula.

Preparation of Nylon

1. Set up a 50-mL reaction beaker and clamp above it a cylindrical paper roll (from toilet paper) or a stick.

2. Add 2.0 mL of 20% NaOH solution and 10 mL of a 5% aqueous solution of hexamethylene diamine.

3. Take 10 mL of 5% adipoyl chloride solution in cyclohexane with a pipet or syringe. Layer the cyclohexane solution slowly on top of the aqueous solution in the beaker. Two layers will form and nylon will be produced at the interface (Fig. 31.1).

4. With a bent wire first scrape off the nylon formed on the walls of the beaker.

Figure 31.1
Preparation of nylon.

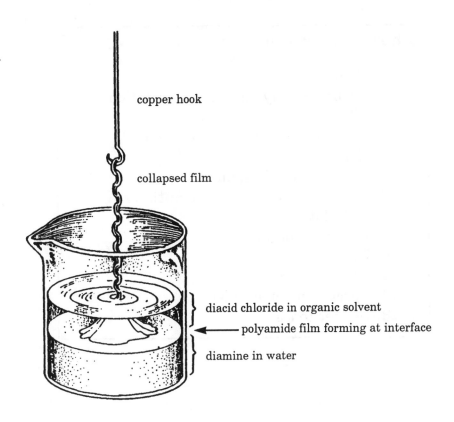

copper hook

collapsed film

diacid chloride in organic solvent

polyamide film forming at interface

diamine in water

5. Slowly lift and pull the film from the center. If you pull it too fast, the nylon rope will break.

6. Wind it around the paper roll or stick two to three times. Do not touch it with your hands.

7. Slowly rotate the roll or the stick and wind at least a 1-meter nylon rope.

8. Cut the rope and transfer the wound rope into a beaker filled with water (or 50% ethanol). Watch as the thickness of the rope collapses. Dry the rope between two filter papers.

9. There are still monomers left in the beaker. Mix the contents vigorously with a glass rod. Observe the beads of nylon that have formed.

10. Pour the mixture into a cold water bath and wash it. Dry the nylon between two filter papers. Note the consistency of your products.

11. Dissolve a small amount of nylon in 80% formic acid. Place a few drops of the solution onto a microscope slide and evaporate the solvent under the hood.

12. Compare the appearance of the solvent cast nylon film with that of the polystyrene.

Chemicals and Equipment

1. Styrene
2. 5% hexamethylene diamine solution
3. 5% adipoyl chloride solution
4. 20% sodium hydroxide solution
5. Xylene
6. 80% formic acid solution
7. *t*-butyl peroxide benzoate initiator
8. Hot plate
9. Paper roll or stick
10. Bent wires
11. 10-mL pipets
12. Spectroline pipet filler
13. Beaker tongs

Experiment 31

PRE-LAB QUESTIONS

1. Why should you *not* expose *t*-butyl peroxide to direct heat?

2. When preparing nylon, what is the purpose of adding NaOH to the polymerization mixture?

3. Write the reaction for the polymerization of tetrafluoroethene. Show the repeating unit of the resulting polymer (Teflon).

4. Write the structure of the monomers and that of the repeating unit in nylon 6-10.

5. What is a free radical? How is it generated?

Experiment 31

REPORT SHEET

1. Describe the appearance of polystyrene and nylon.

2. Can you distinguish between polystyrene and nylon on the basis of solubility?

3. Is there any difference in the appearance of the solvent cast films of nylon and polystyrene?

POST-LAB QUESTIONS

1. A polyester is made of sebacoyl chloride and ethylene glycol,

$$Cl-\underset{\underset{O}{\|}}{C}-(CH_2)_8-\underset{\underset{O}{\|}}{C}-Cl \quad \text{and} \quad \underset{|}{CH_2OH} \\ CH_2OH$$

a. Draw the structure of the polyester formed.

b. What molecules have been eliminated in this condensation reaction?

2. If instead of styrene one would use propene (propylene), $CH_3 - CH = CH_2$, as a monomer in a free radical polymerization, write the reaction in which the free radical of the monomer will be generated using t-butyl benzoyl peroxide as an initiator.

3. Since the polymerization of styrene is an exothermic reaction, why do you need to heat the mixture to 140°C?

Experiment 32

● Preparation of acetyl salicylic acid (aspirin)

One of the most widely used non-prescription drugs is aspirin. In the United States, more than 15,000 pounds are sold each year. It is no wonder there is such wide use when one considers the medicinal applications for aspirin. It is an effective analgesic (pain killer) that can reduce the mild pain of headaches, toothache, neuralgia (nerve pain), muscle pain and joint pain (from arthritis and rheumatism). Aspirin behaves as an antipyretic drug (it reduces fever) and an anti-inflammatory agent capable of reducing the swelling and redness associated with inflammation. It is an effective agent in preventing strokes and heart attacks due to its ability to act as an anti-coagulant.

Early studies showed the active agent that gave these properties to be salicylic acid. However, salicylic acid contains the phenolic and the carboxylic acid groups. As a result, the compound was too harsh to the linings of the mouth, esophagus, and stomach. Contact with the stomach lining caused some hemorrhaging. The Bayer Company in Germany patented the ester acetylsalicylic acid and marketed the product as "aspirin" in 1899. Their studies showed that this material was less of an irritant; the acetylsalicylic acid was hydrolyzed in the small intestine to salicylic acid, which then was absorbed into the bloodstream. The relationship between salicylic acid and aspirin is shown in the following formulas:

Salicylic acid   Acetylsalicylic acid (Aspirin)

Aspirin still has its side effects. Hemorrhaging of the stomach walls can occur even with normal dosages. These side effects can be reduced through the addition of coatings or through the use of buffering agents. Magnesium hydroxide, magnesium carbonate, and aluminum glycinate, when mixed into the formulation of the aspirin (e.g., Bufferin), reduce the irritation.

This experiment will acquaint you with a simple synthetic problem in the preparation of aspirin. The preparative method uses acetic anhydride and an acid catalyst, like sulfuric or phosphoric acid, to speed up the reaction with salicylic acid.

salicylic acid   acetic anhydride   aspirin   acetic acid

If any salicylic acid remains unreacted, its presence can be detected with a 1% iron(III) chloride solution. Salicylic acid has a phenol group in the molecule. The iron(III) chloride gives a violet color with any molecule possessing a phenol group (see Experiment 27). Notice the aspirin no longer has the phenol group. Thus a pure sample of aspirin will not give a purple color with 1% iron(III) chloride solution.

Objectives

1. To illustrate the synthesis of the drug aspirin.
2. To use a chemical test to determine the purity of the preparation.

Procedure

Preparation of aspirin

1. Prepare a bath using a 400-mL beaker filled about half-way with water. Heat to boiling.

2. Take 2.0 g of salicylic acid and place it in a 125-mL Erlenmeyer flask. Use this quantity of salicylic acid to calculate the theoretical or expected yield of aspirin (1). Carefully add 3 mL of acetic anhydride to the flask and, while swirling, add 3 drops of concentrated phosphoric acid.

CAUTION!

Acetic anhydride will irritate your eyes. Phosphoric acid will cause burns to the skin. Use gloves with these reagents. Handle both chemicals with care. Dispense in the hood.

3. Mix the reagents and then place the flask in the boiling water bath; heat for 15 min. (Fig. 32.1). The solid will completely dissolve. Swirl the solution occasionally.

Figure 32.1

Assembly for the
synthesis of aspirin.

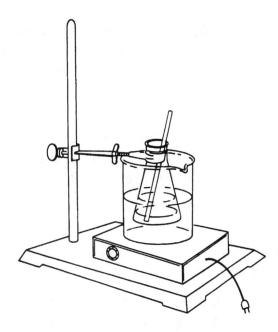

4. Remove the Erlenmeyer flask from the bath and let it cool to approximately
room temperature. Then, slowly pour the solution into a 150-mL beaker containing 20 mL of ice water, mix thoroughly, and place the beaker in an ice
bath. The water destroys any unreacted acetic anhydride and will cause the
insoluble aspirin to precipitate from solution.

5. Collect the crystals by filtering under suction with a Büchner funnel. The
assembly is shown in Fig. 32.2. (Also see Fig. 28.1, p. 313.)

Figure 32.2

Filtering using the
Büchner funnel.

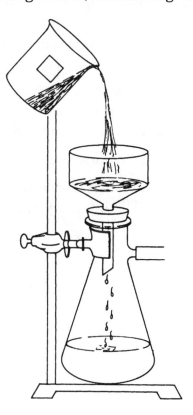

6. Obtain a 250-mL filter flask and connect the side-arm of the filter flask to a water aspirator with heavy wall vacuum rubber tubing. (The thick walls of the tubing will not collapse when the water is turned on and the pressure is reduced.)

7. The Büchner funnel is inserted into the filter flask through either a filtervac, a neoprene adapter, or a one-hole rubber stopper, whichever is available. Filter paper is then placed into the Büchner funnel. Be sure that the paper *lies flat* and *covers all the holes*. Wet the filter paper with water.

8. Turn on the water aspirator to maximum water flow. Pour the solution into the Büchner funnel.

9. Wash the crystals with two 5 mL portions of cold water, followed by one 10 mL portion of cold ethanol.

10. Continue suction through the crystals for several minutes to help dry them. Disconnect the rubber tubing from the filter flask before turning off the water aspirator.

11. Using a spatula, place the crystals between several sheets of paper toweling or filter paper and press-dry the solid.

12. Weigh a 50-mL beaker (2). Add the crystals and reweigh (3). Calculate the weight of crude aspirin (4). Determine the percent yield (5).

Determine the purity of the aspirin

1. The aspirin you prepared is not pure enough for use as a drug and is *not* suitable for ingestion. The purity of the sample will be tested with 1% iron(III) chloride solution and compared with a commercial aspirin and salicylic acid.

2. Label three test tubes (100 × 13 mm) 1, 2, and 3; place a few crystals of salicylic acid into test tube no. 1, a small sample of your aspirin into test tube no. 2, and a small sample of a crushed commercial aspirin into test tube no. 3. Add 5 mL of distilled water to each test tube and swirl to dissolve the crystals.

3. Add 10 drops of 1% aqueous iron(III) chloride to each test tube.

4. Compare and record your observations. The formation of a purple color indicates the presence of salicylic acid. The intensity of the color qualitatively tells how much salicylic acid is present.

Chemicals and Equipment

1. Acetic anhydride
2. Concentrated phosphoric acid, H_3PO_4
3. Commercial aspirin tablets
4. 95% ethanol
5. 1% iron(III) chloride
6. Salicylic acid
7. Boiling chips
8. Büchner funnel, small
9. 250-mL filter flask
10. Filter paper
11. Filtervac or neoprene adaptor
12. Hot plate

Experiment 32

PRE-LAB QUESTIONS

1. A person took aspirin because it is an antipyretic. What was the person trying to accomplish?

2. Draw the structure of aspirin. Should this compound test positive with 1% iron(III) chloride solution? Explain your answer.

3. Aspirin can irritate the stomach. What is usually done in the formulation of the drug that reduces this side effect?

4. What is the active ingredient that gives aspirin its therapeutic properties?

NAME ______________________ SECTION ____________ DATE ____________

PARTNER ______________________ GRADE ______________________

Experiment 32

REPORT SHEET

1. Theoretical yield:

$$_____ \text{g salicylic acid} \times \frac{180 \text{ g aspirin}}{1 \text{ mole}} \times \frac{1 \text{ mole}}{138 \text{ g salicylic acid}}$$

$$= __________ \text{ g aspirin}$$

2. Weight of 50-mL beaker ____________ g

3. Weight of your aspirin and beaker ____________ g

4. Weight of your aspirin: (3) − (2) ____________ g

5. Percent yield: [(4)/(1)] × 100 ____________ %

6. Iron(III) chloride test

No.	Sample	Color	Intensity
1	Salicylic acid		
2	Your aspirin		
3	Commercial aspirin		

POST-LAB QUESTIONS

1. What is the purpose of the concentrated phosphoric acid in the preparation of aspirin? Could some other acid be used?

2. What would happen to your percent yield if in step no. 11 of the procedure you failed to dry completely your aspirin preparation by omitting the drying between filter paper?

3. A student expected 10.0 g of acetylsalicylic acid but obtained only 6.3 g. What is the percentage yield?

4. Tylenol also is an analgesic often taken by people who are allergic to aspirin. The active ingredient is acetaminophen.

$$\text{HO} - \langle \bigcirc \rangle - \overset{\overset{\displaystyle O}{\overset{\|}{}}}{\text{NHCCH}_3}$$

Acetaminophen

Would acetaminophen give a positive phenol test? Explain your answer. Circle the amide functional group in the structure.

Experiment 33

• Measurement of the active
ingredient in aspirin pills

Background

Medication delivered in the form of a pill contains an active ingredient or ingredients. Beside the the drug itself, the pill also contains fillers. The task of the filler is many fold. Sometimes it is there to mask the bitter or otherwise unpleasant taste of the drug. Other times the filler is necessary because the prescribed dose of the drug is so small in mass that it would be difficult to handle. Drugs that have the same generic name contain the same active ingredient. The dosage of the active ingredient must be listed as specified by law. On the other hand, neither the quantity of the filler nor its chemical nature appears on the label. That does not mean that the fillers are completely inactive. They usually effect the rate of drug delivery. In order to deliver the active ingredient, the pill must fall apart in the stomach. For this reason, many fillers are polysaccharides, for example, starch, that either are partially soluble in stomach acid or swell, allowing the drug to be delivered in the stomach or in the intestines.

In the present experiment, we measure the amount of the active ingredient, acetylsalicylic acid (see also Experiment 32), in common aspirin pills. Companies use different fillers and in different amounts, but the active ingredient, acetylsalicylic acid, must be the same in every aspirin tablet. We separate the acetylsalicylic acid from the filler based on their different solubilities. Acetylsalicylic acid is very soluble in ethanol, while neither starch, nor other polysaccharides, or even mono- and disaccharides used as a fillers, are soluble in ethanol. Some companies may use inorganic salts as fillers but these too are not soluble in ethanol. On the other hand, some specially formulated aspirin tablets may contain small amounts of ethanol-soluble substances such as stearic acid or vegetable oil. Thus the ethanol extracts of aspirin tablets may contain small amounts of substances other than acetylsalicylic acid.

Objectives

1. To appreciate the ratio of filler to active ingredients in common aspirin tablets.
2. To learn techniques of quantitative separations.

Procedure

1. Weigh approximately 10 g of aspirin tablets. Record the actual weight on your Report Sheet (1). Count the number of tablets and record it on your Report Sheet (2).

2. Place the weighed aspirin tablets in a mortar of approximately 100 mL capacity. Before starting to grind, place the mortar on a white sheet of paper and loosely cover it with a filter paper. The purpose of this procedure is to catch small fragments of the tablets that may fly out of the mortar during the grinding process. Break up the aspirin tablets by gently hammering them with the pestle. Recover and place back in the mortar any fragments that flew out during the hammering. With a twisting motion of your wrist, grind the aspirin pieces into a fine powder with the aid of the pestle.

3. Add 10 mL of 95% ethanol to the mortar and continue to grind for 2 min. Place a filter paper (Whatmann no. 2, 7 cm) in a funnel and place the funnel in a 250-mL Erlenmeyer flask. With the aid of a glass rod, transfer the supernatant liquid from the mortar to the filter paper. After a few minutes when about 1 mL of clear filtrate has been collected in the Erlenmeyer flask, lift the funnel and allow a drop of the filtrate to fall on a clean microscope slide. Replace the funnel in the Erlenmeyer flask and allow the filtration to continue. The drop on the microscope slide will rapidly evaporate leaving behind crystals of acetylsalicylic acid. This is a qualitative test showing that the extraction of the active ingredient is successful. Report what you see on the microscope slide on your Report Sheet (3).

4. Add another 10 mL of 95% ethanol and repeat the procedure from no. 3.

5. Repeat procedure no. 4 two more times; you will use a total of 40 mL of ethanol in the four extractions. Report after each extraction if the extract carries acetylsalicylic acid. Enter these observations on your Report Sheet (4), (5), and (6).

6. When the filtration is completed and only the white moist solid is left in the filter, transfer the filter paper with its contents into a 100-mL beaker and place the beaker into a drying oven set at 110°C. Dry for 10 min.

7. Carefully remove the beaker from the oven. (**CAUTION!** The beaker is hot.) Allow it to come to room temperature. Weigh a clean and dry 25-mL beaker on your balance. Report the weight on your Report Sheet (7). With the aid of a spatula, **carefully** transfer the dried filler from the filter paper into the 25-mL beaker. Make sure that you do not spill any of the powder. Some of the dried filler may stick to the paper a bit, and you may have to scrape the paper with the spatula. Weigh the 25-mL beaker with its contents on your balance. Report the weight on your Report Sheet (8).

8. Test the dried filler with a drop of Hanus iodine solution. A blue coloration will indicate that it contains starch. Report your findings on your Report Sheet (14).

Chemicals and Equipment

1. Aspirin tablets
2. Mortar and pestle (100 mL capacity)
3. 95% ethanol
4. Filter paper (Whatman no. 2, 7 cm)
5. Balance
6. Drying oven at 110°C
7. Hanus iodine solution.
8. Microscope slides
9. 100-mL beaker
10. 25-ml beaker

EXPERIMENT 33

PRE-LAB QUESTIONS

1. Consider the statement: "Like dissolves like." Explain why acetylsalicylic acid is soluble in ethanol.

2. Consider the same statement again. Explain why starch and cellulose are not soluble in ethanol.

3. The normal adult aspirin tablet contains 5.4 grains of aspirin. If one grain is 64.8 mg, how many mg of aspirin are in one tablet?

Experiment 33

REPORT SHEET

1. Weight of aspirin tablets _____ g

2. Number of aspirin tablets in your sample _____

3. Does your first extract contain acetylsalicylic acid? _____

4. Does your second extract contain acetylsalicylic acid? _____

5. Does your third extract contain acetylsalicylic acid? _____

6. Does your fourth extract contain acetylsalicylic acid? _____

7. Weight of the empty 25-mL beaker _____ g

8. Weight of the 25-mL beaker and filler _____ g

9. Weight of the filler: (8) − (7) _____ g

10. Percent of filler in tablets: [(9)/(1)] × 100 _____ %

11. Weight of one tablet: (1)/(2) _____ g

12. Weight of filler per tablet: (11) × [(10)/100] _____ g

13. Weight of acetylsalicylic acid per tablet: (11) − (12) _____ g

14. Does your filler contain starch? _____

POST-LAB QUESTIONS

1. According to your calculations, does your aspirin tablet contain more, the same, or less active ingredients than the average adult dosage (5.4 grains)?

2. If your ethanol extract contained a filler in addition to the active ingredient, acetylsalicylic acid, how would that effect your calculations of the dosage of the aspirin tablet?

3. If you did not dry your filler to sufficient dryness in the drying oven, how would that effect your calculations of the dosage of the aspirin tablet?

4. On the basis of the Hanus iodine test performed, what can you say about the nature of the filler in your aspirin tablet?

5. On the basis of your observations of residual crystals on the microscope slide, what can you say about the completeness of the extraction procedure?

Experiment 34

● Isolation of caffeine from tea leaves

Background

Many organic compounds are obtained from natural sources through extraction. This method takes advantage of the solubility characteristics of a particular organic substance with a given solvent. In the experiment here, caffeine is readily soluble in hot water and is thus separated from the tea leaves. Caffeine is one of the main substances that make up the water solution called tea. Besides being found in tea leaves, caffeine is present in coffee, kola nuts, and cocoa beans. As much as 5% by weight of the leaf material in tea plants consists of caffeine.

The caffeine structure is shown below. It is classed as an alkaloid, meaning that with the nitrogen present, the molecule has base characteristics (alkali-like). In addition, the molecule has the purine ring system, a framework which plays an important role in living systems.

Caffeine is the most widely used of all the stimulants. Small doses of this chemical (50 to 200 mg) can increase alertness and reduce drowsiness and fatigue. The popular "No-Doz" tablet lists caffeine as the main ingredient. In addition, it affects blood circulation since the heart is stimulated and blood vessels are relaxed (vasodilation). It also acts as a diuretic. There are side effects. Large doses of over 200 mg can result in insomnia, restlessness, headaches, and muscle tremors ("coffee nerves"). Continued, heavy use may bring on physical dependence. (How many of you know somebody who cannot function in the morning until they have that first cup of coffee?)

Tea leaves consist primarily of cellulose; this is the principle structural material of all plant cells. Fortunately, the cellulose is insoluble in water, so that by using a hot water extraction, more soluble caffeine can be separated. Also dissolved in water are complex substances called tannins. These are colored phenolic compounds of high molecular weight (500 to 3000) that have acidic behavior. If a basic salt such as Na_2CO_3 is added to the water solution, the tannins can react to form a salt. These salts are insoluble in organic solvents, such as chloroform or methylene chloride, but are soluble in water.

Although caffeine is soluble in water (2 g/100 g of cold water), it is more soluble in the organic solvent methylene chloride (14 g/100 g). Thus caffeine can be extracted from the basic tea solution with methylene chloride, but the sodium salts of the tannins remain behind in the aqueous solution. Evaporation of the methylene chloride yields crude caffeine; the crude material can be purified by sublimation (see Experiment 4).

Objectives

1. To demonstrate the isolation of a natural product.
2. To learn the techniques of extraction.
3. To use sublimation as a purification technique.

Procedure

The isolation of caffeine from tea leaves follows the scheme below:

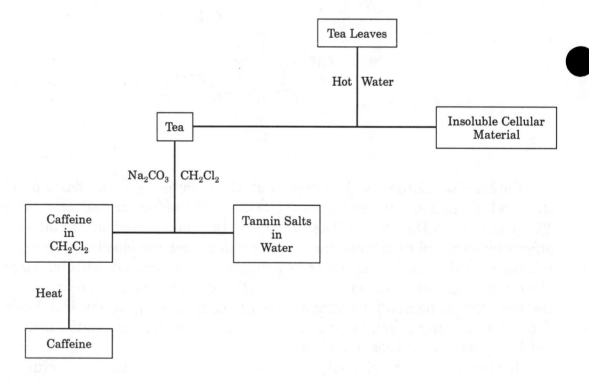

1. Carefully open two commercial tea bags (try not to tear the paper) and weigh the contents to the nearest 0.001 g. Record this weight (1). Place the tea leaves back into the bags, close, and secure the bags with staples.

2. Into a 150-mL beaker, place the tea bags so that they lie flat on the bottom. Add 30 mL of distilled water and 2.0 g of anhydrous Na_2CO_3; heat the contents with a hot plate, keeping a *gentle* boil, for 20 min. While the mixture is boiling, keep a watch glass on the beaker. Hold the tea bags under water by occasionally pushing them down with a glass rod.

3. Decant the hot liquid into a 50-mL Erlenmeyer flask. Wash the tea bags with 10 mL of hot water, carefully pressing the tea bag with a glass rod; add this wash water to the tea extract. (If any solids are present in the tea extract, filter them by gravity to remove.) Cool the combined tea extract to room temperature. The tea bags may be discarded.

4. Transfer the cool tea extract to a 125-mL separatory funnel that is supported on a ring stand with a ring clamp.

5. Carefully add 5.0 mL of methylene chloride to the separatory funnel. Stopper the funnel and lift it from the ring clamp; hold the funnel with two hands as shown in Fig. 34.1. By holding the stopper in place with one hand, invert the funnel. *Make certain the stopper is held tightly and no liquid is spilled; open the stop-cock, being sure to point the opening away from you and your neighbors.* Built-up pressure caused by gases accumulating inside will be released. Now, close the stop-cock and gently mix the contents by inverting the funnel two or three times. Again, release any pressure by opening the stop-cock as before.

Figure 34.1

Using the separatory funnel.

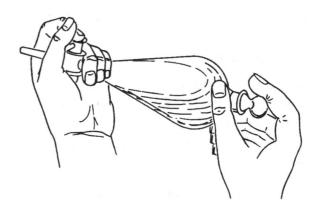

6. Return the separatory funnel to the ring clamp, remove the stopper, and allow the aqueous layer to separate from the methylene chloride layer (Fig. 34.2). You should see two distinct layers form after a few minutes with the methylene chloride layer at the bottom. Sometimes an emulsion may form at the juncture of the two layers. The emulsion often can be broken by gently swirling the contents or by gently stirring the emulsion with a glass rod.

Figure 34.2

Separation of the aqueous layer
and the methylene chloride
layer in the separatory funnel.

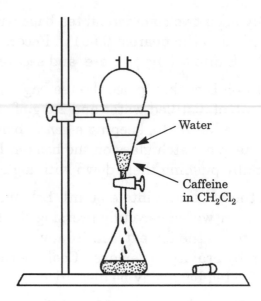

Water

Caffeine
in CH_2Cl_2

7. Carefully drain the lower layer into a 25-mL Erlenmeyer flask. Try not to include any water with the methylene chloride layer; careful manipulation of the stop-cock will prevent this.

8. Repeat the extraction with an additional 5.0 mL of methylene chloride. Combine the separated bottom layer with the methylene chloride layer obtained from step no. 7.

9. Add 0.5 g of anhydrous Na_2SO_4 to the combined methylene chloride extracts. Swirl the flask. The anhydrous salt is a drying agent and will remove any water that may still be present.

10. Weigh a 25-mL side-arm filter flask containing one or two boiling stones. Record this weight (2). By means of a gravity filtration, filter the methylene chloride–salt mixture into the pre-weighed flask. Rinse the salt on the filter paper with an additional 2.0 mL of methylene chloride.

11. Remove the methylene chloride by evaporation *in the hood*. Be careful not to overheat the solvent, since it may foam over. The solid residue which remains after the solvent is gone is the crude caffeine. Reweigh the cooled flask (3). Calculate the weight of the crude caffeine by subtraction (4) and determine the percent yield (5).

12. Take a melting point of your solid. First, scrape the caffeine from the bottom and sides of the flask with a microspatula and collect a sample of the solid in a capillary tube (review Experiment 4 for the technique). Pure caffeine melts at 238°C. Compare your melting point (6) to the literature value.

Optional

13. At the option of your instructor, the caffeine may be purified further. The caffeine may be sublimed directly from the flask with a cold finger condenser (Fig. 34.3). Carefully insert the cold finger condenser into a no. 2 neoprene

adapter (use a drop of glycerine as a lubricant). Adjust the tip of the cold finger to 1 cm from the bottom of the flask. Clean any glycerine remaining on the cold finger with a Kimwipe and acetone; the cold finger surface must be clean and dry. Connect the cold finger to a faucet by latex tubing (water *in* the upper tube; water *out* the lower tube). Connect the side-arm filter flask to a water aspirator with vacuum tubing, installing a trap between the aspirator and the sublimation set-up (Fig. 34.3). When you turn the water on, press the cold finger into the filter flask until a good seal is made. Gently heat the bottom of the filter flask which holds the caffeine with a microburner (hold the *base* of the microburner); move the flame back and forth and along the sides of the flask. **Do not allow the sample to melt.** If the sample melts, *stop* heating and allow to cool before continuing. When the sublimation is complete, disconnect the heat and allow the system to cool; leave the aspirator connected and the water running.

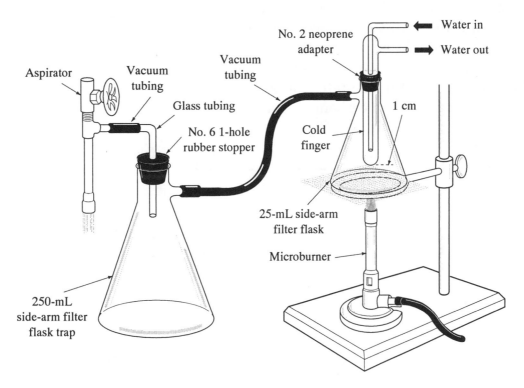

Figure 34.3 • Sublimation apparatus connected to an aspirator.

14. When the system has reached room temperature, carefully disconnect the aspirator from the side-arm filter flask by removing the vacuum tubing from the side-arm. Turn off the water to the cold finger. *Carefully* remove the cold finger from the flask along with the neoprene adapter without dislodging any crystals. Scrape the sublimed caffeine onto a pre-weighed piece of weighing paper (7). Reweigh (8); determine the weight of caffeine (9). Calculate the percent recovery (10). Determine the melting point (11).

15. Collect the caffeine in a sample vial, and submit it to your instructor.

Chemicals and Equipment

1. Boiling chips
2. Cold finger condenser
3. Filter paper, (Whatman no. 7.0) fast flow
4. Hot plate
5. 125-mL separatory funnel with stopper
6. Melting point capillaries
7. No. 2 neoprene adapter
8. 25-mL side-arm filter flask
9. Small sample vials
10. Tea bags
11. Tubing: latex, 2 ft.; vacuum, 2 ft.
12. 250-mL trap: 250-mL side-arm filter flask fitted with a no. 6 one-hole rubber stopper containing a piece of glass tubing (10 cm long × 7 mm OD)
13. Anhydrous sodium sulfate, Na_2SO_4
14. Anhydrous sodium carbonate, Na_2CO_3
15. Methylene chloride, CH_2Cl_2
16. Stapler

Experiment 34

PRE-LAB QUESTIONS

1. Name some sources of caffeine.

2. What part of the caffeine structure gives the alkaloid properties?

3. What are some of the ways caffeine affects an individual?

4. How can you test the purity of a sample of caffeine?

Experiment 34

REPORT SHEET

1. Weight of tea in 2 tea bags   _____ g

2. Weight of 25-mL side-arm filter flask and boiling stones   _____ g

3. Weight of flask, boiling stones, and crude caffeine   _____ g

4. Weight of caffeine: (3) − (2)   _____ g

5. Percent yield: $[(4)/(1)] \times 100$ _____ %

6. Melting point of your crude caffeine   _____ °C

7. Weight of weighing paper   _____ g

8. Weight of sublimed caffeine and paper   _____ g

9. Weight of caffeine: (8) − (7)   _____ g

10. Percent recovery: $[(9)/(4)] \times 100$ _____ %

11. Melting point of sublimed caffeine   _____ °C

POST-LAB QUESTIONS

1. Compare the melting points of the crude and sublimed samples of caffeine. Account for the differences between the melting points. How do they compare to the literature value?

2. What other compounds were extracted along with caffeine in the hot water extract (tea)? How were these compounds separated from the caffeine?

3. In the procedure, why was the caffeine eventually recovered from the methylene chloride extract and not directly from the water?

4. How do you know that methylene chloride has a greater density than water?

Carbohydrates

Carbohydrates are polyhydroxy aldehydes, ketones, or compounds that yield poly-hydroxy aldehydes or ketones upon hydrolysis. Rice, potatoes, bread, corn, candy, and fruits are rich in carbohydrates. A carbohydrate can be classified as a mono-saccharide (glucose or fructose); a disaccharide (sucrose or lactose), which consists of two joined monosaccharides; or a polysaccharide (starch or cellulose), which consists of thousands of monosaccharide units linked together. Monosaccharides exist mostly as cyclic structures containing hemiacetal (or hemiketal) groups. These structures in solutions are in equilibrium with the corresponding open chain structures bearing aldehyde or ketone groups. Glucose, blood sugar, is an example of a polyhydroxy aldehyde (Fig. 35.1).

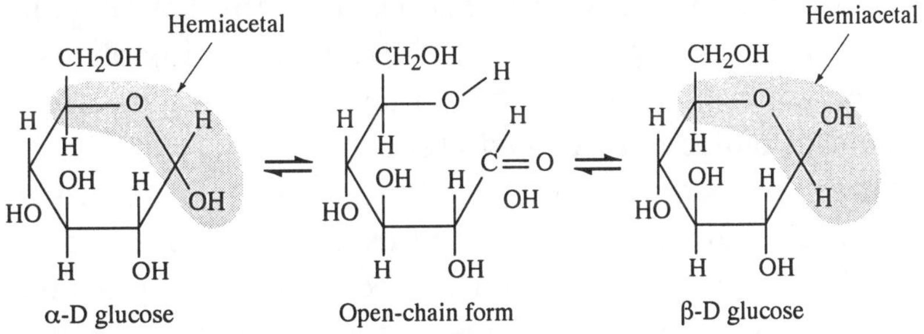

Figure 35.1 • The structures of D-glucose.

Disaccharides and polysaccharides exist as cyclic structures containing functional groups such as hydroxyl groups, acetal (or ketal), and hemiacetal (or hemiketal). Most of the di-, oligo-, or polysaccharides have two distinct ends. The one end which has a hemiacetal (or hemiketal) on its terminal is called the reducing end, and the one which does not contain a hemiacetal (or hemiketal) terminal is the non-reducing end. The name "reducing" is given because hemiacetals (and to a lesser extent hemiketals) can reduce an oxidizing agent such as Benedict's reagent.

Fig. 35.2 is an example:

Figure 35.2

The structure of maltose, a disaccharide.

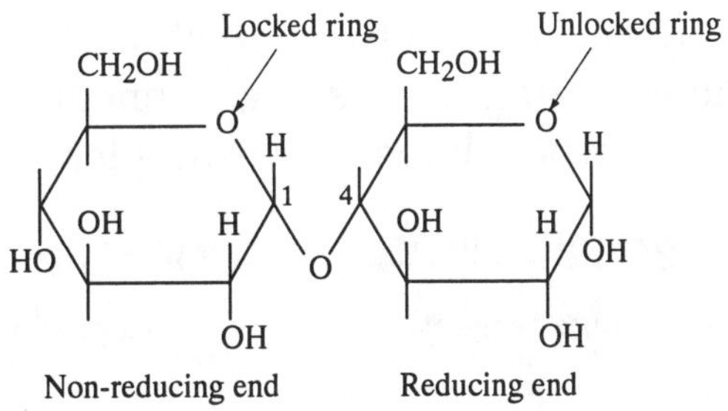

Not all disaccharides or polysaccharides contain a reducing end. An example is sucrose, which does not have a hemiacetal (or hemiketal) group on either of its ends (Fig. 35.3).

Figure 35.3

The structure of sucrose.

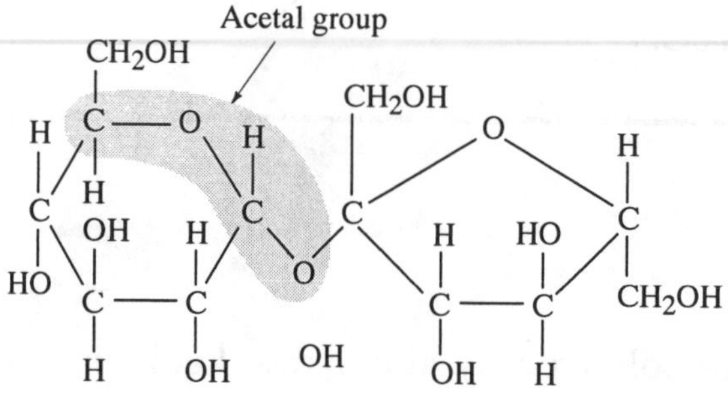

Polysaccharides, such as amylose or amylopectin, do have a hemiacetal group on one of their terminal ends, but practically they are non-reducing substances because there is only one reducing group for every 2,000–10,000 monosaccharidic units. In such a low concentration, the reducing group does not give a positive test with Benedict's or Fehling's reagent.

On the other hand, when a non-reducing disaccharide (sucrose) or a polysaccharide such as amylose is hydrolyzed the glycosidic linkages (acetal) are broken and reducing ends are created. Hydrolyzed sucrose (a mixture of D-glucose and D-fructose) will give a positive test with Benedict's or Fehling's reagent as well as hydrolyzed amylose (a mixture of glucose and glucose containing oligosaccharides). The hydrolysis of sucrose or amylose can be achieved by using a strong acid such as HCl or with the aid of biological catalysts (enzymes).

Starch can form an intense, brilliant, dark blue-, or violet-colored complex with iodine. The straight chain component of starch, the amylose, gives a blue color while the branched component, the amylopectin, yields a purple color. In the presence of iodine, the amylose forms helixes inside of which the iodine molecules assemble as long polyiodide chains. The helix forming branches of amylopectin are much shorter than those of amylose. Therefore, the polyiodide chains are also much shorter in the amylopectin-iodine complex than in the amylose-iodine complex. The result is a different color (purple). When starch is hydrolyzed and broken down to small carbohydrate units, the iodine will not give a dark blue (or purple) color. The iodine test is used in this experiment to indicate the completion of the hydrolysis.

In this experiment, you will investigate some chemical properties of carbohydrates in terms of their functional groups.

1. *Reducing and non-reducing properties of carbohydrates*

   a. **Aldoses (polyhydroxy aldehydes).** All aldoses are reducing sugars because they contain free aldehyde functional groups. The aldehydes are oxidized by mild oxidizing agents (e.g., Benedict's or Fehling's reagent) to the corresponding carboxylates. For example,

$$R\text{—CHO} + 2Cu^{2+} \xrightarrow{NaOH} R\text{—COO}^-Na^+ + Cu_2O\downarrow$$

   (from Fehling's reagent)   red precipitate

b. Ketoses (polyhydroxy ketones). All ketoses are reducing sugars because they have a ketone functional group next to an alcohol functional group. The reactivity of this specific ketone (also called α-hydroxyketone) is attributed to its ability to form an α-hydroxyaldehyde in basic media according to the following equilibrium equations:

$$\begin{array}{ccccc}
CH_2OH & & CHOH & & CHO \\
| & & \| & & | \\
C=O & \underset{base}{\rightleftharpoons} & C-OH & \underset{base}{\rightleftharpoons} & H-C-OH \\
| & & | & & | \\
H-C-OH & & H-C-OH & & H-C-OH \\
| & & | & & | \\
\end{array}$$

$$\begin{array}{ccccc}
\text{ketose} & & \text{enediol} & & \text{aldose}
\end{array}$$

c. Hemiacetal functional group (potential aldehydes). Carbohydrates with hemiacetal functional groups can reduce mild oxidizing agents such as Benedict's reagent because hemiacetals can easily form aldehydes through the following equilibrium equation:

$$\begin{array}{ccc}
\underset{R}{\overset{H}{\diagdown}}C\underset{OH}{\overset{OR'}{\diagup}} & \rightleftharpoons & \underset{R}{\overset{H}{\diagdown}}C=O + R'OH
\end{array}$$

Sucrose is, on the other hand, a nonreducing sugar because it does not contain a hemiacetal functional group. Although starch has a hemiacetal functional group at one end of its molecule, it is, however, considered as a nonreducing sugar because the effect of the hemiacetal group in a very large starch molecule becomes insignificant to give a positive Benedict's test.

2. *Hydrolysis of acetal groups.* Disaccharides and polysaccharides can be converted into monosaccharides by hydrolysis. The following is an example:

$$C_{12}H_{22}O_{11} + H_2O \xrightarrow{\text{catalyst}} C_6H_{12}O_6 + C_6H_{12}O_6$$

$$\begin{array}{ccc}
\text{lactose} & \text{glucose} & \text{galactose} \\
\text{(milk sugar)} & &
\end{array}$$

Objectives

1. To become familiar with the reducing or nonreducing nature of carbohydrates.
2. To experience the enzyme-catalyzed and acid-catalyzed hydrolysis of acetal groups.

Procedure

Reducing or nonreducing carbohydrates

Place approximately 2 mL (40 drops) of Fehling's solution (20 drops each of solution part A and solution part B) into each of five labeled tubes. Add 10 drops of each of the following carbohydrates to the corresponding test tubes as shown in the following table.

Test tube no.	Name of carbohydrate
1	Glucose
2	Fructose
3	Sucrose
4	Lactose
5	Starch

Place the test tubes in a boiling water bath for 5 min. A 600-mL beaker containing about 200 mL of tap water with a few boiling chips is used as the bath. Record your results on your Report Sheet. Which of those carbohydrates are reducing carbohydrates?

Hydrolysis of carbohydrates

Hydrolysis of sucrose (Acid versus base catalysis)

Place 3 mL of 2% sucrose solution in each of two labeled test tubes. To the first test tube (no. 1) add 3 mL of water and 3 drops of dilute sulfuric acid solution (3 M H_2SO_4). To the second test tube (no. 2), add 3 mL of water and 3 drops of dilute sodium hydroxide solution (3 M NaOH). Heat the test tubes in a boiling water bath for about 5 min. Cool both solutions to room temperature. To the contents of test tube no. 1, add dilute sodium hydroxide solution (3 M NaOH) (about 10 drops) until red litmus paper turns blue. Test a few drops of each of the two solutions (test tubes nos. 1 and 2) with Fehling's reagent as described before. Record your results on your Report Sheet.

Hydrolysis of starch (Enzyme versus acid catalysis)

Place 2 mL of 2% starch solution in each of two labeled test tubes. To the first test tube (no. 1), add 2 mL of your own saliva. (Use a 10-mL graduated cylinder to collect your saliva.) To the second test tube (no. 2), add 2 mL of dilute sulfuric acid (3 M H_2SO_4). Place both test tubes in a water bath that has been previously heated to 45°C. Allow the test tubes with their contents to stand in the warm water bath for 30 min. Transfer a few drops of each solution into separate depressions of a spot plate or two separately labeled microtest tubes. (Use two clean, separate medicine droppers for transferring.) To each sample (in microtest tubes or on a spot plate), add 2 drops of iodine solution. Record the color of the solutions on your Report Sheet.

Acid catalyzed hydrolysis of starch

Place 5.0 mL of starch solution in a 15 × 150 mm test tube and add 1.0 mL of dilute sulfuric acid (3 M H_2SO_4). Mix it by gently shaking the test tube. Heat the solution in a boiling water bath for about 5 min. Using a clean medicine dropper, transfer about 3 drops of the starch solution into a spot plate or a microtest tube and then add 2 drops of iodine solution. Observe the color of the solution. If the solution gives a positive test with iodine solution (the solution should turn blue), continue heating. Transfer about 3 drops of the boiling solution at 5-min. intervals for an iodine test. (Note: Rinse the medicine dropper very thoroughly before each test.) When the solution no longer gives a blue color with iodine solution, stop heating and record the time needed for the completion of hydrolysis.

Chemicals and Equipment

1. Bunsen burner
2. Medicine droppers
3. Microtest tubes or a white spot plate
4. Boiling chips
5. Fehling's reagent
6. 3 M NaOH
7. 2% starch solution
8. 2% sucrose
9. 2% fructose
10. 2% glucose
11. 2% lactose
12. 3 M H_2SO_4
13. 0.01 M iodine in KI

Experiment 35

PRE-LAB QUESTIONS

1. Circle the hemiacetal functional group and acetal functional group of the
following carbohydrates:

a. sucrose

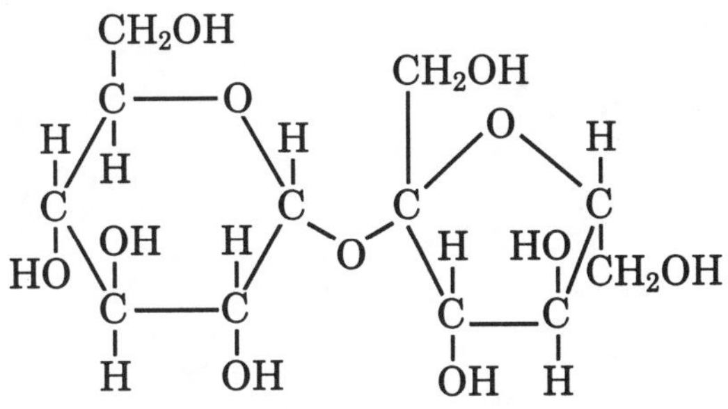

b. lactose

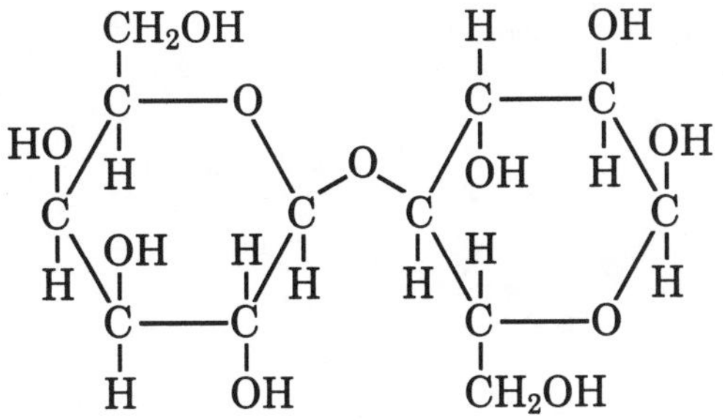

2. Which carbohydrate in question no. 1 is a reducing sugar?

3. The structure of D-fructose is given below. Write the stucture of the corresponding aldose to which D-fructose is converted in basic media.

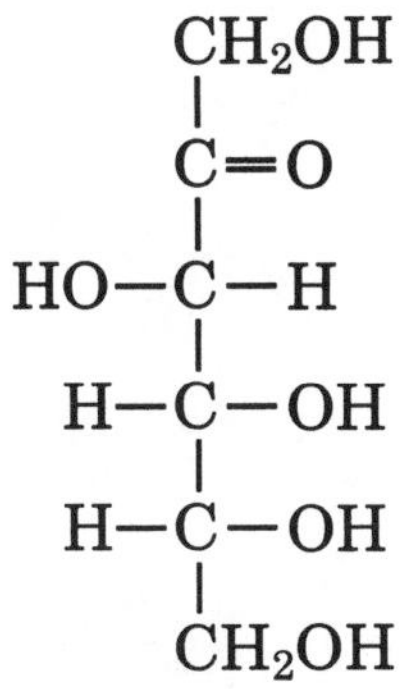

$$
\begin{array}{c}
CH_2OH \\
| \\
C=O \\
| \\
HO-C-H \\
| \\
H-C-OH \\
| \\
H-C-OH \\
| \\
CH_2OH
\end{array}
$$

NAME SECTION DATE

PARTNER GRADE

Experiment 35

REPORT SHEET

Reducing or nonreducing carbohydrates

Test tube no.	Substance	Reducing or nonreducing carbohydrates
1	Glucose	
2	Fructose	
3	Sucrose	
4	Lactose	
5	Starch	

Hydrolysis of carbohydrates

Hydrolysis of sucrose (Acid versus base catalysis)		
Sample	Condition of hydrolysis	Fehling's reagent (positive or negative)
1	Acidic (H_2SO_4)	
2	Basic (NaOH)	

Hydrolysis of starch (Enzyme versus acid catalysis)		
Sample	Condition of hydrolysis	Iodine test (positive or negative)
1	Enzymatic (saliva)	
2	Acidic (H_2SO_4)	

Acid catalyzed hydrolysis of starch		
Test tube no.	Heating time (min.)	Iodine test (positive or negative)
1	5	
2	10	
3	15	
4	20	

POST-LAB QUESTIONS

1. What chemical test is performed to indicate the completion of the hydrolysis of starch by acid?

2. Which hydrolysis of starch is faster: the acid or the enzyme catalyzed reaction?

3. The structure of amylose is depicted in your textbook. Amylose gives a positive iodine test and a negative test with Fehling's reagent. On the basis of the structure of amylose, explain the results of the two tests.

4. In an unusual disaccharide, two α-D-glucose units are linked together in an $\alpha(1 \longrightarrow 1)$ glycosidic linkage. Is this a reducing or non-reducing disaccharide? Explain.

Experiment 36

Preparation and properties of a soap

Background

A soap is the sodium or potassium salt of a long-chain fatty acid. The fatty acid usually contains 12 to 18 carbon atoms. Solid soaps usually consist of sodium salts of fatty acids, whereas liquid soaps consist of the potassium salts of fatty acids.

A soap such as sodium stearate consists of a nonpolar end (the hydrocarbon chain of the fatty acid) and a polar end (the ionic carboxylate).

$$CH_3CH_2CH_2CH_2CH_2CH_2CH_2CH_2CH_2CH_2CH_2CH_2CH_2CH_2CH_2CH_2CH_2-\overset{\overset{\displaystyle O}{\|}}{C}-O^-Na^+$$

nonpolar
(dissolves in oils)

polar
(dissolves in water)

Because "like dissolves like," the nonpolar end (hydrophobic or water-hating part) of the soap molecule can dissolve the greasy dirt, and the polar or ionic end (hydrophilic or water-loving part) of the molecule is attracted to water molecules. Therefore, the dirt from the surface being cleaned will be pulled away and suspended in water. Thus soap acts as an *emulsifying agent*, a substance used to disperse one liquid (oil molecules) in the form of finely suspended particles or droplets in another liquid (water molecules).

Treatment of fats or oils with strong bases such as lye (NaOH) or potash (KOH) causes them to undergo hydrolysis (saponification) to form glycerol and the salt of a long-chain fatty acid (soap).

$$\begin{array}{l} CH_2-O-\overset{\overset{\displaystyle O}{\|}}{C}-C_{17}H_{35} \\ | \\ CH-O-\overset{\overset{\displaystyle O}{\|}}{C}-C_{17}H_{35} \\ | \\ CH_2-O-\overset{\overset{\displaystyle O}{\|}}{C}-C_{17}H_{35} \end{array} + 3NaOH \xrightarrow{\Delta} \begin{array}{l} CH_2OH \\ | \\ CHOH \\ | \\ CH_2OH \end{array} + 3C_{17}H_{35}\overset{\overset{\displaystyle O}{\|}}{C}-O^-Na^+$$

tristearin

glycerol

sodium stearate
(a soap)

Because soaps are salts of strong bases and weak acids, they should be weakly alkaline in aqueous solution. However, a soap with free alkali can cause damage to skin, silk, or wool. Therefore, a test for basicity of the soap is quite important.

Soap has been largely replaced by synthetic detergents during the last two decades, because soap has two serious drawbacks. One is that soap becomes ineffective in hard water; this is water which contains appreciable amounts of Ca^{2+} or Mg^{2+} salts.

$$2C_{17}H_{35}COO^-Na^+ + M^{2+} \longrightarrow [C_{17}H_{35}COO^-]_2\,M^{2+}\downarrow + 2Na^+$$

soap $\qquad\qquad\qquad\qquad\qquad$ scum

$$M = (Ca^{2+} \text{ or } Mg^{2+})$$

The other is that, in an acidic solution, soap is converted to free fatty acid and therefore loses its cleansing action.

$$C_{17}H_{35}COO^-Na^+ + H^+ \longrightarrow C_{17}H_{35}COOH\downarrow + Na^+$$

soap $\qquad\qquad\qquad\qquad\qquad$ fatty acid

Objectives

1. To prepare a simple soap.
2. To investigate some properties of a soap.

Procedure

Preparation of a soap

Measure 23 mL of a vegetable oil into a 250-mL Erlenmeyer flask. Add 20 mL of ethyl alcohol (to act as a solvent) and 20 mL of 25% sodium hydroxide solution (25% NaOH). While stirring the mixture constantly with a glass rod, the flask with its contents is heated gently in a boiling water bath. A 600-mL beaker containing about 200 mL of tap water and a few boiling chips can serve as a water bath (Fig. 36.1).

Figure 36.1

Experimental set-up for soap preparation.

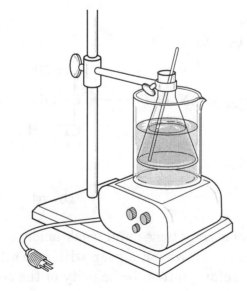

After being heated for about 20 min, the odor of alcohol will disappear, indicating the completion of the reaction. A pasty mass containing a mixture of the soap, glycerol, and excess sodium hydroxide is obtained. Use an ice-water bath to cool the flask with its contents. To precipitate or "salt out" the soap, add 150 mL of a saturated sodium chloride solution to the soap mixture while stirring vigorously. This process increases the density of the aqueous solution; therefore, soap will float out from the aqueous solution. Filter the precipitated soap with the aid of suction and wash it with 10 mL of ice cold water. Observe the appearance of your soap and record your observation on the Report Sheet.

Properties of a soap

1. *Emulsifying Properties.* Shake 5 drops of mineral oil in a test tube containing 5 mL of water. A temporary emulsion of tiny oil droplets in water will be formed. Repeat the same test, but this time add a small piece of the soap you have prepared before shaking. Allow both solutions to stand for a short time. Compare the appearance and the relative stabilities of the two emulsions. Record your observations on the Report Sheet.

2. *Hard Water Reactions.* Place about one-third spatula full of the soap you have prepared in a 50-mL beaker containing 25 mL of water. Warm the beaker with its contents to dissolve the soap. Pour 5 mL of the soap solution into each of 5 test tubes (nos. 1, 2, 3, 4, and 5). Test no. 1 with 2 drops of a 5% solution of calcium chloride (5% $CaCl_2$), no. 2 with 2 drops of a 5% solution of magnesium chloride (5% $MgCl_2$), no. 3 with 2 drops of a 5% solution of iron(III) chloride (5% $FeCl_3$), and no. 4 with tap water. The no. 5 tube will be used for a basicity test, which will be performed later. Record your observations on the Report Sheet.

3. *Alkalinity* (Basicity). Test soap solution no. 5 with a wide-range pH paper. What is the approximate pH of your soap solution? Record your answer on the Report Sheet.

Chemicals and Equipment

1. Hot plate
2. Ice cubes
3. Büchner funnel in no. 7 one-hole rubber stopper
4. 500-mL filter flask
5. Filter paper, 7 cm diameter
6. pHydrion paper
7. Boiling chips
8. 95% ethanol
9. Saturated sodium chloride solution
10. 25% NaOH
11. Vegetable oil
12. 5% $FeCl_3$
13. 5% $CaCl_2$
14. Mineral oil
15. 5% $MgCl_2$

Experiment 36

PRE-LAB QUESTIONS

1. Define the following terms:

    a. Hydrophobic

    b. Hydrophilic

    c. Emulsifying agent

2. Consult Table 17.2 of your textbook. If corn oil is used to make soap, what is the chemical formula of the *most abundant* soap you formed?

3. How would you convert this soap to the corresponding fatty acid?

4. Why is the use of a soap in hard water impractical? Explain with a chemical equation.

NAME _____ SECTION _____ DATE _____

PARTNER _____ GRADE _____

Experiment 36

REPORT SHEET

Preparation

Appearance of your soap _____

Properties

Emulsifying Properties

Which mixture, oil-water or oil-water-soap, forms a more stable emulsion?

Hard Water Reaction

No. 1 + $CaCl_2$ _____

No. 2 + $MgCl_2$ _____

No. 3 + $FeCl_3$ _____

No. 4 + tap water _____

Alkalinity

pH of your soap solution (no. 5) _____

POST-LAB QUESTIONS

1. When you made soap, first you dissolved vegetable oil in ethanol. What happened to the ethanol during the reaction?

2. What are the two main disadvantages of soaps versus detergents?

3. Soaps that have a pH above 8.0 tend to irritate some sensitive skins. Was your soap good enough to compete with commercial preparations?

Experiment 37

● Preparation of a hand cream

Hand creams are formulated to carry out a variety of cosmetic functions. Among these are softening the skin and preventing dryness; elimination of natural waste products (oils) by emulsification; cooling the skin by radiation thus helping to maintain body temperature. In addition, hand creams must have certain ingredients that aid spreadibility and provide body. In many cases added fragrance improves the odor, and in some special cases medications combat assorted ills.

The basic hand cream formulations all contain water to provide moisture and lanolin which helps its absorption by the skin. The latter is a yellowish wax. Chemically, wax is made of esters of long chain fatty acids and long chain alcohols. Lanolin is usually obtained from sheep wool; it has the ability to absorb 25–30% of its own weight of water and to form a fine emulsion. Mineral oil, which consists of high molecular weight hydrocarbons, provides spreadibility. In order to allow nonpolar substances, such as lanolin and mineral oil, to be uniformly dispersed in a polar medium, water, one needs strong emulsifying agents. An emulsifying agent must have nonpolar, hydrophobic portions to interact with the oil and also polar, hydrophilic portions to interact with water. A mixture of stearic acid and triethanolamine, through acid-base reaction, yields the salt that has the requirements to act as an emulsifying agent.

Besides the above five basic ingredients, some hand creams also contain alcohols such as propylene glycol (1,2-propanediol), and esters such as methyl stearate, to provide the desired texture of the hand cream.

In this experiment you will prepare four hand creams using the combination of ingredients as shown in Table 37.1.

Objectives

1. To learn the method of preparing a hand cream.
2. To appraise the function of the ingredients in the hand cream.

Preparation of the Hand Creams

For each sample in Table 37.1, assemble the ingredients in two beakers. Beaker 1 contains the polar ingredients, and beaker 2 contains the nonpolar contents.

Table 37.1	Recipes to Prepare Hand Creams				
Ingredients	**Sample 1**	**Sample 2**	**Sample 3**	**Sample 4**	
Water	25 mL	25 mL	25 mL	25 mL	
Triethanolamine	1 mL	1 mL	1 mL	—	Beaker 1
Propylene glycol	0.5 mL	0.5 mL	—	0.5 mL	
Stearic acid	5 g	5 g	5 g	5 g	
Methyl stearate	0.5 g	0.5 g	—	0.5 g	
Lanolin	4 g	4 g	4 g	4 g	Beaker 2
Mineral oil	5 mL	—	5 mL	5 mL	

1. To prepare sample 1, put the nonpolar ingredients in a 50-mL beaker (beaker 2) and heat it in a water bath. The water bath can be a 400-mL beaker half-filled with tap water and heated with a Bunsen burner (Fig. 37.1). Carefully hold the beaker with crucible tongs in the boiling water until all ingredients melt.

Figure 37.1
Heating ingredients.

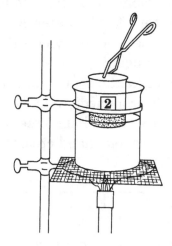

2. In the same water bath, heat the 100-mL beaker (beaker 1) containing the polar ingredients for about 5 min. Remove the beaker and set it on the bench top.

3. Into the 100-mL beaker containing polar ingredients, pour slowly the contents of the 50-mL beaker that holds the molten nonpolar ingredients (Fig. 37.2). Stir the mixture for 5 min. until you have a smooth uniform paste.

4. Repeat the same procedure in preparing the other three samples.

Figure 37.2
Mixing hand cream
ingredients.

Characterization of the Hand Cream Preparations

1. Test the pH of the hand creams prepared using a wide-range pH paper.

2. Rubbing a small amount of the hand cream between your fingers, test for smoothness and homogeneity. Also note the appearance. Record your observations on the Report Sheet.

Chemicals and Equipment

1. Bunsen burner
2. Lanolin
3. Stearic acid
4. Methyl stearate
5. Mineral oil
6. Triethanolamine
7. Propylene glycol
8. pHydrion paper

Experiment 37

PRE-LAB QUESTIONS

1. Write the formula of propylene glycol (1,2-propanediol).

2. The emulsifying agent was prepared from stearic acid and triethanolamine. Give the name of this salt. Write its formula.

3. What functional groups of the emulsifying agent provide the hydrophilic character?

NAME SECTION DATE

PARTNER GRADE

Experiment 37

REPORT SHEET

Characterization of the Hand Cream Samples

Properties	Sample 1	Sample 2	Sample 3	Sample 4
pH				
Smoothness				
Homogeneity				
Appearance				

POST-LAB QUESTIONS

1. In comparing the properties of the hand creams you produced, ascertain the function of each of the missing ingredients in the hand cream:

(a) Mineral oil

(b) Triethenolamine

(c) Methyl stearate and propylene glycol

2. A hand cream appears smooth and uniform after you prepared it, but in a week of storage most of the water settles on the bottom and most of the oil separates on the top. What do you think may have gone wrong with the hand cream preparation?

3. Commercial hand creams contain many more ingredients than you had in the recipe. These additional compounds enhance the effect of one or another ingredient in your formulation. Considering the functional groups of the following commercial ingredients, pair them up with one of your ingredients whose function they may enhance.

    (a)   glycerol: $CH_2-CH-CH_2$
$$\underset{OH\quad OH\quad OH}{|\quad\ |\quad\ |}$$

    (b)   hexadecane: $CH_3(CH_2)_{14}CH_3$

    (c)   sodium hydroxide: $NaOH$

Experiment 38

Extraction and identification of fatty acids from corn oil

Background

Fats are esters of glycerol and fatty acids. Liquid fats are often called oils. Whether a fat is solid or liquid depends on the nature of the fatty acids. Solid animal fats contain mostly saturated fatty acids, while vegetable oils contain high amounts of unsaturated fatty acids. To avoid arteriosclerosis, hardening of the arteries, diets which are low in saturated fatty acids as well as in cholesterol are recommended.

Note that even solid fats contain some unsaturated fatty acids, and oils contain saturated fatty acids as well. Besides the degree of unsaturation, the length of the fatty acid chain also influences whether a fat is solid or liquid. Short chain fatty acids, such as found in coconut oil, convey liquid consistency in spite of the low unsaturated fatty acid content. Two of the unsaturated fatty acids, linoleic and linolenic acids, are essential fatty acids because the body cannot synthesize them from precursors; they must be included in the diet.

The four unsaturated fatty acids most frequently found in vegetable oils are:

Oleic acid: $CH_3(CH_2)_7CH = CH(CH_2)_7COOH$

Linoleic acid: $CH_3(CH_2)_4CH = CHCH_2CH = CH(CH_2)_7COOH$

Linolenic acid: $CH_3CH_2CH = CHCH_2CH = CHCH_2CH = CH(CH_2)_7COOH$

Arachidonic acid:
$CH_3(CH_2)_4CH = CHCH_2CH = CHCH_2CH = CHCH_2CH = CH(CH_2)_3COOH$

All the $C = C$ double bonds in the unsaturated fatty acids are cis double bonds, which interrupt the regular packing of the aliphatic chains, and thereby convey a liquid consistency at room temperature. This physical property of the unsaturated fatty acid is carried over to the physical properties of triglycerides (oils).

In order to extract and isolate fatty acids from corn oil, first, the ester linkages must be broken. This is achieved in the saponification reaction in which a triglyceride is converted to glycerol and the potassium salt of its fatty acids:

$$
\begin{array}{ccc}
& O & \\
& \| & \\
CH_2-O-C-C_{17}H_{35} & & \\
| & O & \\
& \| & \\
CH-O-C-C_{17}H_{35} + 3KOH & \longrightarrow & \\
| & O & \\
& \| & \\
CH_2-O-C-C_{17}H_{35} & &
\end{array}
$$

$$
\begin{array}{cc}
CH_2OH & O \\
| & \| \\
CHOH + & 3C_{17}H_{35}C-O^-K^+ \\
| & \\
CH_2OH &
\end{array}
$$

triglyceride   glycerol   potassium stearate

In order to separate the potassium salts of fatty acids from glycerol, the products of the saponification mixture must be acidified. Subsequently, the fatty acids can be extracted by petroleum ether. To identify the fatty acids that were isolated, they must be converted to their respective methyl ester by a perchloric acid catalyzed reaction:

$$C_{17}H_{35}COOH + CH_3OH \xrightarrow{HClO_4} C_{17}H_{35}\overset{\overset{\displaystyle O}{\|}}{C}\text{-O-CH}_3 + H_2O$$

The methyl esters of fatty acids can be separated by thin layer chromatography (TLC). They can be identified by comparison of their rate of migration (R_f values) to the R_f values of authentic samples of methyl esters of different fatty acids (Fig. 38.1).

Figure 38.1

TLC chromatogram.

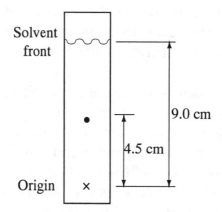

R_f = distance travelled by fatty acid/distance travelled by the solvent front.

Objectives

1. To extract fatty acids from neutral fats.
2. To convert them to their methyl esters.
3. To identify them by thin-layer chromatography.

Procedure

Part A. Extraction of fatty acids

1. Weigh a 50-mL Erlenmeyer flask and record the weight on your Report Sheet (1).

2. Add 2 mL of corn oil and weigh it again. Record the weight on your Report Sheet (2).

3. Add 5 mL of 0.5 M KOH in ethanol to the Erlenmeyer flask. Stopper it. Place the flask in a water bath at 55°C for 20 min.

CAUTION!

Strong acid; use gloves with concentrated HCl.

---

4. When the saponification is completed, add 2.5 mL of the concentrated HCl. Mix it by swirling the Erlenmeyer flask. Transfer the contents into a 50-mL separatory funnel. Add 5 mL of petroleum ether. Mix it thoroughly (see Fig. 34.1). Drain the lower aqueous layer into a flask and the upper petroleum ether layer into a glass-stoppered test tube. Repeat the process by adding back the aqueous layer into the separatory funnel and extracting it with another portion of 5 mL of petroleum ether. Combine the extracts.

Part B. Preparation of methyl esters

1. Place a plug of glass wool (the size of a pea) into the upper stem of a funnel, fitting it loosely. Add 10 g of anhydrous Na_2SO_4. Rinse the salt on to the glass wool with 5 mL of petroleum ether; discard the wash. Pour the combined petroleum ether extracts into the funnel and collect the filtrate in an evaporating dish. Add another portion (2 mL) of petroleum ether to the funnel and collect this wash, also in the evaporating dish.

2. Evaporate the petroleum ether under the hood by placing the evaporating dish on a water bath at 60°C. (Alternatively, if dry N_2 gas is available, the evaporation could be achieved by bubbling nitrogen through the extract. This also must be done under the hood.)

3. When dry, add 10 mL of the $CH_3OH:HClO_4$ mixture (95:5). Place the evaporating dish in the water bath at 55°C for 10 min.

Part C. Identification of fatty acids

1. Transfer the methyl esters prepared above into a separatory funnel. Extract twice with 5 mL of petroleum ether. Combine the extracts.

2. Prepare another funnel with anhydrous Na_2SO_4 on top of the glass wool. Filter the combined petroleum ether extracts through the salt into a dry, clean evaporating dish. Evaporate the petroleum ether on the water bath at 60°C, as before. When dry, add 0.2 mL of petroleum ether and transfer the solution to a clean and dry test tube.

3. Take a 15 × 6.5 cm TLC plate. Make sure you do not touch the TLC plate with your fingers. Preferably use plastic gloves, or handle the plate by holding it only at the edges. This precaution must be observed throughout the whole operation because your fingers may contaminate the sample. With a pencil, lightly draw a line parallel to the 6.5 edge about 1 cm from the edge. Mark the positions of the five spots, equally spaced, where you will spot your samples (Fig. 38.2).

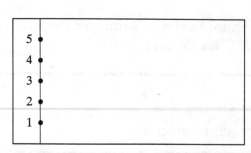

Figure 38.2
Spotting.

5
4
3
2
1

4. For spots no. 1 and no. 5, use your isolated methyl esters obtained from corn oil. For spot no. 2, use methyl oleate; for spot no. 3, methyl linoleate; and for spot no. 4, methyl palmitate. For each sample use a separate capillary tube. In spotting, apply each sample in the capillary to the plate until it spreads to a spot of 1 mm diameter. Dry the spots with a heat lamp. Pour about 15 mL of solvent (hexane:diethyl ether; 4:1) into a 500-mL beaker. Place the spotted TLC plate diagonally for ascending chromatography. Make certain that the spots applied are **above** the surface of the eluting solvent. Cover the beaker lightly with aluminum foil to avoid excessive solvent evaporation.

5. When the solvent front has risen to about 1–2 cm from the top edge, remove the plate from the beaker. Mark the advance of the solvent front with a pencil. Dry the plate with a heat lamp under the hood. Place the dried plate in a beaker containing a few iodine crystals. Cover the beaker tightly with aluminum foil. Place the beaker in a 110°C oven for 3–4 min. Remove the beaker and let it cool to room temperature. **This part is essential to avoid inhaling iodine vapors.** Remove the TLC plate from the beaker and mark the spots with a pencil.

6. Record the distance the solvent front advanced on your Report Sheet (4). Record on your Report Sheet (5) the distance of each iodine-stained spot from its origin. Calculate the R_f values of your samples (6).

Chemicals and Equipment

1. Corn oil
2. Methyl palmitate
3. Methyl oleate
4. Methyl linoleate
5. Petroleum ether (b. p. 30–60°C)
6. 0.5 M KOH in ethanol
7. Concentrated HCl
8. Anhydrous Na_2SO_4
9. Methanol:perchloric acid mixture (95:5)
10. Hexane:diethyl ether mixture (4:1)
11. Iodine crystals, I_2
12. Aluminum foil
13. Polyethylene gloves
14. 15 × 6.5 cm silica gel TLC plate
16. Capillary tubes open on both ends
17. Heat lamp
18. Water bath
19. Ruler
20. Drying oven, 110°C.

Experiment 38

PRE-LAB QUESTIONS

1. Write the schematic structures of palmitic and oleic acids. Indicate the single bonds of the aliphatic chain with zig-zag drawing and the cis double bonds with horizontal double lines.

2. Write the formulas of the reaction, converting linolenic acid to its methyl ester.

3. The R_f on a chromatogram is the ratio of the distance a particular spot traveled divided by the distance the solvent front advanced. A particular compound has an R_f value of 0.45. How far did the spot travel on a chromatogram in which the solvent front advanced 13 cm?

4. Why do you have to cool the iodine chamber (the beaker containing the chromatogram and iodine vapor) from 110°C to room temperature?

Experiment 38

REPORT SHEET

1. Weight of beaker   _____ g

2. Weight of beaker and oil   _____ g

3. Weight of oil   _____ g

Distances on the chromatogram in cm

4. The solvent front   _____

5. Spot no. 1 a, b, c, d, e   a___ b___ c___ d___ e___

6. Spot no. 2   _____

7. Spot no. 3   _____

8. Spot no. 4   _____

9. Spot no. 5 a, b, c, d, e   a___ b___ c___ d___ e___

Calculated R_f values

10. For spot no.1 [(5)/(4)] a, b, c, d, e   a___ b___ c___ d___ e___

11. For spot no. 2 [(6)/(4)]   _____

12. For spot no. 3 [(7)/(4)]   _____

13. For spot no. 4 [(8)/(4)]   _____

14. For spot no. 5 [(9)/(4)] a, b, c, d, e   a___ b___ c___ d___ e___

15. How many fatty acids were present in your corn oil?

16. How many fatty acids could you identify? Name the identifiable fatty acids in the corn oil.

POST-LAB QUESTIONS

1. Which of the identifiable fatty acids of your corn oil was a saturated fatty acid?

2. Judging from the iodine spots of samples 2, 3, and 4, which fatty acid reacts most strongly with iodine? Why?

3. Does the intensity of the iodine spots in the corn oil sample indicate their abundance? Explain.

4. Given two saturated fatty acids, one a short chain of 10 carbons and the other a long chain of 20 carbons, which would move faster on the TLC plate? Explain.

Analysis of lipids

Lipids are chemically heterogeneous mixtures. The only common property they have is their insolubility in water. We can test for the presence of various lipids by analyzing their chemical constituents. Foods contain a variety of lipids, most important among them are fats, complex lipids, and steroids. Fats are triglycerides, esters of fatty acids and glycerol. Complex lipids also contain fatty acids, but their alcohol may be either glycerol or sphingosine. They also contain other constituents such as phosphate, choline, or ethanolamine or mono- to oligo-saccharides. An important representative of this group is lecithin, a glycerophospholipid, containing fatty acids, glycerol, phosphate, and choline. The most important steroid in foods is cholesterol. Different foods contain different proportions of these three groups of lipids.

Structurally, cholesterol contains the steroid nucleus that is the common core of all steroids.

steroid nucleus   cholesterol

There is a special colorimetric test, the Lieberman-Burchard reaction, which uses acetic anhydride and sulfuric acid as reagents, that gives a characteristic green color in the presence of cholesterol. This color is due to the -OH group of cholesterol and the unsaturation found in the adjacent fused ring. The color change is gradual: first it appears as a pink coloration, changing later to lilac, and finally to deep green.

When lecithin is hydrolyzed in acidic medium, both the fatty-acid ester bonds and the phosphate ester bonds are broken and free fatty acids and inorganic phosphate are released.

Using a molybdate test, we can detect the presence of phosphate in the hydrolysate by the appearance of a purple color. Although this test is not specific for lecithin (other phosphate containing lipids will give a positive molybdate test), it differentiates clearly between fat and cholesterol, on the one hand (negative test), and phospholipid, on the other (positive test).

A second test that differentiates between cholesterol and lecithin is the acrolein reaction. When lipids containing glycerol are heated in the presence of potassium hydrogen sulfate, the glycerol is dehydrated, forming acrolein, which has an unpleasant odor. Further heating results in polymerization of acrolein, which is indicated by the slight blackening of the reaction mixture. Both the pungent smell and the black color indicate the presence of glycerol, and thereby fat and/or lecithin. Cholesterol gives a negative acrolein test.

$$
\begin{array}{c}
CH_2OH \\
| \\
CHOH \\
| \\
CH_2OH
\end{array}
\xrightarrow{\Delta}
\begin{array}{c}
CH_2 \\
\| \\
CH \\
| \\
C=O \\
| \\
H
\end{array}
+ 2H_2O
$$

Objectives

To investigate the lipid composition of common foods such as corn oil, butter, and egg yolk.

Procedure

Use six samples for each test: (1) pure cholesterol, (2) pure glycerol, (3) lecithin preparation, (4) corn oil, (5) butter, (6) egg yolk.

Phosphate test

CAUTION!

6 M nitric acid is a strong acid. Handle it with care. Use gloves.

1. Take six clean and dry test tubes. Label them. Add about 0.2 g of sample to each test tube. Hydrolyze the compounds by adding 3 mL of 6 M nitric acid to each test tube.

2. Prepare a water bath by boiling about 100 mL of tap water in a 250-mL beaker on a hot plate. Place the test tubes in the boiling water bath for 5 min. Do not inhale the vapors. Cool the test tubes. Neutralize the acid by adding 3 mL of

6 M NaOH. Mix. During the hydrolysis a precipitate may form, especially in the egg yolk sample. The samples in which a precipitate appeared must be filtered. Place a piece of cheese cloth on top of a 25-mL Erlenmeyer flask. Pour the turbid hydrolysate in the test tube on the cheese cloth and filter it.

3. Transfer 2 mL of each neutralized (and filtered) sample into clean and labeled test tubes. Add 3 mL of a molybdate solution to each test tube and mix the contents. **(Be careful. The molybdate solution contains sulfuric acid.)** Heat the test tubes in a boiling water bath for 5 min. Cool them to room temperature.

4. Add 0.5 mL of an ascorbic acid solution and mix the contents thoroughly. Wait 20 min. for the development of the purple color. Record your observations on the Report Sheet. While you wait, you can perform the rest of the colorimetric tests.

The acrolein test for glycerol

1. Place 1 g of potassium hydrogen sulfate, $KHSO_4$, in each of seven clean and dry test tubes. Label them. Add a few grains of your pure preparations, lecithin and cholesterol, to two of the test tubes. Add a drop, about 0.1 g, from each, glycerol, corn oil, butter, and egg yolk to the other four test tubes. To the seventh test tube add a few crystals of sucrose.

2. Set up your Bunsen burner in the hood. **It is important that this test be performed under the hood because of the pungent odor of the acrolein.**

3. Gently heat each test tube, one at a time, over the Bunsen burner flame, shaking it continuously from side to side. When the mixture melts it slightly blackens, and you will notice the evolution of fumes. Stop the heating. Smell the test tubes by moving them sideways under your nose or waft the vapors. Do not inhale the fumes directly. A pungent odor, resembling burnt hamburgers, is the positive test for glycerol. Sucrose in the seventh test tube also will be dehydrated and will give a black color. However, its smell is different, and thus is not a positive test for acrolein. Do not overheat the test tubes, for the residue will become hard, making it difficult to clean the test tubes. Record your observations on the Report Sheet.

Lieberman-Burchard test for cholesterol

1. Place a few grains of your cholesterol and lecithin preparations in labeled clean and dry test tubes. Similarly, add about 0.1 g samples of glycerol, corn oil, butter, and egg yolk to the other four labeled clean and dry test tubes. **(The next step should be done in the hood.)**

2. Transfer 3 mL of chloroform and 1 mL of acetic anhydride to each test tube. Finally, add 1 drop of concentrated sulfuric acid to each mixture. Mix the contents and record the color changes, if any. Wait 5 min. Record again the color of your solutions. Record your observations on the Report Sheet.

Chemicals and Equipment

1. 6 M NaOH
2. 6 M HNO_3
3. Molybdate reagent
4. Ascorbic acid solution
5. $KHSO_4$
6. Chloroform
7. Acetic anhydride
8. Sulfuric acid, H_2SO_4
9. Cholesterol
10. Lecithin
11. Glycerol
12. Corn oil
13. Butter
14. Egg yolk
15. Hot plate
16. Cheese cloth

Experiment 39

PRE-LAB QUESTIONS

1. Cephalins are also glycerophospholipids present in foods. They differ from lecithins by having ethanolamine or serine instead of choline in their structure. Could you differentiate between lecithins and cephalins on the basis of the three tests to be performed in this experiment?

2. Cholesterol has an alcohol group. One could also dehydrate cholesterol (removing one water molecule by heating). Show what kind of structure you would expect from the dehydration of cholesterol.

3. List all the functional groups in (a) acrolein and (b) choline.

4. List the acid hydrolysis products of glycerol trioleate.

Experiment 39

REPORT SHEET

Tests	Cholesterol	Lecithin	Glycerol	Corn oil	Butter	Egg yolk	Sucrose
1. Phosphate **a.** Color							
b. Conclusions							
2. Acrolein **a.** Odor							
b. Color							
c. Conclusions							
3. Lieberman-Burchard **a.** Initial color							
b. Color after 5 min.							
c. Conclusion							

POST-LAB QUESTIONS

1. What is your overall conclusion regarding the composition of your corn oil? Was it pure triglyceride?

2. Based on the intensity of color developed in your test for cholesterol, which food contained the most and which contained the least cholesterol?

3. Assume that the lecithin still contained all its ester linkages. Would you get a positive acrolein test? Explain.

4. A positive acrolein test is indicated by its odor as well as by its color. Which comes first? Explain.

5. When sucrose is dehydrated by heating it with $KHSO_4$, you can observe the black residue (carbon) and water. This is the origin of the name *carbohydrate*. Can you detect the presence of acrolein by its smell in the dehydration of sucrose?

TLC separation of amino acids

Background

Amino acids are the building blocks of peptides and proteins. They possess two functional groups—the carboxylic acid group gives the acidic character, and the amino group provides the basic character. The common structure of all amino acids is

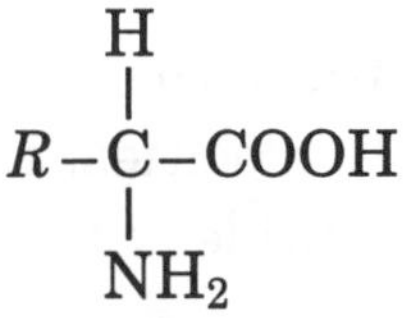

$$R-\underset{\underset{NH_2}{|}}{\overset{\overset{H}{|}}{C}}-COOH$$

The R represents the side chain that is different for each of the amino acids that are commonly found in proteins. However, all 20 amino acids have a free carboxylic acid group and a free amino (primary amine) group, except proline which has a cyclic side chain and a secondary amino group.

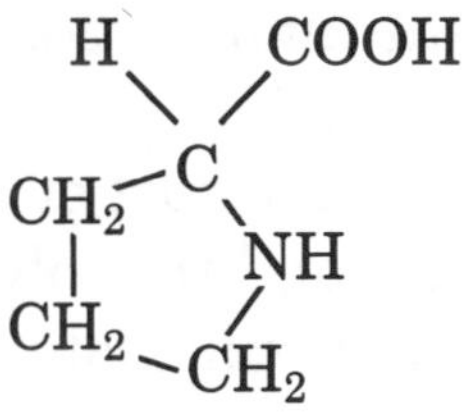

proline

We use the properties provided by these groups to characterize the amino acids. The common carboxylic acid and amino groups provide the acid-base nature of the amino acids. The different side chains, and the solubilities provided by these side chains, can be utilized to identify the different amino acids by their rate of migration in thin-layer chromatography.

In this experiment, we use thin-layer chromatography to identify aspartame, an artificial sweetener, and its hydrolysis products from certain foods.

Aspartame

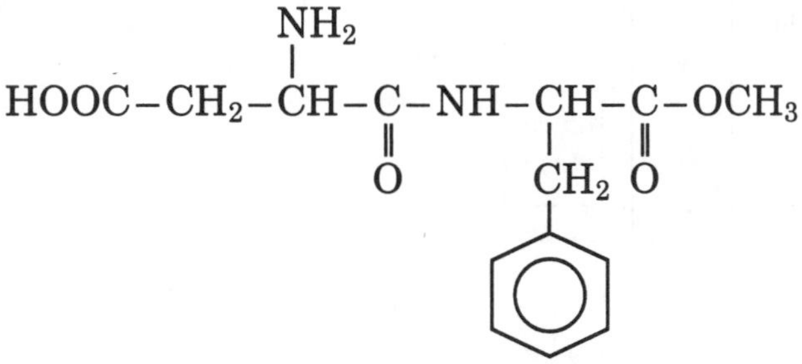

Aspartame is the methyl ester of the dipeptide aspartylphenylalanine. Upon hydrolysis with HCl it yields aspartic acid, phenylalanine, and methyl alcohol. When this artificial sweetener was approved by the Food and Drug Administration, opponents of aspartame claimed that it is a health hazard, because aspartame would be hydrolyzed and would yield poisonous methyl alcohol in soft drinks that are stored over long periods of time. The Food and Drug Administration ruled, however, that aspartame is sufficiently stable and fit for human consumption. Only a warning must be put on the labels containing aspartame. This warning is for patients suffering from phenylketonurea who cannot tolerate phenylalanine.

To run a thin-layer chromatography experiment, we use silica gel in a thin layer on a plastic or glass plate. We apply the sample (aspartame or amino acids) as a spot to a strip of a thin-layer plate. The plate is dipped into a mixture of solvents. The solvent moves up the thin gel by capillary action and carries the sample with it. Each amino acid may have a different migration rate depending on the solubility of the side chain in the solvent. Amino acids with similar side chains are expected to move with similar, though not identical rates; those that have quite different side chains are expected to migrate with different velocities. Depending on the solvent system used, almost all amino acids and dipeptides can be separated from each other by thin-layer chromatography (TLC).

We actually do not measure the rate of migration of an amino acid or a dipeptide, but rather, how far a particular amino acid travels in the thin silica gel layer relative to the migration of the solvent. This ratio is called the R_f value. In order to calculate the R_f values, one must be able to visualize the position of the amino acid or dipeptide. This is done by spraying the thin layer silica gel plate with a ninhydrin solution that reacts with the amino group of the amino acid. A purple color is produced when the gel is heated. (The proline not having a primary amine gives a yellow color with ninhydrin.) For example, if the purple spot of an amino acid appears on the TLC plate 4.5 cm away from the origin and the solvent front migrates 9.0 cm (Fig. 40.1), the R_f value for the amino acid is calculated

$$R_f = \frac{\text{distance traveled by the amino acid}}{\text{distance traveled by the solvent front}} = \frac{4.5 \text{ cm}}{9.0 \text{ cm}} = 0.50$$

In the present experiment you will determine the R_f values of three amino acids: phenylalanine, aspartic acid, and leucine. You will also measure the R_f value of aspartame.

Figure 40.1

TLC chromatogram.

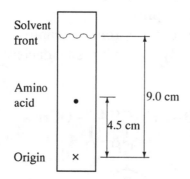

The aspartame you will analyze is actually a commercial sweetener, Equal by the NutraSweet Co., that contains silicon dioxide, glucose, cellulose, and calcium phosphate in addition to the aspartame. None of these other ingredients of Equal will give a purple or any other colored spot with ninhydrin. Occasionally, some sweeteners may contain a small amount of leucine which can be detected by the ninhydrin test. You will also hydrolyze aspartame using HCl as a catalyst to see if the hydrolysis products will prove that the sweetener is truly aspartame. Finally, you will analyze some commercial soft drinks supplied by your instructor. The analysis of the soft drink can tell you if the aspartame was hydrolyzed at all during the processing and storing of the soft drink.

Objectives

1. To separate amino acids and a dipeptide by TLC.
2. To identify hydrolysis products of aspartame.
3. To analyze the state of aspartame in soft drinks.

Procedure

1. Dissolve about 10 mg of the sweetener Equal in 1 mL of 3 M HCl in a test tube. Heat it with a Bunsen burner to a boil for 30 sec., but make sure that the liquid does not completely evaporate. Cool the test tube and label it as "Hydrolyzed Aspartame."

2. Label five additional small test tubes, respectively, for aspartic acid, phenylalanine, leucine, aspartame, and Diet Coca-Cola. Place about 0.5 mL samples in each test tube.

3. Take two 15 × 6.5 cm TLC plates. With a pencil, lightly draw a line parallel to the 6.5 cm edge and about 1 cm from the edge. Mark the positions of five spots on each plate, spaced equally, where you will spot your samples (Fig. 40.2). **You must make sure that you don't touch the plates with your fingers.**

Figure 40.2
Spotting.

Either use plastic gloves or handle the plates by holding them only at their edges. This precaution must be observed throughout the whole operation, because amino acids from your fingers will contaminate the plate.

On plate A you will spot samples of (1) phenylalanine, (2) aspartic acid, (3) leucine, (4) aspartame in Equal, and (5) the hydrolyzed aspartame you prepared in step no. 1. On plate B you will spot samples of Diet Coca-Cola on lanes (1) and (4), aspartic acid on lane (2), aspartame in Equal on lane (3), and the hydrolyzed aspartame you prepared previously on lane (5).

4. First spot plate A. For each sample use a separate capillary tube. Apply the sample to the plate until it spreads to a spot of 1 mm diameter. Dry the spots. (If a heat lamp is available, use it for drying.) Pour about 15 mL of solvent mixture (butanol:acetic acid:water) into a large (500-mL or 1-L) beaker and place your spotted plate diagonally for an ascending chromatography. Make certain that the spots applied to the plate are above the surface of the eluting solvent. Cover the beaker with aluminum foil to avoid the evaporation of the solvent mixture.

5. Spot plate B. For aspartic acid, lane (2), and for the hydrolyzed and non-hydrolyzed aspartame, lanes (3) and (5), use one spot as before. For Coca-Cola [lanes (1) and (4)] multiple spotting is needed. Apply the capillary tube 12–15 times to the same spot, making certain that between each application the previous sample has been dried. Also, try to control the size of the spots that they should not spread too much, not more than 2 mm in diameter. Dry the spots as before. Place the plate in a large beaker containing the eluting solvent as before. Cover the beaker with aluminum foil. Allow about 50–60 min. for the solvent front to advance.

6. When the solvent front nears the edge of the plate, about 1–2 cm from the edge, remove the plate from the beaker. You must not allow the solvent front to advance up to or beyond the edge of the plate. Mark immediately *with a pencil* the position of the solvent front. Under a hood dry the plates with the aid of a heat lamp or hair dryer. Using polyethylene gloves, spray the dry plates with ninhydrin solution. *Be careful not to spray ninhydrin on your hand and not to touch the sprayed areas with bare hands. If the ninhydrin spray touches your skin (which contains amino acids) your fingers will be discolored for a few days.* Place the sprayed plates into a drying oven at 105–110°C for 2–3 min.

7. Remove the plates from the oven. Mark the center of the spots and calculate the R_f values of each spot. Record your observations on the Report Sheet.

Chemicals and Equipment

1. 0.1% solutions of aspartic acid, phenylalanine, and leucine
2. 0.5% solution of aspartame (Equal)
3. Diet Coca-Cola
4. 3 M HCl
5. 0.2% ninhydrin spray
6. Butanol:acetic acid:water–solvent mixture
7. Equal sweetener
8. Aluminum foil
9. 15×6.5 cm silica gel TLC plates
10. Ruler
11. Polyethylene gloves
12. Capillary tubes open on both ends
13. Heat lamp or hair dryer
14. Drying oven, 110°C

Experiment 40

PRE-LAB QUESTIONS

1. If an amino acid moved 2.1 cm on a TLC plate and the solvent moved 7.0 cm, what is the R_f value of the amino acid?

2. Why must you use a pencil rather than ink to mark the origin of your spot on the TLC plate?

3. What happens if you don't use gloves and your finger comes in contact with the ninhydrin spray?

4. Write the structure of the mono- and di-methyl ester of aspartic acid.

Experiment 40

REPORT SHEET

1.

Plate A	Distance traveled (mm)	Solvent front (mm)	R_f
Phenylalanine			
Aspartic acid			
Leucine			
Aspartame			
Hydrolyzed aspartame			

Plate B	Distance traveled (mm)	Solvent front (mm)	R_f
Phenylalanine			
Aspartic acid			
Aspartame			
Coca-Cola			
Hydrolyzed aspartame			

2. Identification

(a) Name the amino acids you found in the hydrolysate of the sweetener Equal.

(b) How many spots were stained with ninhydrin (1) in Equal and (2) in Diet Coca-Cola samples?

POST-LAB QUESTIONS

1. Can you separate aspartame, a dipeptide, from its constituent amino acids using the TLC technique? Which sample migrates the slowest?

2. In testing the hydrolysate of aspartame, you forgot to mark the position of the solvent front on your TLC plate. Could you

   (a) determine how many amino acids were in the aspartame;

   (b) identify those amino acids?

3. Do you have any evidence that the aspartame was hydrolyzed during the processing and storage of the Diet Coca-Cola sample? Explain.

4. The difference between aspartic acid and phenylalanine is twofold. Aspartic acid has a polar, acidic side chain, while phenylalanine has a nonpolar side chain. The molecular weight of aspartic acid is smaller than the molecular weight of phenylalanine. Based on the R_f values you obtained for these two amino acids in the solvent employed, which property influenced the rate of migration?

5. If the R_f values of two amino acids are 0.3 and 0.35, respectively, what length should the TLC plate be in order that the two spots should be separated by 1 cm?

Acid-base properties of amino acids

Background

In the body, amino acids exist as zwitterions.

$$\begin{array}{c} H \\ | \\ R-C-COO^- \\ | \\ NH_3{}^+ \end{array}$$

This is an amphoteric compound because it behaves as both an acid and a base in the Brønsted definition. As an acid, it can donate an H^+ and becomes the conjugate base:

$$\begin{array}{c} H \\ | \\ R-C-COO^- \\ | \\ NH_3{}^+ \end{array} + OH^- \rightleftharpoons \begin{array}{c} H \\ | \\ R-C-COO^- \\ | \\ NH_2 \end{array} + H_2O$$

$$\text{acid} \qquad \text{base} \qquad \text{conj. base} \qquad \text{conj. acid}$$

As a base, it can accept an H^+ ion and becomes the conjugate acid:

$$\begin{array}{c} H \\ | \\ R-C-COO^- \\ | \\ NH_3{}^+ \end{array} + H_3O^+ \rightleftharpoons \begin{array}{c} H \\ | \\ R-C-COOH \\ | \\ NH_3{}^+ \end{array} + H_2O$$

$$\text{base} \qquad \text{acid} \qquad \text{conj. acid} \qquad \text{conj. base}$$

To study the acid-base properties, one can perform a simple titration. We start our titration with the amino acid being in its acidic form at a low pH:

$$\begin{array}{c} H \\ | \\ R-C-COOH \\ | \\ NH_3{}^+ \end{array} \quad (I)$$

As we add a base, OH^-, to the solution, the pH will rise. We record the pH of the solution by a pH meter after each addition of the base. To obtain the titration curve, we plot the milliliters of NaOH added against the pH of the solution (Fig. 41.1).

Figure 41.1

The titration curve of an amino acid.

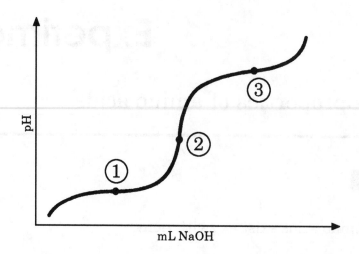

Note that there are two flat portions (called legs) on the titration curve where the pH does not increase appreciably with the addition of NaOH. The midpoint of the first leg, ①, is when half of the original acidic amino acid (I) has been titrated and it becomes a zwitterion (II).

$$R-\overset{\displaystyle H}{\underset{\displaystyle NH_3^+}{\overset{|}{\underset{|}{C}}}}-COO^- \quad (II)$$

The point of inflection, ②, occurs when the amino acid is entirely in the zwitterion form (II). At the midpoint of the second leg, ③, half of the amino acid is in the zwitterion form and half is in the basic form (III).

$$R-\overset{\displaystyle H}{\underset{\displaystyle NH_2}{\overset{|}{\underset{|}{C}}}}-COO^- \quad (III)$$

From the pH at the midpoint of the first leg we obtain the pK value of the carboxylic acid group, since this is the group that is titrated with NaOH at this stage (the structure going from I to II). The pH of the midpoint of the second leg, ③, is equal to the pK of the $-NH_3^+$, since this is the functional group that donates its H^+ at this stage of the titration. The pH at the inflection point, ②, is equal to the isoelectric point. At the isoelectric point of a compound, the positive and negative charges balance each other. This occurs at the inflection point when all the amino acids are in the zwitterion form.

You will obtain a titration curve of an amino acid with a neutral side chain such as glycine, alanine, phenylalanine, leucine, or valine. If pH meters are available, you read the pH directly from the instrument after each addition of the base. If a pH meter is not available, you can obtain the pH with the aid of indicator papers. From the titration curve obtained, you can determine the pK values and the isoelectric point.

Procedure

1. Pipet 20 mL of 0.1 M amino acid solution (glycine, alanine, phenylalanine, leucine, or valine) that has been acidified with HCl to a pH of 1.5 into a 100-mL beaker.

2. If a pH meter is available, insert the clean and dry electrode of the pH meter into a standard buffer solution with known pH. Turn the knob of the meter to the pH mark and adjust it to read the pH of the buffer. Turn the knob of the pH meter to "Standby" position. Remove the electrode from the buffer, wash it with distilled water, and dry it. Insert the dry electrode into the amino acid solution. Turn the knob of the meter to "pH" position and record the pH of the solution. Fill a buret with 0.25 M NaOH solution. Add the NaOH solution from the buret in 1.0 mL increments to the beaker. After each increment, stir the contents with a glass rod and then read the pH of the solution. Record these on your Report Sheet. Continue the titration as described until you reach pH 12. Turn off your pH meter, wash the electrode with distilled water, wipe it dry, and store it in its original buffer.

3. If a pH meter is not available, perform the titration as above, but use pH indicator papers. After the addition of each increment and stirring, withdraw a drop of the solution with a Pasteur pipet. Touch the end of the pipet to a dry piece of the pH indicator paper. Compare the color of the indicator paper with the color on the charts supplied. Read the corresponding pH from the chart and record it on your Report Sheet.

4. Draw your titration curve. From the graph, determine your pK values and the isoelectric point of your amino acid. Record these on your Report Sheet.

Chemicals and Equipment

1. 0.1 M amino acid solution (glycine, alanine, leucine, phenylalanine, or valine)
2. 0.25 M NaOH solution
3. pH meter and standard buffer (or pH indicator paper and Pasteur pipet)
4. 50-mL buret
5. 20-mL pipet
6. Spectroline pipet filler

Experiment 41

PRE-LAB QUESTIONS

1. Write the acidic, the amphoteric (zwitterion), and the basic form of alanine.

2. If the equilibrium constant, K_a, for the ionization of the carboxylic acid group is 1×10^{-4}, what is the pK_a?

3. If in a solution of alanine, the number of negatively charged carboxylate groups, $-COO^-$, is 1×10^{-19} at the isoelectric point, what is the number of the positively charged amino groups, $-NH_3^+$?

Experiment 41

REPORT SHEET

1. Amino acid used for titration _____

mL of 0.25 M NaOH added	pH	mL of 0.25 M NaOH added	pH
0		7.0	
1.0		8.0	
2.0		9.0	
3.0		10.0	
4.0		11.0	
5.0		12.0	
6.0			

2. Plot your data below to get the titration curve.

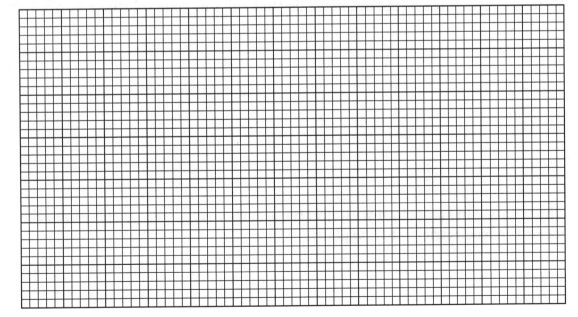

3. a. Indicate the positions of the midpoints of each leg and the position of the inflection point on your graph.

  b. Record the pK values for the carboxylic acid group _____, for the amino group _____.

  c. Record the pH of the isoelectric point _____.

POST-LAB QUESTIONS

1. The isoelectric point of an amino acid is an intensive property.

  (a) Knowing that, would you expect to find your inflection point at a different pH value, if you had titrated 0.5 M solution of the same amino acid instead of the 0.1 M solution? Explain.

  (b) Would your result be different if you had used 50 mL of amino acid solution instead of 20 mL? Explain.

2. Write the structures of the amino acids that exist in your solutions at pH 4 and at pH 12.

3. Which data can you obtain with greater accuracy from your graph—the pK values or the isoelectric point? Explain.

Experiment 42

- Isolation and identification of casein

Background

Casein is the most important protein in milk. It functions as a storage protein, fulfilling nutritional requirements. Casein can be isolated from milk by acidification to bring it to its isoelectric point. At the isoelectric point, the number of positive charges on a protein equal the number of negative charges. Proteins are least soluble in water at their isoelectric points because they tend to aggregate by electrostatic interaction. The positive end of one protein molecule attracts the negative end of another protein molecule, and the aggregates precipitate out of solution.

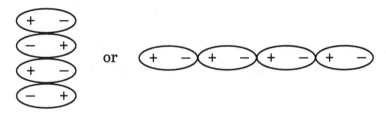

On the other hand, if a protein molecule has a net positive (at low pH or acidic condition) or a net negative charge (at high pH or basic condition), its solubility in water is increased.

$$\overset{+}{N}H_3 \sim\!\!\sim COOH \xleftarrow[\text{low pH}]{H^+} \overset{+}{N}H_3 \sim\!\!\sim\!\!\sim COO^- \xrightarrow[\text{high pH}]{OH^-} NH_2 \sim\!\!\sim COO^- + H_2O$$

more soluble   least soluble   more soluble
              (at isolelectric pH)

In the first part of this experiment, you are going to isolate casein from milk which has a pH of about 7. Casein will be separated as an insoluble precipitate by acidification of the milk to its isoelectric point (pH = 4.6). The fat that precipitates along with casein can be removed by dissolving it in alcohol.

In the second part of this experiment, you are going to prove that the precipitated milk product is a protein. The identification will be achieved by performing a few important chemical tests.

1. *The Biuret Test.* This is one of the most general tests for proteins. When a protein reacts with copper(II) sulfate, a positive test is the formation of a copper complex which has a violet color.

protein   blue color   protein-copper complex
(violet color)

This test works for any protein or compound that contains two or more of the following groups:

$$-\underset{\underset{O}{\parallel}}{C}-NH-, \quad -\underset{\underset{O}{\parallel}}{C}-NH_2, \quad -CH_2-NH_2, \quad -\underset{\underset{NH}{\parallel}}{C}-NH_2, \quad -\underset{\underset{S}{\parallel}}{C}-NH_2$$

2. *The Ninhydrin Test.* Amino acids with a free $-NH_2$ group and proteins containing free amino groups react with ninhydrin to give a purple-blue complex.

amino acid   ninhydrin

$$+ \; RCHO + CO_2 + 3H_2O$$

purple-blue complex

3. *Heavy Metal Ions Test.* Heavy metal ions precipitate proteins from solution. The ions that are most commonly used for protein precipitation are Zn^{2+}, Fe^{3+}, Cu^{2+}, Sb^{3+}, Ag^+, Cd^{2+}, and Pb^{2+}. Among these metal ions, Hg^{2+}, Cd^{2+}, and Pb^{2+} are known for their notorious toxicity. They can cause serious damage to proteins (especially the enzymes) by denaturing them. This can result in death. The precipitation occurs because proteins become cross-linked by heavy metals as shown below:

$$2\ NH_2 \text{~~~~~} \overset{\overset{\displaystyle O}{\|}}{C}-O^- + Hg^{2+} \longrightarrow \left\{ \begin{array}{c} H_2N \searrow \quad \swarrow O-C \\ Hg \quad \overset{\|}{O} \\ \nearrow \quad \nwarrow \\ \overset{\overset{\displaystyle C-O}{\|}}{\underset{O}{}} \quad NH_2 \end{array} \right\}$$

insoluble precipitate

Victims swallowing Hg^{2+} or Pb^{2+} ions are often treated with an antidote of a food rich in proteins, which can combine with mercury or lead ions in the victim's stomach and, hopefully, prevent absorption! Milk and raw egg white are used most often. The insoluble complexes are then immediately removed from the stomach by an emetic.

4. *The Xanthoprotein Test.* This is a characteristic reaction of proteins that contain phenyl rings

Concentrated nitric acid reacts with the phenyl ring to give a yellow-colored aromatic nitro compound. Addition of alkali at this point will deepen the color to orange.

$$HO-\hspace{-4pt}\left\langle \rule{0pt}{12pt}\right\rangle\hspace{-4pt}-CH_2-\overset{\overset{\displaystyle NH_2}{|}}{\underset{\underset{\displaystyle H}{|}}{C}}-COOH + HNO_3 \longrightarrow HO-\hspace{-4pt}\overset{NO_2}{\left\langle \rule{0pt}{12pt}\right\rangle}\hspace{-4pt}-CH_2-\overset{\overset{\displaystyle NH_2}{|}}{\underset{\underset{\displaystyle H}{|}}{C}}-COOH + H_2O$$

tyrosine   colored compound

The yellow stains on the skin caused by nitric acid are the result of the xanthoprotein reaction.

Objectives

1. To isolate the casein from milk under isoelectric conditions.
2. To perform some chemical tests to identify proteins.

Part A: Isolation of Casein

1. To a 250-mL Erlenmeyer flask, add 50.00 g of milk and heat the flask in a water bath (a 600-mL beaker containing about 200 mL of tap water; see Fig. 42.1). Stir the solution constantly with a stirring rod. When the bath temperature has reached about 40°C, remove the flask from the water bath, and add about 10 drops of glacial acetic acid while stirring. Observe the formation of a precipitate.

2. Filter the mixture into a 100-mL beaker by pouring it through a cheese cloth which is fastened with a rubber band over the mouth of the beaker (Fig. 42.2). Remove most of the water from the precipitate by squeezing the cloth gently. Discard the filtrate in the beaker. Using a spatula, scrape the precipitate from the cheese cloth into the empty flask.

3. Add 25 mL of 95% ethanol to the flask. After stirring the mixture for 5 min., allow the solid to settle. Carefully decant (pour off) the liquid that contains fats into a beaker. Discard the liquid.

Figure 42.1

Precipitation of casein.

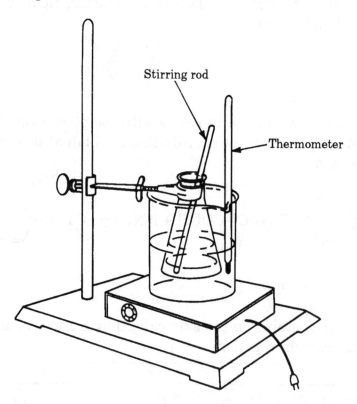

Stirring rod

Thermometer

Figure 42.2
Filtration of casein.

Rubber band

4. To the residue, add 25 mL of a 1:1 mixture of ether-ethanol. After stirring the resulting mixture for 5 min., collect the solid by vacuum filtration.

5. Spread the casein on a paper towel and let it dry. Weigh the dried casein and calculate the percentage of casein in the milk. Record it on your Report Sheet.

$$\% \text{ casein} = \frac{\text{weight of solid (casein)}}{50.00 \text{ g of milk}} \times 100$$

Part B: Chemical Analysis of Proteins

1. *Biuret Test*. Place 15 drops of each of the following solutions in five clean, labeled test tubes.

   a. 2% glycine

   b. 2% gelatin

   c. 2% albumin

   d. Casein prepared in Part A (one-quarter of a full spatula) + 15 drops of distilled water

   e. 1% tyrosine

   To each of the test tubes, add 5 drops of 10% NaOH solution and 2 drops of a dilute $CuSO_4$ solution while swirling. The development of a purplish-violet color is evidence of the presence of proteins. Record your results on the Report Sheet.

2. *The Ninhydrin Test*. Place 15 drops of each of the following solutions in five clean, labeled test tubes.

   a. 2% glycine

   b. 2% gelatin

   c. 2% albumin

   d. Casein prepared in Part A (one-quarter of a full spatula) + 15 drops of distilled water

   e. 1% tyrosine

To each of the test tubes, add 5 drops of ninhydrin reagent and heat the test tubes in a boiling water bath for about 5 min. Record your results on the Report Sheet.

3. *Heavy Metal Ions Test.* Place 2 mL of milk in each of three clean, labeled test tubes. Add a few drops of each of the following metal ions to the corresponding test tubes as indicated below:

a. Pb^{2+} as $Pb(NO_3)_2$ in test tube no. 1

b. Hg^{2+} as $Hg(NO_3)_2$ in test tube no. 2

c. Na^+ as $NaNO_3$ in test tube no. 3

Record your results on the Report Sheet.

CAUTION!

The following test will be performed by your instructor.

---

4. *The Xanthoprotein Test.* (Perform the experiment under the hood.) Place 15 drops of each of the following solutions in five clean, labeled test tubes:

a. 2% glycine

b. 2% gelatin

c. 2% albumin

d. Casein prepared in Part A (one-quarter of a full spatula) + 15 drops of distilled water

e. 1% tyrosine

To each test tube, add 10 drops of concentrated HNO_3 while swirling. Heat the test tubes carefully in a warm water bath. Observe any change in color. Record the results on your Report Sheet.

Chemicals and Equipment

1. Hot plate
2. Büchner funnel in a no. 7 one-hole rubber stopper
3. 500-mL filter flask
4. Filter paper, (Whatman no. 2) 7 cm
5. Cheese cloth
6. Rubber band
7. Boiling chips
8. 95% ethanol
9. Ether-ethanol mixture
10. Regular milk
11. Glacial acetic acid
12. Concentrated nitric acid
13. 2% albumin
14. 2% gelatin
15. 2% glycine
16. 5% copper(II) sulfate
17. 5% lead(II) nitrate
18. 5% mercury(II) nitrate
19. Ninhydrin reagent
20. 10% sodium hydroxide
21. 1% tyrosine
22. 5% sodium nitrate

Experiment 42

PRE-LAB QUESTIONS

1. Casein has an isoelectric point at pH 4.6. What kind of charges will be on the casein in its native environment, that is, in milk?

2. If a protein is void of tyrosine, could it give a positive xanthoprotein reaction? Explain.

3. Biuret is the most common test for proteins. Would an amino acid, such as proline, give a positive biuret test?

4. What are the three most toxic heavy metal ions?

NAME ______________________ SECTION ____________ DATE ____________

PARTNER ______________________ GRADE ______________________

Experiment 42

REPORT SHEET

Isolation of casein

1. Weight of milk ____________ g

2. Weight of dried casein ____________ g

3. Percentage of casein in milk ____________ %

Chemical analysis of proteins

Biuret test

Substance	Color formed
2% glycine	
2% gelatin	
2% albumin	
casein + H_2O	
1% tyrosine	

Which of these chemicals gives a positive test with this reagent? ____________

Ninhydrin test

Substance	Color formed after heating
2% glycine	
2% gelatin	
2% albumin	
casein + H_2O	
1% tyrosine	

Which of these chemicals gives a positive test with this reagent? ____________

Heavy metal ion test

Substance	Precipitates formed
$Pb(NO_3)_2$	
$Hg(NO_3)_2$	
$NaNO_3$	

Which of these metal ions gives a positive test with casein in milk? _____

Xanthoprotein test

Substance	Color formed before or after heating
2% glycine	
2% gelatin	
2% albumin	
casein + H_2O	
1% tyrosine	

Which of these chemicals gives a positive test with this reagent? _____

POST-LAB QUESTIONS

1. Explain why casein precipitates when acetic acid is added to it.

2. In the isolation of casein following the acidification, you removed the precipitate by filtering through a cheese cloth and squeezing the cloth. If you did not squeeze out all the liquids, would your yield of casein be different? Explain.

3. What functional group(s) will give a positive reaction with the ninhydrin reagent?

4. If by mistake (*don't try it*) your finger touches nitric acid and you observe a yellow color on your fingers, what functional group(s) in your skin is (are) responsible for this reaction?

5. Why is milk or raw egg used as an antidote in cases of heavy metal ion poisoning?

6. Which of the following solutions would give an insoluble precipitate with a casein solution: (a) KNO_3; (b) $Cd(NO_3)_2$; (c) $MgCl_2$?

Isolation and identification of DNA from yeast

Background

Hereditary traits are transmitted by genes. Genes are parts of giant deoxyribonucleic acid (DNA) molecules. In lower organisms, such as bacteria and yeast, both DNA and RNA (ribonucleic acid) occur in the cytoplasm. In higher organisms, most of the DNA is inside the nucleus, and the RNA is outside the nucleus in other organelles and in the cytoplasm.

In this experiment, we will isolate DNA molecules from yeast cells. The first task is to break up the cells. This is achieved by a combination of different techniques and agents. Grinding up the cells with sand disrupts them and the cytoplasm of many yeast cells is spilled out. However, this is not a complete process. The addition of a detergent, hexadecyltrimethylammonium bromide, CTAB, accomplishes two functions: (1) it helps to solubilize cell membranes and thereby further weakens the cell structure, and (2) it helps to inactivate the nucleic acid-degrading enzymes, nucleases, that are present. The addition of a chelating agent, ethylenediamine tetraacetate, EDTA, also inactivates these enzymes. EDTA removes the di- and tri-valent cations necessary for the activity of nucleases. Without this inhibition, the nucleases would degrade the nucleic acids to their constituent nucleotides. The final assault on the yeast cell is the osmotic shock. This is provided by a hypotonic saline-EDTA solution. The already weakened cells (by grinding and treatment with CTAB) will burst in the hypotonic medium and spill their contents, nucleic acids, among them.

Once the nucleic acids are in solution, they must be separated from the other constituents of the cell. First, the protein molecules must be removed. Many of the proteins of the cell are strongly associated with nucleic acids. The addition of sodium perchlorate ($NaClO_4$) dissociates the proteins from nucleic acids. When the mixture is shaken with the organic solvent, chloroform-isoamyl alcohol, the proteins are denatured, and they precipitate at the interface. At the same time, the lipid components of the cells are dissolved in the organic solvent. Thus the aqueous layer will contain nucleic acids, small water-soluble molecules, and even some proteins as contaminants.

The addition of ethanol precipitates the large molecules (DNA, RNA and proteins) and leaves the small molecules in solution. DNA, being the largest fibrous molecule, forms thread-like precipitates that can be spooled off onto a rod. The protein and RNA form a gelatinous precipitate that cannot be picked up by winding them on a glass rod. Thus, the spooling separates DNA from RNA.

After the isolation of DNA we will probe its identity by the diphenylamine test. The blue color of this test is specific for deoxyribose and the appearance of a blue color can be used to identify the deoxyribose containing DNA molecule.

Flow diagram of the DNA isolation process

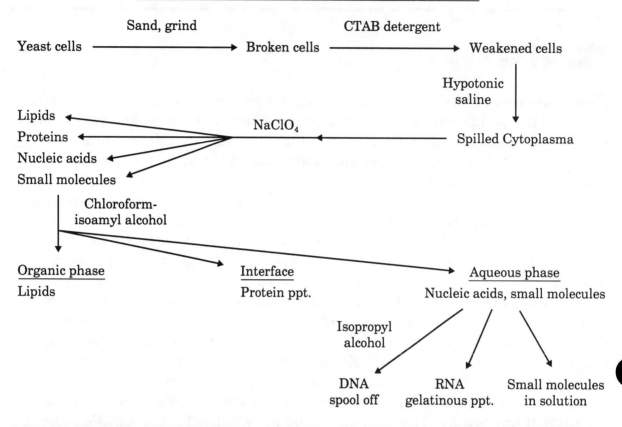

Objectives

To demonstrate the separation of DNA molecules from other cell constituents and to prove their identity.

Procedure

1. Cool a mortar in ice water. Add 2 to 3 g of baker's yeast and twice as much acid-washed sand. Grind the yeast and the sand vigorously with a pestle for 5–10 min. to disrupt the cells. (Two groups can work together in grinding; then divide the product.)

2. Preheat 25 mL of hexadecyltrimethylammonium bromide (CTAB) isolation buffer (2% CTAB, 0.15 M NaCl, 0.2% 2-mercaptoethanol, 20 mM EDTA, and 100 mM Tris-HCl at pH 8.0) in a 100-mL beaker in a 60°C water bath.

3. Add the ground yeast and sand to the saline-CTAB solution. Mix the solution with the sand. Let it stand for 20 min., with occasional swirling, while maintaining the temperature at 60°C.

4. Decant the cell suspension into a 250-mL Erlenmeyer flask leaving the sand behind. Cool the solution to room temperature. Add 5 mL of 6 M $NaClO_4$ solution and mix well. Transfer 40 mL of the chloroform-isoamyl alcohol mixture into the flask. Stopper the flask with a cork. Shake it for 10 min., sloshing the contents from side to side once every 15 sec. A frothy emulsion will form. After 10 min., let the emulsion settle.

5. Break up the emulsion by gently swirling with a glass rod that reaches into the interface. The complete separation into two distinct layers is not possible without centrifugation. (If desk top centrifuges are available, it is preferable to separate the layers by centrifuging at 1600 × gravity for 5 min.) However, one can proceed without centrifugation as well. When a sufficient amount (20–30 mL) of the top aqueous layer is cleared, remove this with a Pasteur pipet and transfer it to a graduated cylinder. Measure the volume and pour the contents into a 250-mL beaker. Pay attention that none of the brownish precipitate, droplets of emulsion, is transferred.

6. To the viscous DNA containing aqueous solution, add slowly twice its volume of cold isopropyl alcohol, taking care that the alcohol flows along the side of the beaker settling on top of the aqueous solution. With a flame sterilized glass rod, gently stir the DNA-isopropyl alcohol solution. This procedure is **critical**. The DNA will form a thread-like precipitate. **Rotating (not stirring) the glass rod** spools all the DNA precipitate onto the glass rod. As the DNA is wound on the rod, squeeze out the excess liquid by pressing the rod against the wall of the beaker. Transfer the spooled DNA on the rod into a test tube containing 95% ethanol.

7. Discard the alcohol solution left in the beaker and the chloroform-isoamyl alcohol solution left in the Erlenmeyer flask into specially labeled waste jars. Do not pour them down the sink.

8. Remove the rod and the spooled DNA from the test tube. Dry the DNA with a clean filter paper. Note its appearance. Dissolve the isolated crude DNA in 2 mL of citrate buffer (0.15 M NaCl, 0.015 M sodium citrate). Set up four dry and clean test tubes. Add 2 mL each of the following into the test tubes:

Test tube	Solution
1	1% glucose
2	1% ribose
3	1% deoxyribose
4	crude DNA solution

Add 5 mL diphenylamine reagent to each test tube. Mix the contents of the test tubes. Heat the test tubes in boiling water bath for 10 min. Record the color on your Report Sheet.

Chemicals and Equipment

1. Baker's yeast
2. Sand
3. Saline-hexadecyltrimethylammonium bromide (CTAB) isolation buffer
4. $NaClO_4$ solution
5. Chloroform-isoamyl alcohol solvent
6. Citrate buffer
7. Isopropyl alcohol (2-propanol)
8. Glucose solution
9. Ribose solution
10. Deoxyribose solution
11. Diphenylamine reagent
12. 95% ethanol
13. Mortar and pestle
14. Desk top clinical centrifuges (optional)

Experiment 43

PRE-LAB QUESTIONS

Consult Chapter 23 of your textbook to answer the structural questions.

1. Draw the structure of the purine and pyrimidine bases that are part of the DNA molecule.

2. Draw the structures of ribose and deoxyribose.

3. Why must you handle the diphenylamine reagent with great care?

4. The most demanding part of this experiment is grinding the yeast cells with sand. What does this process accomplish? Would you be able to isolate DNA without this grinding?

Experiment 43

REPORT SHEET

1. Describe the appearance of the crude DNA preparation.

2. Diphenylamine test.

Solution	Color
1% glucose	_____
1% ribose	_____
1% deoxyribose	_____
crude DNA sample	_____

Did the diphenylamine test confirm the identity of DNA?

3. How did you separate the DNA from proteins?

4. After mixing the aqueous extract with chloroform-isoamyl alcohol mixture, which layer contained the RNA (aqueous or organic)?

5. What compounds were left behind in the isopropyl alcohol solution after spooling the DNA?

POST-LAB QUESTIONS

1. Can the diphenylamine reagent distinguish between ribose and deoxyribose, and between DNA and RNA?

2. If you forgot to add $NaClO_4$ solution to your extract, how would the omission affect the purity of your DNA preparation?

3. Why can we isolate DNA from the precipitate, which also contains RNA and proteins, by the simple "spooling" procedure?

Experiment 44

Viscosity and secondary structrue of DNA

Background

In 1953, Watson and Crick proposed a three-dimensional structure of DNA which is a cornerstone in the history of biochemistry and molecular biology. The double helix they proposed for the secondary structure of DNA gained immediate acceptance, partly because it explained all known facts about DNA, and partly because it provided a beautiful model for DNA replication.

In the DNA double helix, two polynucleotide chains run in opposite directions. This means that at each end of the double helix there is one 5'-OH and one 3'-OH terminal. The sugar phosphate backbone is on the outside, and the bases point inward. These bases are paired so that for each adenine (A) on one chain a thymine (T) is aligned opposite it on the other chain. Each cytosine (C) on one chain has a guanine (G) aligned with it on the other chain. The AT and GC base pairs form hydrogen bonds with each other. The AT pair has two hydrogen bonds; the GC pair has three hydrogen bonds (Fig. 44.1).

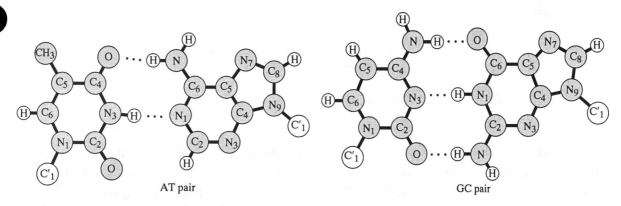

Figure 44.1 • Hydrogen bonding between base pairs.

Most of the DNA in nature has the double helical secondary structure. The hydrogen bonds between the base pairs provide the stability of the double helix. Under certain conditions the hydrogen bonds are broken. During the replication process itself this happens and parts of the double helix unfold. Under other conditions the whole molecule unfolds, becomes single stranded and assumes a random coil conformation. This can happen in denaturation processes aided by heat, extreme acidic or basic conditions, etc. Such a transformation is often referred to as helix to coil transition. There are a number of techniques that can monitor such a transition. One of the most sensitive is the measurement of viscosity of DNA solutions.

Viscosity is the resistance to flow of a liquid. Honey has a high viscosity and gasoline a low viscosity, at room temperature. In a liquid flow, the molecules must slide past each other. The resistance to flow comes from the interaction between the molecules as they slide past each other. The stronger this interaction, i.e., hydrogen bonds vs. London dispersion forces, the greater the resistance and the higher the viscosity. Even more than the nature of the intermolecular interaction, the size and the shape of the molecules influence their viscosity. A large molecule has greater surface over which it interacts with other molecules than a small molecule. Therefore, its viscosity is greater than that of a small molecule. If two molecules have the same size and the same interaction forces but have different shapes, their viscosity will be different. For example, a needle-shaped molecule has a greater surface of interaction than a spherical molecule of the same molecular weight (Fig. 44.2). The needle-shaped molecule will have a higher viscosity than the spherical molecule. The DNA double helix is a rigid structure held together by hydrogen bonds. Its long axis along the helix exceeds by far its short axis perpendicular to it. Thus the DNA double helix has large surface area and consequently high viscosity. When the hydrogen bonds are broken and the DNA molecule becomes single stranded, it assumes a random coil shape which has much lower surface area and lower viscosity. Thus a helix to coil transition is accompanied by a drop in viscosity.

Figure 44.2

Surface area of interaction between molecules of different shapes.

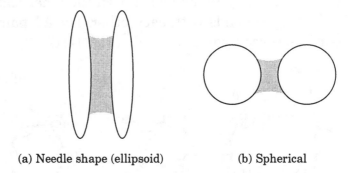

(a) Needle shape (ellipsoid) (b) Spherical

In practice, we can measure viscosity by the efflux time of a liquid in a viscometer (Fig. 44.3). The capillary viscometer is made of two bulbs connected by a tube in which the liquid must flow through a capillary tube. The capillary tube provides a laminar flow in which concentric layers of the liquid slide past each other. Originally, the liquid is placed in the storage bulb (A). By applying suction above the capillary, the liquid is sucked up past the upper calibration mark. With a stopwatch in hand, the suction is released and the liquid is allowed to flow under the force of gravity. The timing starts when the meniscus of the liquid hits the upper calibration mark. The timing ends when the meniscus of the liquid hits the lower calibration mark of the viscometer. The time elapsed between these two marks is the efflux time.

Figure 44.3.

Ostwald capillary viscometer.

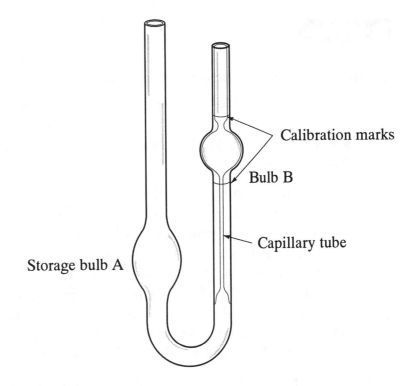

Calibration marks

Bulb B

Capillary tube

Storage bulb A

With dilute solutions, such as the DNA in this experiment, the viscosity of the solution is compared to the viscosity of the solvent. The efflux time of the solvent, aqueous buffer, is t_o and that of the solution is t_s. The relative viscosity of the solution

$$\eta_{rel} = t_s/t_o.$$

The viscosity of a solution also depends on the concentration; the higher the concentration the higher the viscosity. In order to make the measurement independent of concentration, a new viscometric parameter is used which is called intrinsic viscosity, $[\eta]$. This number is

$$[\eta] = (\log\eta_{rel})/c$$

which is almost a constant for a particular solute (DNA in our case) in very dilute solutions.

In this experiment, we follow the change in the viscosity of a DNA solution when we change the pH of the solution from the very acidic (pH 2.0) to very basic (pH 12.0). At extreme pH values, we expect that the hydrogen bonds will break and the double helix will become single-stranded random coils. A change in the viscosity will tell at what pH this happens. We shall also determine whether two acid-denatured single-stranded DNA molecules can refold themselves into a double helix when we neutralize the denaturing acid.

Objectives

1. To demonstrate helix to coil to helix transitions.
2. To learn how to measure viscosity.

Because of the cost of viscometers the students may work in groups of 5–6.

1. To 3 mL of a buffer solution, add 1 drop of 1.0 M HCl using a Pasteur pipet. Measure its pH with a universal pH paper. If the pH is above 2.5, add another drop of 1 M HCl. Measure the pH again. Record the pH on your Report Sheet (1).

2. Clamp one clean and dry viscometer on a stand. Pipet 3 mL of your acidified buffer solution into bulb A of your viscometer. Using a suction bulb of a Spectroline pipet filler, raise the level of the liquid in the viscometer above the upper calibration mark. Release the suction by removing the suction bulb and time the efflux time between the two calibration marks. Record this as t_o on your Report Sheet (2). Remove all the liquid from your viscometer by pouring the liquid out from the wide arm. Then apply pressure with the suction bulb on the capillary arm of the viscometer and blow out any remaining liquid into the storage bulb (A); pour out this residual liquid.

3. Take 3 mL of the prepared DNA solution. Add the same amount of 1 M HCl as above (1 or 2 drops). Mix it thoroughly by shaking the solution. Test the pH of the solution with a universal pH paper and record the pH (3) and the DNA concentration of the prepared solution on your Report Sheet (4).

4. Pour the acidified DNA solution into the wide arm (bulb A) of your viscometer. Using a suction bulb, raise the level of your liquid above the upper calibration mark. Release the suction by removing the suction bulb and measure and record the efflux time of the acidified DNA solution (5).

5. Add the same amount (1 or 2 drops) as above of neutralizing 1 M NaOH solution to the liquid in the wide arm of your viscometer. With the suction bulb on the capillary arm blow a few air bubbles through the solution to mix the ingredients. Repeat the measurement of the efflux time and record it on your Report Sheet (6). For the next 100 min. or so, repeat the measurement of the efflux times every 20 min. and record the results on your Report Sheet (7–11).

6. While the efflux time measurements in viscometer 1 are repeated every 20 min., another dry and clean viscometer will be used for establishing the pH dependence of the viscosity of DNA solutions. First, measure the pH of the buffer solution with a universal pH paper. Record it on your Report Sheet (12). Second, transfer 3 mL of the buffer into the viscometer 2 and measure and record its efflux time (13). Empty the viscometer as above. Test the pH of the DNA solution with a universal pH paper (14) and transfer 3 mL into the viscometer. Measure its efflux time and record it on your Report Sheet (15). Empty your viscometer.

7. Repeat the procedure described in step no. 6, but this time, with the aid of a Pasteur pipet, add one drop of 0.1 M HCl both to the 3 mL buffer solution, as well as to the 3 mL DNA solution. Measure the pH and the efflux times of both buffer and DNA solutions and record them (16–19) on your Report Sheet. *Make sure that you empty the viscometer after each viscosity measurement.*

8. Repeat the procedure described in step no. 6, but this time add one drop of 0.1 M NaOH solution to both the 3 mL buffer and 3 mL DNA solutions. Measure their pH and efflux times and record them on your Report Sheet (20–23).

9. Repeat the procedure described in step no. 6, but this time add 2 drops of 1 M NaOH to both buffer and DNA solutions (3 mL of each solution). Measure and record their pH and efflux times on your Report Sheet (24–27).

10. If time allows, you may repeat the procedure at other pH values; for example, by adding two drops of 1 M HCl (28–31), or two drops of 0.1 M HCl (32–35), or two drops of 0.1 M NaOH (36–39) to the separate samples of buffer and DNA solutions.

Chemicals and Equipments

1. Viscometers, 3-mL capacity
2. Stopwatch or watch with a seconds hand
3. Stand with utility clamp
4. Pasteur pipets
5. Buffer at pH 7.0
6. Prepared DNA solution
7. 1 M HCl
8. 0.1 M HCl
9. 1 M NaOH
10. 0.1 M NaOH
11. Spectroline pipet fillers

Experiment 44

PRE-LAB QUESTIONS

1. Which functional groups in the AT base pair form hydrogen bonds?

2. Write an equation for the reaction between an amine, $-NH_2$, and an acid, H_3O^+.

3. Write an equation for the reaction between an ammonium cation, $-NH_3^+$, and a base, $-OH^-$.

4. Which will have a greater surface of interaction—DNA in a double helix or the same DNA denatured, single-stranded random coil? Justify your answer with a diagram of helix to coil transition.

NAME ______________________ SECTION ____________ DATE ____________

PARTNER ______________________ GRADE ______________________

Experiment 44

REPORT SHEET

1. pH of acidified buffer ____________
2. Efflux time of acidified buffer ____________ sec.
3. pH of acidified DNA solution ____________
4. Concentration of DNA solution ____________
5. Efflux time of acidified DNA solution ____________ sec.
6. Efflux time of neutralized DNA solution at time of neutralization ____________ sec.
7. 20 min. later ____________ sec.
8. 40 min. later ____________ sec.
9. 60 min. later ____________ sec.
10. 80 min. later ____________ sec.
11. 100 min. later ____________ sec.
12. pH of neutral buffer ____________
13. Efflux time of neutral buffer ____________ sec.
14. pH of DNA solution in neutral buffer ____________
15. Efflux time of DNA in neutral buffer ____________ sec.

After addition of 1 drop of 0.1 M HCl

16. pH of buffer ____________
17. Efflux time of buffer ____________ sec.
18. pH of DNA solution ____________
19. Efflux time of DNA solution ____________ sec.

After addition of 1 drop of 0.1 M NaOH

20. pH of buffer _____

21. Efflux time of buffer _____ sec.

22. pH of DNA solution _____

23. Efflux time of DNA solution _____ sec.

After addition of 2 drops of 1 M NaOH

24. pH of buffer _____

25. Efflux time of buffer _____ sec.

26. pH of DNA solution _____

27. Efflux time of DNA solution _____ sec.

After addition of 2 drops of 1 M HCl

28. pH of buffer _____

29. Efflux time of buffer _____ sec.

30. pH of DNA solution _____

31. Efflux time of DNA solution _____ sec.

After addition of 2 drops of 0.1 M HCl

32. pH of buffer _____

33. Efflux time of buffer _____ sec.

34. pH of DNA solution _____

35. Efflux time of DNA solution _____ sec.

After addition of 2 drops of 0.1 M NaOH

36. pH of buffer _____

37. Efflux time of buffer _____ sec.

38. pH of DNA solution _____

39. Efflux time of DNA solution _____ sec.

Tabulate your data on the pH dependence of relative viscosity.

pH	η_{rel}
(3)_____	(5)/(2)_____
(14)_____	(15)/(13)_____
(18)_____	(19)/(17)_____
(22)_____	(23)/(21)_____
(26)_____	(27)/(25)_____
(30)_____	(31)/(29)_____
(34)_____	(35)/(33)_____
(38)_____	(39)/(37)_____

POST-LAB QUESTIONS

1. Plot your tabulated data—relative viscosity on the Y axis and pH on the X axis.

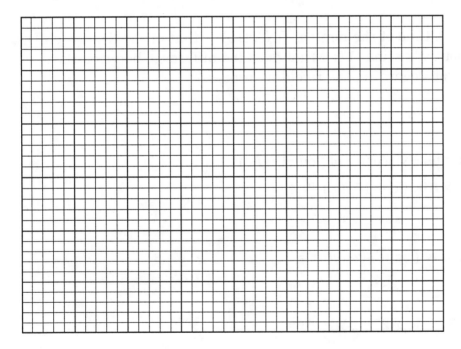

2. At what pH values did you observe helix to coil transitions?

3. Plot your data on the refolding of DNA double helix (5)–(11).
Plot the time on the X axis and the efflux times on the Y axis.

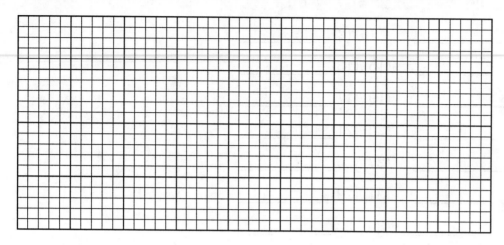

4. Was there any indication that, upon neutralization of the denaturing acid, the DNA did refold into a double helix? Explain.

5. Compare the efflux time of the neutral DNA (15) to that of the denatured DNA 100 min. after neutralization (11). What does the difference between these two efflux times tell you regarding the refolding process?

6. Calculate the intrinsic viscosity of your DNA at

(a) neutral pH = 2.3 × {log[(15)/(13)]}/(4) =

(b) acidic pH = 2.3 × {log[(5)/(2)]}/(4) =

(c) basic pH = 2.3 × {log[(27)/(25)]}/(4) =

(d) neutralized pH 100 min. after neutralization = 2.3 × {log[(11)/(13)]}/(4) =

7. A high intrinsic viscosity implies a double helix, a low intrinsic viscosity means a random coil. What do you think is the shape of the DNA after acid denaturation and subsequent neutralization? [See 6(d).]

Kinetics of urease—catalyzed decomposition of urea

Enzymes speed up the rates of reactions by forming an enzyme-substrate complex. The reactants can undergo the reaction on the surface of the enzyme, rather than finding each other by collision. Thus the enzyme lowers the energy of activation of the reaction.

Urea decomposes according to the following equation:

$$H_2N-\overset{\overset{\displaystyle O}{\|}}{C}-NH_2 + H_2O \rightleftharpoons CO_2 + 2\,NH_3 \qquad (1)$$

This reaction is catalyzed by a highly specific enzyme, urease. Urease is present in a number of bacteria and plants. The most common source of the enzyme is jack bean or soy bean. Urease was the first enzyme that was crystallized. Sumner, in 1926, proved unequivocally that enzymes are protein molecules.

Urease is a -SH group (thiol) containing enzyme. The cysteine residues of the protein molecule must be in the reduced -SH form in order for the enzyme to be active. Oxidation of these groups will form -S-S-, disulfide bridges and the enzyme loses its activity. Reducing agents such as cysteine or glutathione can reactivate the enzyme.

Heavy metals such as Ag^+, Hg^{2+}, or Pb^{2+}, which form complexes with the -SH groups, also inactivate the enzyme. For example, the poison phenylmercuric acetate is a potent inhibitor of urease.

$$\text{enzyme}-SH + CH_3-\overset{\overset{\displaystyle O}{\|}}{C}-O-Hg-C_6H_5 \rightleftharpoons CH_3\overset{\overset{\displaystyle O}{\|}}{C}-OH + \text{enzyme}-S-Hg-C_6H_5 \quad (2)$$

active   phenylmercuric acetate   acetic acid   inactive

In this experiment, we study the kinetics of the urea decomposition. As shown in equation (1), the products of the reaction are carbon dioxide, CO_2, and ammonia, NH_3. Ammonia, being a base, can be titrated with an acid, HCl, and in this way we can determine the amount of NH_3 that is produced.

$$NH_3\,(aq) + HCl\,(aq) \rightleftharpoons NH_4Cl\,(aq) \qquad (3)$$

For example, a 5 mL aliquot of the reaction mixture is taken before the reaction starts. We use this as a blank. We titrate this with 0.05 N HCl to an end point. The amount of acid used was 1.5 mL. This blank then must be subtracted from all subsequent titration values. Next, we take a 5 mL sample of the reaction mixture after the reaction has proceeded for 10 min. We titrate this with 0.05 N HCl and, let's assume, get a value of 5.0 mL HCl. Therefore, $5.0 - 1.5 = 3.5$ mL of 0.05 N HCl was used to neutralize the NH_3 produced in a 10 min. reaction time. This means that

$$(3.5 \text{ mL} \times 0.05 \text{ moles HCl})/1000 \text{ mL} = 1.75 \times 10^{-4} \text{ moles HCl was used up.}$$

According to reaction (3), one mole of HCl neutralizes 1 mole of NH_3, therefore, the titration indicates that in our 5 mL sample, 1.75×10^{-4} moles of NH_3 was produced in 10 min. Equation (1) also shows that for each mole of urea decomposed, 2 moles of NH_3 are formed. Therefore, in 10 min.

$$(1 \text{ mole urea} \times 1.75 \times 10^{-4} \text{ moles } NH_3)/2 \text{ moles } NH_3$$
$$= 0.87 \times 10^{-4} \text{ moles urea or } 8.7 \times 10^{-5} \text{ moles of urea}$$

were decomposed. Thus the rate was 8.7×10^{-6} moles of urea per min. This is the result we obtained using a 5 mL sample in which 1 mg of urease was dissolved. This rate of reaction corresponds to 8.7×10^{-6} moles urea/mg enzyme-min.

A *unit of activity* of urease is defined as the micromoles (1×10^{-6} moles) of urea decomposed in 1 min. Thus the enzyme in the preceding example had an activity of 8.7 units per mg enzyme.

In this experiment we also study the rate of the urease-catalyzed decomposition in the presence of an inhibitor. We use a dilute solution of phenylmercuric acetate to inhibit but not completely inactivate urease.

CAUTION!

Mercury compounds are poisons. Take extra care to avoid getting the mercuric salt solution in your mouth or swallowing it.

Many of the enzymes in our body are also -SH containing enzymes, and these will be inactivated if we ingest such compounds. As a result of mercury poisoning, many body functions will be inhibited.

Objectives

1. To demonstrate how to measure the rate of an enzyme-catalyzed reaction.
2. To investigate the effect of an inhibitor on the rate of reaction.
3. To calculate urease activity.

Enzyme Kinetics in the Absence of Inhibitor

1. Prepare a 37°C water bath in a 250-mL beaker. Maintain this temperature by occasionally adding hot water to the bath. To a 100-mL Erlenmeyer flask, add 20 mL of 0.05 M Tris buffer and 20 mL of 0.3 M urea in a Tris buffer. Mix the two solutions, and place the corked Erlenmeyer flask into the water bath for 5 min. This is your *reaction vessel*.

2. Set up a buret filled with 0.05 N HCl. Place into a 100-mL Erlenmeyer flask 3 to 4 drops of a 1% $HgCl_2$ solution. This will serve to stop the reaction, once the sample is pipetted into the titration flask. Add a few drops of methyl red indicator. This Erlenmeyer flask will be referred to as the *titration vessel*.

3. Take the *reaction vessel* from the water bath. Add 10 mL of urease solution to your reaction vessel. The urease solution contains a specified amount of enzyme (e.g., 20 mg enzyme in 10 mL of solution). Note the time of adding the enzyme solution as zero reaction time. Immediately pipet a 5 mL aliquot of the urea mixture into your *titration vessel*. Stopper the reaction vessel, and put it back into the 37°C bath.

4. Titrate the contents of the titration vessel with 0.05 N HCl to an end point. The end point is reached when the color changes from yellow to pink and stays that way for 10 sec. Record the amount of acid used. This is your blank.

5. Wash and rinse your titration vessel after each titration and reuse it for subsequent titrations.

6. Take a 5 mL aliquot from the reaction vessel **every 10 min.** Pipet these aliquots into the cleaned titration vessel into which methyl red indicator and $HgCl_2$ inhibitor were already placed similar to the procedure in step no. 2 that you used in your first titration (blank). Record the time you placed the aliquots into the titration vessels and titrate them with HCl to an end point. Record the amount of HCl used in your titration. Use five samples over a period of 50 min.

Enzyme Kinetics in the Presence of Inhibitor

> **CAUTION!**
>
> Be careful with the phenylmercuric acetate solution. Do not get it in your mouth or eyes.

1. Use the same water bath as in the first experiment. Maintain the temperature at 37°C. To a new 100-mL reaction vessel, add 19 mL of 0.05 M Tris buffer, 20 mL of 0.3 M urea solution, and 1 mL of phenylmercuric acetate (1×10^{-3} M). Mix the contents, and place the reaction vessel into the water bath for 5 min.

2. Ready the *titration vessel* as before by adding a few drops of $HgCl_2$ and methyl red indicator. To the *reaction vessel*, add 10 mL of urease solution. Note the time of addition as zero reaction time. Mix the contents of the *reaction vessel*. Transfer immediately a 5 mL aliquot into the *titration vessel*. This will serve as your blank.

3. Titrate it as before. Record the result. **Every 10 min.** take a 5 mL aliquot for titration. The duration of this experiment should be 40 min.

Chemicals and Equipment

1. Tris buffer
2. 0.3 M urea
3. 0.05 N HCl
4. 1×10^{-3} M phenylmercuric acetate
5. 1% $HgCl_2$
6. Methyl red indicator
7. Urease solution
8. 50-mL buret
9. 10-mL graduated pipets
10. 5-mL volumetric pipets
11. 10-mL volumetric pipets
12. Buret holder
13. Spectroline pipet filler

Experiment 45

PRE-LAB QUESTIONS

1. To what enzyme group does urease belong?

2. What groups are in the active site of urease?

3. Why is phenylmercuric acetate such a dangerous poison?

4. How do we measure the concentration of a product from the urease-catalyzed reaction?

Experiment 45

REPORT SHEET

Enzyme kinetics in the absence of inhibitor

Reaction time (min.)	Buret readings before titration (A)	Buret readings after titration (B)	mL acid titrated (B) − (A)	mL 0. 05 N HCl used up in the reaction (B) − (A) − blank
0 (blank)				
10				
20				
30				
40				
50				

Enzyme kinetics in the presence of Hg salt inhibitor

Reaction time (min.)	Buret readings before titration (A)	Buret readings after titration (B)	mL acid titrated (B) − (A)	mL 0. 05 N HCl used up in the reaction (B) − (A) − blank
0 (blank)				
10				
20				
30				
40				

1. Present the preceding data in the graphical form by plotting column 1 on the x-axis and column 5 on the y-axis for both reactions.

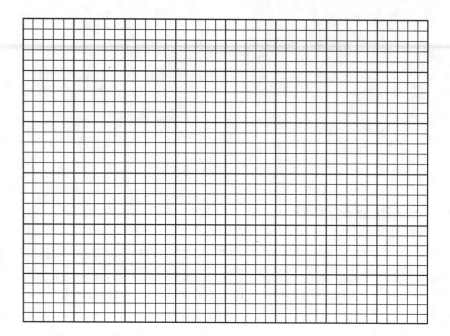

2. Calculate the urease activity only for the reaction without the inhibitor. Use the titration data from the first 10 min. of reaction (initial slope).

Urease activity

$$= \frac{\text{X mL HCl consumed} \times 0.05 \text{ moles HCl} \times 1 \text{ mole NH}_3 \times 1 \text{ mole urea} \times 50 \text{ mL sol.}}{10 \text{ min.} \times 5 \text{ mL sol} \times 1000 \text{ mL HCl} \times 1 \text{ mole HCl} \times 2 \text{ moles NH}_3 \times 20 \text{ mg urease}}$$

= Z units activity/mg enzyme

POST-LAB QUESTIONS

1. What would be the urease activity if you used the slope between 40 and 50 min. instead of the initial slope from your diagram?

2. Your instructor will provide the activity of urease as it was specified by the manufacturer. Compare this activity with the one you calculated. Can you account for the difference? (Enzymes usually lose their activity in long storage.)

3. What was the purpose of adding $HgCl_2$ to the titration vessel?

4. If you did not add a few drops of $HgCl_2$ to the titration vessel, would the titration values of HCl used be higher or lower than the one you obtained?

5. In studying enzyme reactions, you must work at constant temperature and pH. What steps were taken in your experiment to satisfy these requirements?

6. Your lab ran out of 0.05 N HCl. You found a bottle labeled 0.05 N H_2SO_4. Could you use it for your titration? If you do, would your calculation of urease activity be different?

Experiment 46

Isocitrate dehydrogenase—an enzyme of the citric acid cycle

Background

The citric acid cycle is the first unit of the common metabolic pathway through which most of our food is oxidized to yield energy. In the citric acid cycle, the partially fragmented food products are broken down further. The carbons of the C_2 fragments are oxidized to CO_2, released as such, and expelled in the respiration. The hydrogens and the electrons of the C_2 fragments are transferred to the coenzyme, nicotinamide adenine dinucleotide, NAD^+, or to flavin adenine dinucleotide, FAD, which in turn become $NADH + H^+$ or $FADH_2$, respectively. These enter the second part of the common pathway, oxidative phosphorylation, and yield water and energy in the form of ATP.

The first enzyme of the citric acid cycle to catalyze both the release of one carbon dioxide and the reduction of NAD^+ is isocitrate dehydrogenase. The overall reaction of this step is as follows:

Isocitrate   α-Ketoglutarate

The reduction of the NAD^+ itself is given by the equation:

NAD$^+$ NADH

The enzyme has been isolated from many tissues, the best source being a heart muscle or yeast. The isocitrate dehydrogenase requires the presence of cofactors Mg^{2+} or Mn^{2+}. As an allosteric enzyme, it is regulated by a number of modulators. ADP, adenosine diphosphate, is a positive modulator and therefore stimulates enzyme activity. The enzyme has an optimum pH of 7.0. As is the case with all enzymes of the citric acid cycle, isocitrate dehydrogenase is found in the mitochondria.

In the present experiment, you will determine the activity of isocitrate dehydrogenase extracted from pork heart muscle. The commercial preparation comes in powder form and it uses $NADP^+$ rather than NAD^+ as a coenzyme. The basis of the measurement of the enzyme activity is the absorption spectrum of NADPH. This reduced coenzyme has an absorption maximum at 340 nm. Therefore, an increase in the absorbance at 340 nm indicates an increase in NADPH concentration, hence the progress of the reaction. We define the unit of isocitrate dehydrogenase activity as one that causes an increase of 0.01 absorbance per min. at 340 nm.

For example, if a 10 mL solution containing isocitrate and isocitrate dehydrogenase and $NADP^+$ exhibits a 0.04 change in the absorbance in two min., the enzyme activity will be

$$\frac{0.04 \text{ abs.}}{2 \text{ min.} \times 10 \text{ mL}} \times \frac{1 \text{ unit}}{0.01 \text{ abs.}/1 \text{ min.}} = 0.2 \text{ units/mL}$$

If the 10 mL test solution contained 1 mL of isocitrate dehydrogenase solution with a concentration of 1 mg powder/1 mL of enzyme solution, then the activity will be

(0.2 units/mL test soln.) × (10 mL test soln./1 mL enzyme soln.) × (1 mL enzyme soln./1 mg enzyme powder)

= 2 units/mg enzyme powder.

Objectives

To measure the activity of an enzyme of the citric acid cycle, isocitrate dehydrogenase, and the effect of enzyme concentration on the rate of reaction.

Procedure

1. Turn *on* the spectrophotometer and let it warm up for a few minutes. Turn the wavelength control knob to read 340 nm. With *no sample tube in the sample compartment*, adjust the amplifier control knob so that 0% transmittance or infinite absorbance is read.

2. Prepare a cocktail of reactants in the following manner: In a 10-mL test tube, mix 2.0 mL phosphate buffer, 1.0 mL $MgCl_2$ solution, 1.0 mL 15 mM isocitrate solution, and 5 mL distilled water.

3. To prepare a **Blank** for the spectrophotometric reading, take a sample tube and add to it 1.0 mL of reagent cocktail (prepared as above), 0.2 mL NADP$^+$ solution, and 1 mL of distilled water. Mix the solutions by shaking the sample tube. *Be careful to pipet exactly 0.2 mL NADP$^+$.*

4. Insert the sample tube with the **Blank** solution into the spectrophotometer. Adjust the reading to 100% transmittance (or 0 absorbance). This zeroing must be performed every 10 min. before each enzyme activity run because some instruments have a tendency to drift. The instrument is now ready to measure enzyme activity.

5. Prepare one sample tube for the **enzyme activity measurements.** Add 1.0 mL of reagent cocktail and 0.7 of mL of distilled water. Next, add 0.2 mL NADP$^+$ solutions. *Be careful to pipet exactly 0.2 mL NADP$^+$.* Mix the contents of the sample tube. Readjust the spectrophotometer with the "Blank" (prepared in step no. 3) to read 0.00 absorbance or 100.00% transmission. Remove the "Blank" and save it for future readjustments. *In the next step the timing is very important.* Take a watch, and *at a set time* (for example 2 hr. 15 min. 00 sec.) add *exactly* 0.3 mL enzyme solution to the sample tube. Mix it thoroughly and quickly by shaking the tube. Insert the sample tube into the spectrophotometer and take a first reading 1 min. after the mixing time (i.e., 2 hr. 16 min. 00 sec.). Record the absorbancies on your Report Sheet in column 1. Thereafter, take a reading of the spectrophotometer every 30 sec. and record the readings for 5–6 min. on your Report Sheet in column 1.

6. Repeat the experiment exactly as in step no. 5: Preparing the sample solution, readjusting the instrument with the "Blank," and reading the sample solution every 30 sec. for 5 min. Record the spectrophotometric readings on your Report Sheet in column 2.

7. Prepare a new sample tube with the following contents: 1.0 mL reagent cocktail, 0.8 mL distilled water, and 0.2 mL NADP$^+$ solution. *Be careful to pipet exactly 0.2 mL NADP$^+$.* Mix it thoroughly. Readjust the spectrophotometer with the "Blank" to zero absorbance (100% transmission). *At a set time* (i.e., 2 hr. 33 min. 00 sec.), add *exactly* 0.2 mL of enzyme solution. Mix the sample tube and insert into the spectrophotometer. Take your first reading 1 min. after the mixing and every 30 sec. for 5 min. thereafter. Record the absorbancies on your Report Sheet in column 3.

8. Prepare a new sample tube with the following contents: 1.0 mL reagent cocktail, 0.6 mL distilled water, and *exactly* 0.2 mL NADP$^+$ solution. Mix it thoroughly. Readjust the spectrophotometer with the "Blank" to zero absorbance (100% transmission). *At a set time* (i.e., 2 hr. 45 min. 00 sec.), add *exactly* 0.4 mL enzyme solution to the sample tube. Mix it thoroughly. Insert the sample tube into the spectrophotometer. Take a first reading 1 min. after mixing and every 30 sec. for 5 min. thereafter. Record the absorbancies on your Report Sheet in column 4.

9. Plot the numerical data you recorded in the four columns on graph paper. Note that somewhere *between 3 and 5 min.* your graphs are linear. Obtain the slopes of these linear portions and record them on your Report Sheet. Calculate the activities of your enzyme first as (a) units per mL sample solution and second as (b) units per mg enzyme powder.

Chemicals and Equipment

1. Phosphate buffer, pH 7.0
2. 0.1 M $MgCl_2$ solution
3. 6.0 mM $NADP^+$ solution
4. 15.0 mM isocitrate solution
5. Isocitrate dehydrogenase (0.2 mg powder/mL solution).
6. Spectrophotometers

Experiment 46

PRE-LAB QUESTIONS

1. Explain the difference between the structures of citrate and isocitrate.

2. What reaction follows the isocitrate oxidation (dehydrogenation) in the citric acid cycle?

3. How do we measure isocitrate dehydrogenase activity?

4. What is the difference between NAD^+ and $NADP^+$? (Consult your textbook!)

Experiment 46

REPORT SHEET

Time (sec.) after mixing	Absorbance of sample			
	Column 1	2	3	4
60				
90				
120				
150				
180				
210				
240				
270				
300				
330				
360				

1. Plot your data: Absorbance versus time.

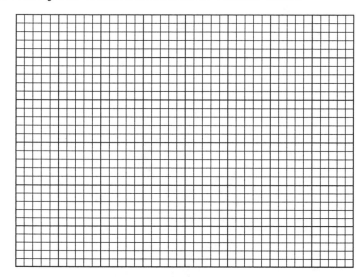

2. Calculate the enzyme activity:

(a) Units of enzyme activity/mL reaction mixture: The slope of the plot is usually a straight line. If so, read the value of change in absorbance per min. Divide it by 0.01. This gives you the number of enzyme activity units/reaction mixture. (One unit of enzyme activity is 0.01 absorbance/min.) Your reaction mixture had a volume of 2.2 mL. Thus dividing by 2.2 will give you the activities in units/mL reaction mixture.

	(1)	(2)	(3)	(4)
Units/reaction mixture				
Units/mL reaction mixture				

(b) Calculate the isocitrate dehydrogenase activity per mg powder extract. For example, your enzyme solution contained 0.2 mg powder extract/mL solution. If you added 0.2 mL of enzyme solution, it contained 0.04 mg powder extract. Dividing the units/reaction mixture (obtained above) by the number of mg of powder extract added gives you the units/mg powder extract.

	(1)	(2)	(3)	(4)
Units of enzyme activity/mg powder extract				

POST-LAB QUESTIONS

1. You ran two experiments with the same enzyme concentration in (1) and (2). Calculate the average activity for this concentration of enzyme in units/mg powder. Does the reproducibility fall within ±5%?

2. Does your enzyme activity (units/mg powder) give you the same number for the different enzyme concentrations employed?

3. If your powder extract contained 80% protein, what would be the average isocitrate dehydrogenase activity per mg protein (enzyme)?

Quantitative analysis of vitamin C contained in foods

Background

Ascorbic acid is commonly known as vitamin C. It was one of the first vitamins that played a role in establishing the relationship between a disease and its prevention by proper diet. The disease *scurvy* has been known for ages, and a vivid description of it was given by Jacques Cartier, a 16th century explorer of the American continent: "...Some did lose their strength and could not stand on their feet.... Others...had their skin spotted with spots of blood...their mouth became stinking, their gums so rotten that all the flesh did fall off...." Prevention of scurvy can be obtained by eating fresh vegetables and fruits. The active ingredient in fruits and vegetables that helps to prevent scurvy is ascorbic acid. It is a powerful biological antioxidant (reducing agent). It helps to keep the iron in the enzyme, prolyl hydroxylase, in the reduced form and, thereby, it helps to maintain the enzyme activity. Prolyl hydroxylase is essential for the synthesis of normal collagen. In scurvy, the abnormal collagen causes skin lesions and broken blood vessels.

Vitamin C cannot be synthesized in the human body and must be obtained from the diet (e.g., citrus fruits, broccoli, turnip greens, sweet peppers, tomatoes) or by taking synthetic vitamin C (e.g., vitamin C tablets, "high-C" drinks, and other vitamin C-fortified commercial foods). The minimum recommended adult daily requirement of vitamin C to prevent scurvy is 60 mg. Some people, among them the late Linus Pauling, twice Nobel Laureate, suggested that very large daily doses (250 to 10,000 mg) of vitamin C could help prevent the common cold, or at least lessen the symptoms for many individuals. No reliable medical data support this claim. At present, the human quantitative requirement for vitamin C is still controversial and requires further research.

In this experiment, the amount of vitamin C is determined quantitatively by titrating the test solution with a water-soluble form of iodine I_3^-:

$$\ddot{I}:\ddot{I}:+:\ddot{I}:^- \longrightarrow \left[:\ddot{I}:\ \ddot{I}::\ddot{I}:\right]^- \text{ (tri-iodode ion)}$$

expanded octets

Vitamin C is oxidized by I_2 (as I_3^-) according to the following chemical reaction:

vitamin C   (MW 254)   oxidized product
(MW 176)   (MW 174)

As vitamin C is oxidized by iodine, I_2 becomes reduced to I^-. When the end point is reached (no vitamin C is left), the excess of I_2 will react with a starch indicator to form a starch-iodine complex which is blackish-blue in color.

$$I_2 + \text{starch} \longrightarrow \text{iodine - starch complex (blackish-blue)}$$

It is worthwhile to know that although vitamin C is very stable when dry, it is readily oxidized by air (oxygen) when in solution; therefore, a solution of vitamin C should not be exposed to air for long periods. The amount of vitamin C can be calculated by using the following conversion factor:

$$1 \text{ mL of } I_2 \text{ (0.01 M)} = 1.76 \text{ mg vitamin C}$$

Objective

To determine the amount of vitamin C that is present in certain commercial food products by the titration method.

Procedure

1. Pour about 60 mL of a fruit drink that you wish to analyze into a clean, dry 100-mL beaker. The fruit drink should be light colored, apple, orange, or grape-fruit, but not dark colored, such as grape. Record the kind of drink on the Report Sheet (1).

2. If the fruit drink is cloudy or contains suspended particles, it can be clarified by the following procedure: Add Celite, used as a filter aid, to the fruit drink (about 0.5 g). After swirling it thoroughly, filter the solution through a glass funnel, bedded with a large piece of cotton. Collect the filtrate in a 50-mL Erlenmeyer flask (Fig. 47.1).

3. Using a 10-mL volumetric pipet and a Spectroline pipet filler, transfer 10.00 mL of the fruit drink into a 125-mL Erlenmeyer flask. Then add 20 mL of distilled water, 5 drops of 3 M HCl (as a catalyst), and 10 drops of 2% starch solution to the flask.

4. Clamp a clean, dry 50-mL buret onto the buret stand. Rinse the buret twice with 5 mL portions of iodine solution. Let the rinses run through the tip of the buret and discard them. Fill the buret slightly above the zero mark with a standardized iodine solution (A dry funnel may be used for easy transfer.) Air bubbles should be removed by turning the stopcock several times to force the air bubbles out of the tip. Record the molarity of standardized iodine solution (2). Record the initial reading of standardized iodine solution to the nearest 0.02 mL (3a).

Figure 47.1
Clarification of fruit drinks.

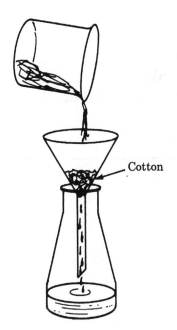

Cotton

5. Place the flask that contains the vitamin C sample under the buret and add the iodine solution dropwise, while swirling, until the indicator just changes to dark blue. This color should persist for at least 20 sec. Record the final buret reading (3b). Calculate the total volume of iodine solution required for the titration (3c), the weight of vitamin C in the sample (4), and percent (w/v) of vitamin C in the drink (5). Repeat this titration procedure twice more, except using 20 and 30 mL portions of the same fruit drink instead of 10 mL. Record the volumes of iodine solution that are required for each titration.

Chemicals and Equipment

1. 50-mL buret
2. Buret clamp
3. Spectroline pipet filler
4. 10-mL volumetric pipet
5. 50-mL Erlenmeyer flask
6. Cotton
7. Filter aid
8. Hi-C apple drink
9. Hi-C orange drink
10. Hi-C grapefruit drink
11. 0.01 M iodine in potassium iodide
12. 3 M HCl
13. 2% starch solution

Experiment 47

PRE-LAB QUESTIONS

1. What are the symptoms of scurvy?

2. What natural foods contain appreciable amounts of vitamin C?

3. What is the minimum daily requirement of vitamin C to prevent scurvy in adults?

4. What enzyme is kept in its active form by the presence of vitamin C in the diet?

Experiment 47

REPORT SHEET

1. The kind of fruit drink _____

2. Molarity of iodine solution _____

3. Titration results

	Sample 1 (10.0 mL)	Sample 2 (20.0 mL)	Sample 3 (30.0 mL)
a. Initial buret reading	_____mL	_____mL	_____mL
b. Final buret reading	_____mL	_____mL	_____mL
c. Total volume of iodine solution used (b − a)	_____mL	_____mL	_____mL

4. The weight of vitamin C in the fruit drink sample [(3c) × 1.76 mg/mL]

_____mg _____mg _____mg

5. Concentration of vitamin C in the fruit drink (mg/100 mL) [(4)/volume of drink] × 100

_____ _____ _____

6. Average concentration of vitamin C in the fruit drink _____ mg/100 mL

POST-LAB QUESTIONS

1. Why is HCl added for the titration of vitamin C?

2. What gives the blue color in your titration?

3. What volume of fruit drink would satisfy your minimum daily vitamin C requirement?

Analysis of vitamin A in margarine

Vitamin A, or retinol, is one of the major fat soluble vitamins. It is present in many foods; the best natural sources are liver, butter, margarine, egg yolk, carrots, spinach, and sweet potatoes. Vitamin A is the precursor of retinal, the essential component of the visual pigment rhodopsin.

Vitamin A (All-*trans*-retinol)   11-*cis*-retinal

When a photon of light penetrates the eye, it is absorbed by the 11-*cis*-retinal. The absorption of light converts the 11-*cis*-retinal to all-*trans*-retinal:

11-*cis*-retinal   All-*trans*-retinal

This isomerization converts the energy of a photon into an atomic motion which in turn is converted into an electrical signal. The electrical signal generated in the retina of the eye is transmitted through the optic nerve into the brain's visual cortex.

Even though part of the all-*trans*-retinal is regenerated in the dark to 11-*cis*-retinal, for good vision, especially for night vision a constant supply of vitamin A is needed. The recommended daily allowance of vitamin A is 750 μg. Deficiency in vitamin A results in night blindness and keratinization of epithelium. The latter compromises the integrity of healthy skin. In young animals, vitamin A is also required for growth. On the other hand, large doses of vitamin A, sometimes recommended in faddish diets, can be harmful. A daily dose above 1500 μg can be toxic.

To analyze the vitamin A content of margarine by spectrophotometric method.

Procedure

The analysis of vitamin A requires a multi-step process. In order that you should be able to follow the step-by-step procedure, a flow chart is provided here:

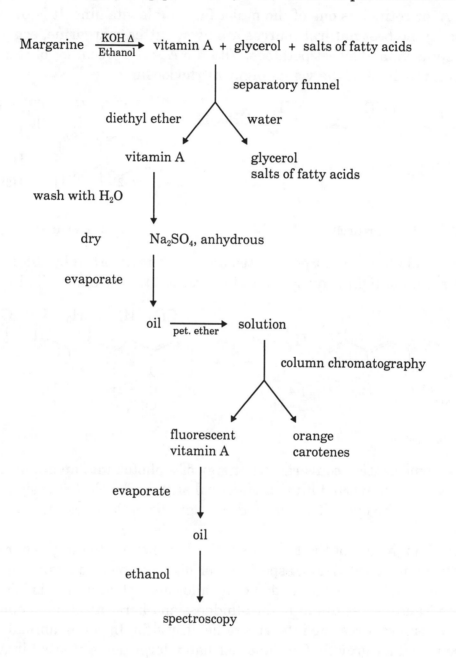

1. Margarine is largely fat. In order to separate vitamin A from the fat in margarine, first the sample must be saponified. This converts the fat to water soluble products, glycerol and potassium salts of fatty acids. Vitamin A can be extracted by diethyl ether from the products of the saponification process. To start, weigh a cover glass to the nearest 0.1 g. Report this weight on the Report Sheet (1). Add approximately 10 g of margarine to the watch glass. Record the weight of watch glass plus sample to the nearest 0.1 g on your Report Sheet (2). Transfer the sample from the watch glass into a 250-mL Erlenmeyer flask with the aid of a glass rod, and wash it in with 75 mL of 95% ethanol. Add 25 mL of 50% KOH solution. Cover the Erlenmeyer flask loosely with a cork and put it on an electric hot plate. Bring it gradually to a boil. Maintain the boiling for 5 min. with an occasional swirling of the flask using tongs. The stirring should aid the complete dispersal of the sample. Remove the Erlenmeyer from the hot plate and let it cool to room temperature (approximately 20 min.).

CAUTION!

50% KOH solution can cause burns on your skin. Handle the solution with care, do not spill it. If a drop gets on your skin, wash it immediately with copious amounts of water.

---

2. While the sample is cooling, prepare a chromatographic column. Take a 25-mL buret. Add a small piece of glass wool. With the aid of a glass rod, push it down near the stopcock. Add 15–16 mL of petroleum ether to the buret. Open the stopcock slowly, and allow the solvent to fill the tip of the buret. Close the stopcock. You should have 12–13 mL of petroleum ether above the glass wool. Weigh about 20 g of alkaline aluminum oxide (alumina) in a 100-mL beaker. Place a small funnel on top of your buret. Pour the alumina slowly, in small increments, into the buret. Allow it to settle to form a 20 cm column. Drain the solvent but do not allow the column to run dry. **Always have at least 0.5 mL clear solvent on top of the column.** If the alumina adheres to the walls of the buret, wash it down with more solvent.

3. Transfer the solution (from your reaction in step no. 1) from the Erlenmeyer flask to a 500-mL separatory funnel. Rinse the flask with 30 mL of distilled water and add the rinsing to the separatory funnel. Repeat the rinsing two more times. Add 100 ml of diethyl ether to the separatory funnel. Close the separatory funnel with the glass stopper. Shake the separatory funnel vigorously. (See Exp. 34 Fig. 34.1 for technique.) Allow it to separate into two layers. Drain the bottom aqueous layer into an Erlenmeyer flask. Add the top (diethyl ether) layer to a second clean 250-mL Erlenmeyer flask. Pour back the aqueous layer into the separatory funnel. Add another 100 mL portion of diethyl ether. Shake and allow it to separate into two layers. Drain again the bottom (aqueous) layer and discard. Combine the first diethyl ether extract with the residual diethyl ether extract in the separatory funnel. Add 100 mL of distilled water to the

combined diethyl ether extracts in the separatory funnel. Agitate it gently and allow the water to drain. Discard the washing.

CAUTION!

Diethyl ether is very volatile and inflammable. Make certain that there are no open flames, not even a hot electrical plate in the vicinity of the operation.

---

4. Transfer the diethyl ether extracts into a clean 300-mL beaker. Add 3–5 g of anhydrous Na_2SO_4 and stir it gently for 5 min. to remove traces of water. Decant the diethyl ether extract into a clean 300-mL beaker. Add a boiling chip or a boiling stick. Evaporate the diethyl ether solvent to about 25 mL volume by placing the beaker **in the hood** on a steam bath. Transfer the sample to a 50-mL beaker and continue to evaporate on the steam bath until an oily residue forms. Remove the beaker from the steam bath. Cool it in an ice bath for one min. Add 5 mL of petroleum ether and transfer the liquid (without the boiling chip) to a 10-mL volumetric flask. Add sufficient petroleum ether to bring it to volume.

5. Add 5 mL of extracts in petroleum ether to your chromatographic column. By opening the stopcock drain the sample into your column, but **take care not to let the column run dry**. (Always have about 0.5 mL liquid on top of the column.) Continue to add solvent to the top of your column. Collect the eluents in a beaker. First you will see the orange colored carotenes moving down the column. With the aid of a UV lamp, you can also observe a fluorescent band following the carotenes. This fluorescent band contains your vitamin A. Allow all the orange color band to move to the bottom of your column and into the collecting beaker. When the fluorescent band reaches the bottom of the column, close the stopcock. By adding petroleum ether on the top of the column continuously, elute the fluorescent band from the column into a 25-mL graduated cylinder. Continue the elution until all the fluorescent band has been drained into the graduated cylinder. Close the the stopcock, and record the volume of the eluate in the graduated cylinder on your Report Sheet (4). Add the vitamin A in the petroleum ether eluate to a dry and clean 50-mL beaker. Evaporate the solvent in the hood on a steam bath. The evaporation is complete when an oily residue appears in the beaker. Add 5 mL of absolute ethanol to the beaker. Transfer the sample into a 10-mL volumetric flask and bring it to volume by adding absolute ethanol.

6. Place your sample in a 1-cm length quartz spectroscopic cell. The control (blank) spectroscopic cell should contain absolute ethanol. Read the absorbance of your sample against the blank, according to the instructions of your spectrophotometer, at 325 nm. Record the absorption at 325 nm on your Report Sheet (5).

7. Calculate the amount of margarine that yielded the vitamin A in the petroleum ether eluate. Remember that you added only half (5 mL) of the extract to the column. Report this value on your Report Sheet (6). Calculate the grams of margarine that would have yielded the vitamin A in 1 mL absolute ethanol by dividing (6)/10 mL. Record it on your Report Sheet (7). Calculate the vitamin A in a lb of margarine by using the following formula:

$$\mu g \text{ Vitamin A/lb of margarine} = \text{Absorption} \times 5.5 \times [454/(7)].$$

Record your value on the Report Sheet (8).

Chemicals and Equipment

1. Separatory funnel (500-mL)
2. Buret (25-mL)
3. UV lamp
4. Spectrophotometer (near uv)
5. Margarine
6. Petroleum ether (30°–60°C)
7. 95% ethanol
8. Absolute ethanol
9. Diethyl ether
10. Glass wool
11. Alkaline aluminum oxide (alumina)

Experiment 48

PRE-LAB QUESTIONS

1. The structure of β-carotene is given below. What is the difference between β-carotene and vitamin A?

β-carotene

2. Why should you not have a lit Bunsen burner in the lab while you are working with diethyl ether?

3. In the saponification process, you hydrolyzed fat in the presence of KOH. Write an equation of a reaction in which the fat is hydrolyzed in the presence of HCl. What is the difference between the products of the saponification and that of the acid hydrolysis?

Experiment 48

REPORT SHEET

1. Weight of watch glass   _____ g

2. Weight of watch glass + margarine   _____ g

3. Weight of margarine: (2) − (1)   _____ g

4. Volume of petroleum ether eluate   _____ mL

5. Absorption at 325 nm   _____

6. Grams margarine in 1 mL of petroleum ether eluate:
   $2 \times [(3)/(4)]$ _____ g

7. Grams of margarine in 1 mL of absolute ethanol: (6)/10 mL   _____ g

8. μg vitamin A/lb margarine:
   $(5) \times 5.5 \times [454/(7)]$ _____

POST-LAB QUESTIONS

1. Looking at the structure of vitamin A, explain why is it not soluble in water.

2. In your separation scheme the fatty acids of the margarine ended up in the aqueous wash, which was discarded. Could you have removed the fatty acids, similarly, if, instead of saponification, you used acid hydrolysis? Explain.

3. Suppose that your petroleum ether eluate had a faint orange color. Vitamin A is colorless. What could have caused the coloration?

4. The label on a commercial margarine sample states that 1 g of it contains 15% of the daily recommended allowance. Was your sample richer or poorer in vitamin A than the above mentioned commercial sample?

Experiment 49

● Measurement of sulfur dioxide
preservative in foods

Many foods contain preservatives that prolong the shelf life and/or combat infestations by insects and microorganisms. Sulfur dioxide is probably one of the oldest
preservatives. For centuries, people found that if the summer harvest of fruits is
to be preserved and stored for the winter months, a drying process can accomplish
the task. Raisins, dates, dried apricots, and prunes are still sun-dried in many
countries. The drying process increases the sugar concentration in such dried
fruits and bacteria and most other microorganisms cannot use the dried fruit as a
carbohydrate source because of the hypertonic (hyperosmotic) conditions.

It was found, by trial and error, that when storage areas are fumigated by
burning sulfur, the dried fruits have a longer shelf life and are mostly void of
insect and mold infestations as well. Sulfur dioxide, the product of the sulfur
fumigation is still used today as a preservative. It is harmless when consumed in
small quantities. The Food and Drug Administration requires the listing of sulfur
dioxide on the labels of food products. You may see such listings on almost every
bottle of wine, on packaged dried fruits, and in some processed meat products.

In the present experiment, we use a colorimetric technique to analyze the SO_2
content of raisins.

Objectives

1. To learn the use of standard curves for analysis.
2. To determine the SO_2 content by colorimetric analysis.

Procedure

Part A: Preparation of Sample

1. Weigh 10 g of raisins and transfer it to a blender containing 290 mL of distilled
   water. Record the weight of the raisins on your Report Sheet (1). This amount
   will be sufficient for a class of 25 students. Cover and blend for 2 min.

2. Each student should prepare two 100-mL volumetric flasks. One will be labeled
   "blank," the other "sample." To each flask add 1 mL of 0.5 N NaOH solution. To
   the "blank" add 10 mL of distilled water. To the "sample" flask add 10 mL of

the raisin extract. *(Use a volumetric pipet to withdraw 10 mL from the bottom portion of the blender.)* Mix both solutions by swirling them for 15-30 sec.

3. Add to each volumetric flask 1 mL of 0.5 N H_2SO_4 solution and 20 mL of mercurate reagent. Add sufficient distilled water to bring both flasks to 100 mL volume.

CAUTION!

Use polyethylene gloves to protect your skin from touching mercurate reagent. Mercurate reagent is toxic and if spills occur you should wash them immediately with copious amounts of water.

Part B: Standard Curve

1. Label five 100-mL volumetric flasks as no. 1, 2, 3, 4, and 5. To each flask add 5 mL of mercurate reagent. Add standard sulfur dioxide solutions to the flasks as labeled (i.e., 1 mL to flask no.1, 2 mL to flask no. 2, etc.). Bring each volumetric flask to 100 mL volume with distilled water.

2. Label five clean and dry test tubes as no. 1, 2, 3, 4, and 5. Transfer to each 5 mL portions of the corresponding samples (i.e, to test tube no. 1 from volumetric flask no. 1, etc.). Add 2.5 mL of rosaniline reagent to each test tube. Add also 5 mL of 0.015% formaldehyde solution to each test tube. Cork the test tubes. Mix the contents by shaking and swirling. Let it stand for 30 min. at room temperature. The SO_2 concentrations in your test tubes will be as follows:

Test tube no.	Concentration in μg/mL
1	10.0
2	20.0
3	30.0
4	40.0
5	50.0

3. At the end of 30 min., read the intensity of the color in each test tube in a spectrophotometer. Your instructor will demonstrate the use of the spectrophotometer. *(For reading absorbance values in spectrophotometers, read the details in Experiment 46, p. 488.)*

4. Construct your standard curve by plotting the absorbance readings on the y-axis and the corresponding concentration readings on the x-axis. Connect the points with the best straight line.

Part C: Measurement of SO$_2$ Content of Raisins

1. Add 2.5 mL of rosaniline reagent to each of four test tubes labeled "blank," no. 1, no. 2, and no. 3. To these test tubes, add the "sample" and "blank" you prepared in Part A using the following scheme:

Test tube	"sample" mL	"blank" mL
Blank	0	2.5
no. 1	0.5	2.0
no. 2	1.0	1.5
no. 3	2.0	0.5

2. To each test tube, add 5 mL of the formaldehyde reagent. Mix by swirling and let it stand at room temperature for 30 min.

3. Read the absorbance of the solutions in your four test tubes and record it on your Report Sheet (2).

4. The "net" absorbance is the absorbance of the "sample" minus the absorbance of the "blank." Record the "net" absorbance on your Report Sheet (3). Using the standard curve obtained in Part B, record on your Report Sheet the SO$_2$ content (in μg/mL) of your test tubes that correspond to your "net" absorbance values (4).

5. Calculate the SO$_2$ content of your raisin sample in each test tube. Record it on your Report Sheet (5). Here is a sample calculation for test tube no. 2:

$$\frac{2.0 \text{ SO}_2 \text{ μg/mL solution from std.curve}}{1 \text{ mL sample/10 mL solution}} \times \frac{100 \text{ mL total sample}}{10 \text{ g dried fruit}} = 200 \text{ μg/g}$$

Average the three values obtained and record it on your Report Sheet (6).

Chemicals and Equipment

1. Raisins
2. Blender
3. 100-mL volumetric flasks
4. 10-mL pipet
5. 0.5 N NaOH solution
6. 0.5 N H$_2$SO$_4$ solution
7. 0.015% formaldehyde solution
8. Rosaniline reagent
9. Mercurate reagent
10. Sulfur dioxide stock solution
11. Spectrophotometers

Experiment 49

PRE-LAB QUESTIONS

1. What is a standard curve?

2. Write the balanced equation that shows that burning sulfur in air produces sulfur dioxide.

3. What is the difference between "blank" and "sample" solutions?

4. Your "standard sulfur dioxide stock solution" is actually sulfur dioxide gas dissolved in water. What is the product of this reaction? Write a balanced equation showing reactants and product.

Experiment 49

REPORT SHEET

1. Weight of raisins _____ g

Standard curve

Test tube no.	Concentration of SO_2 in $\mu g/mL$	Absorbance
1	10.0	
2	20.0	
3	30.0	
4	40.0	
5	50.0	

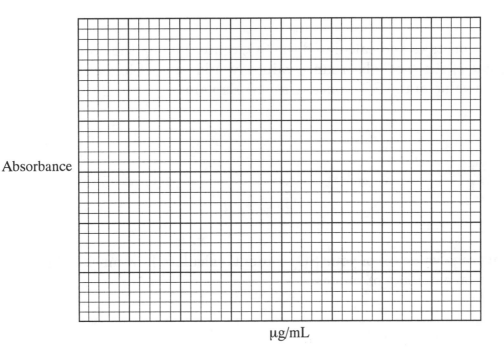

Absorbance

$\mu g/mL$

SO$_2$ content

Test tube	"sample" mL	Absorbance (2)	"Net" absorbance (3)	SO$_2$ μg/mL (4)	SO$_2$ μg/g (5)
Blank	0				
no. 1	0.5				
no. 2	1.0				
no. 3	2.0				

Average SO$_2$ content μg/g raisin _____(6)

POST-LAB QUESTIONS

1. Is your standard curve a straight line going through the origin?

2. If you made a mistake and calculated the SO$_2$ content using the absorbance values instead of the "net" absorbance values, would you get higher or lower than the true value of the SO$_2$ concentration? Explain.

3. From the average value (6) of your SO$_2$ content, calculate how much *sulfur* you ingest when you eat 100 g of raisins.

Background

The kidney is an important organ that filters materials from the blood that are harmful, or in excess, or both. These materials are excreted in the urine. A number of tests are routinely run in clinical laboratories on urine samples. These involve the measurements of glucose or reducing sugars, ketone bodies, albumin, specific gravity, and pH.

Normal urine contains little or no glucose or reducing sugars; the amount varies from 0.05 to 0.15%. Higher concentrations may occur if the diet contains a large amount of carbohydrates or if strenuous work was performed shortly before the test. Patients with diabetes or liver damage have chronically elevated glucose content in the urine. A semiquantitative test of glucose levels can be performed with the aid of test papers such as Clinistix. This is a quick test that uses a paper containing two enzymes—glucose oxidase and peroxidase. In the presence of glucose, the glucose oxidase catalyzes the formation of gluconic acid and hydrogen peroxide. The hydrogen peroxide is decomposed with the aid of peroxidase and yields atomic oxygen.

α-D-glucose   gluconic acid   hydrogen peroxide

In Clinistix, the atomic oxygen reacts with an indicator, o-tolidine, and produces a purple color. The intensity of the purple color is proportional to the glucose concentration.

$$H_2O_2 \xrightarrow{\text{peroxidase}} H_2O + [O]$$

o-tolidine

H_2N- ⬡ ⬡ $-NH_2 + [O] \longrightarrow$ purple oxidation product

(with CH$_3$ groups)

This test is specific for glucose only. No other reducing sugar will give positive results.

Normal urine contains no albumin or only a trace amount of it. In case of kidney failure or malfunction, the protein passes through the glomeruli and is not reabsorbed in the tubule. So, albumin and other proteins end up in the urine. The condition known as proteinuria may be symptomatic of kidney disease. The loss of albumin and other blood proteins will decrease the osmotic pressure of blood. This allows water to flow from the blood into the tissues, and creates swelling (edema). Renal malfunction is usually accompanied by swelling of the tissues. The Albustix test is based on the fact that a certain indicator at a certain pH changes its color in the presence of proteins.

Albustix contains the indicator, tetrabromophenol blue, in a citrate buffer at pH 3. At this pH, the indicator has a yellow color. In the presence of protein, the color changes to green. The higher the protein concentration, the greener the indicator will be. Therefore, the color produced by the Albustix can be used to estimate the concentration of protein in urine.

Three substances that are the products of fatty acid catabolism—acetoacetic acid, β-hydroxybutyric acid, and acetone—are commonly called ketone bodies. These are normally present in the blood in small amounts and can be used as an energy source by the cells. Therefore, no ketone bodies will normally be found in the urine. However, when fats are the only energy source, excess production of ketone bodies will occur. They will be filtered out by the kidney and appear in the urine. Such abnormal conditions of high fat catabolism take place during starvation or in diabetes mellitus when glucose, although abundant, cannot pass through the cell membranes to be utilized inside where it is needed. Acetoacetic acid (CH_3COCH_2COOH), and to a lesser extent acetone (CH_3COCH_3) and β-hydroxybutyric acid ($CH_3CHOHCH_2COOH$), react with sodium nitroprusside {$Na_2[Fe(CN)_5NO]$}·2 H_2O to give a maroon-colored complex. In Ketostix, the test area contains sodium nitroprusside and a sodium phosphate buffer to provide the proper pH for the reaction. The addition of lactose to the mixture in the Ketostix enhances the development of the color.

Some infants are born with a genetic defect known as phenylketonuria (PKU). They lack the enzyme phenylalanine oxidase, which converts phenylalanine to tyrosine. Thus phenylalanine accumulates in the body and it is degraded to phenylpyruvate by transamination:

phenylalanine pyruvate phenylpyruvate alanine

Phenylpyruvate is excreted in the urine. Normal urine does not contain any phenylpyruvate. People suffering from PKU disease have varying amounts of phenylpyruvate in their urine. PKU disease causes severe mental retardation in infants if it is not treated immediately after birth, which is done by restricting the phenylalanine content of the diet. In many states, the law requires that every newborn be tested for phenylpyruvate in the urine. The test is based on the reaction of the iron(III) ion with the phenylpyruvate, producing a gray-green color. Phenistix strips which are coated with $Fe(NH_4)(SO_4)_2$ and a buffer can detect as little as 8 mg of phenylpyruvate in 100 mL of urine. Some drugs such as aspirin produce metabolites (salicylic acid) that are excreted in the urine and give color with an iron(III) ion. However, this produces a deep purple color and not the gray-green of PKU. The purple color that is given by the Phenistix can be used to diagnose an overdose of aspirin. Other drugs, such as phenylthiazine, in an overdose, give a gray-purple color with Phenistix. For PKU diagnosis only the appearance of the gray-green color means a positive test.

Urobilinogen and other bile pigments are normally minor components of urine (2 to 50 μg/100 mL). They are the products of hemoglobin breakdown. Bile pigments are usually excreted in the feces. In case of obstruction of the bile ducts (gallstones, obstructive jaundice), the normal excretion route through the small intestines is blocked and the excess bilirubin is filtered out of the blood by the kidneys and appears in the urine. Urobilistix is a test paper that can detect the presence of urobilinogen, because it is impregnated with p-dimethylaminobenzaldehyde. In strongly acidic media, this reagent gives a yellow-brownish color with urobilinogen.

p-dimethylaminobenzaldehyde

The specific gravity of normal urine may range from 1.008 to 1.030. After excessive fluid intake (like a beer party), the specific gravity may be on the low side; after heavy exercise and perspiration, it may be on the high side. High specific gravity indicates excessive dissolved solutes in the urine.

The pH of normal urine can vary between 4.7 through 8.0. The usual value is about 6.0. High protein diets and fever can lower the pH of urine. In severe acidosis, the pH may be as low as 4.0. Vomiting and respiratory or metabolic alkalosis can raise the pH above 8.0.

Objectives

1. To perform quick routine analytical tests on urine samples.
2. To compare results obtained on "normal" and "pathological" urine samples.

Procedure

Each student must analyze her (his) own urine. A fresh urine sample will be collected prior to the laboratory period. The stockroom will provide paper cups for sample collection. *While handling body fluids, such as urine, plastic gloves should be worn.* The used body fluid will be collected in a special jar and disposed of collectively. The plastic gloves worn will be collected and autoclaved before disposal. In addition, the stockroom will provide one "normal" and two "pathological" urine samples.

Place 5 mL of the urine sample from each source into four different test tubes. These will be tested with the different test papers.

Glucose Test

For the glucose test, use for comparison two test tubes, each half-filled, one with 0.25% and the other with 1% glucose solutions. Take six strips of Clinistix from the bottle. Replace the cap immediately. Dip the test area of the Clinistix into one of the samples. Tap the edge of the strip against a clean, dry surface to remove the excess urine. Compare the test area of the strip to the color chart supplied on the bottle exactly 10 sec. after the wetting. Do not read the color changes that occur after 10 sec. Record your observation: The "light" on the color chart means 0.25%, or less, glucose; the "medium" means 0.4%; and the "dark" means 0.5%, or more. Repeat the test with the other five samples.

Protein Test

For the protein test, take four Albustix strips, one for each of the four urine samples, from the bottle. Replace the cap immediately. Dip the Albustix strips into the test solutions, making certain that the reagent area of the strip is completely immersed. Tap the edge of the strip against a clean, dry surface to remove the excess urine. Compare the color of the Albustix test area with the color chart supplied on the bottle. The time of the comparison is not critical; you can do it immediately or any time within 1 min. after wetting. Read the color from yellow (negative) to different shades of green, indicating trace amounts of albumin, up to 0.1%.

Ketone Bodies

To measure the ketone bodies' concentration in the urine, take four Ketostix strips from the bottle. Replace the cap immediately. Dip a Ketostix in each of the urine samples and remove the strips at once. Tap them against a clean, dry surface to remove the excess urine. Compare the color of the test area of the strips to the color chart supplied on the bottle. Read the colors exactly 15 sec. after wetting. A buff pink color indicates the absence of ketone bodies. A progression to a maroon color indicates increasing concentration of ketone bodies from 50 to 160 mg/L of urine.

Test for PKU

To test for PKU disease, use four Phenistix strips, one for each urine sample. Immerse the test area of the strips into the urine samples and remove them immediately. Remove the excess urine by tapping the strips against a clean, dry surface. Read the color after 30 sec. of wetting, and match them against the color charts provided on the bottle. Record your estimated phenylpyruvate content: negative or 0.015 to 0.1%.

Urobilinogen

To measure the urobilinogen content of the urine samples, use Urobilistix strips, one for each urine sample. Dip them into the samples and remove the excess urine by tapping the strips against a clean, dry surface. Sixty sec. after wetting, compare the color of the test area of the strips to the color chart provided on the bottle. Estimate the urobilinogen content and record it in Ehrlich units.

pH in Urine

To measure the pH of each urine sample, use pH indicator paper, such as pHydrion test paper within the pH range of 3.0 to 9.0.

For the preceding tests, you may use a multipurpose strip such as Labstix that contains test areas for all these tests, except for the test for phenylpyruvate, on one strip. The individual test areas are separated from each other and clearly marked. The time requirements to read the colors are also indicated on the chart. The results of five tests can be read in 1 min.

Specific Gravity of Urine

The specific gravity of your urine samples will be measured with the aid of a hydrometer (urinometer; see Fig. 50.1). Place the bulb in a cylinder. Add sufficient urine to the cylinder to make the bulb float. Read the specific gravity of the sample from the stem of the hydrometer where the meniscus of the urine intersects the calibration lines. Be sure the hydrometer is freely floating and does not touch the walls of the cylinder. In order to use as little urine as possible, the instructor may read the normal and 2 pathological urine samples for the whole class. If so, you will measure the specific gravity of your own urine sample only.

Figure 50.1

A urinometer.

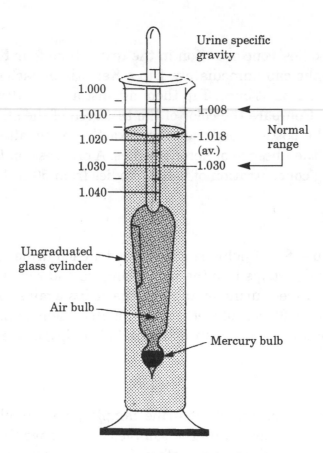

Urine specific gravity

1.000

1.010 ———— 1.008

1.020 ———— 1.018 (av.)

1.030 ———— 1.030

1.040 ————

Normal range

Ungraduated glass cylinder

Air bulb

Mercury bulb

Chemicals and Equipment

1. 0.25 and 1% glucose solutions
2. Clinistix
3. Albustix
4. Ketostix
5. Urobilistix
6. pH paper in the 3.0 to 9.0 range
7. A multipurpose Labstix (instead of these test papers)
8. Phenistix
9. 3 M NaOH
10. Hydrometer (urinometer)

NAME _____   SECTION _____   DATE _____

PARTNER _____   GRADE _____

Experiment 50

PRE-LAB QUESTIONS

1. Why should you wear gloves when dealing with urine samples?

2. In what tests do we use the following reagents?

    a. Fe^{3+}

    b.

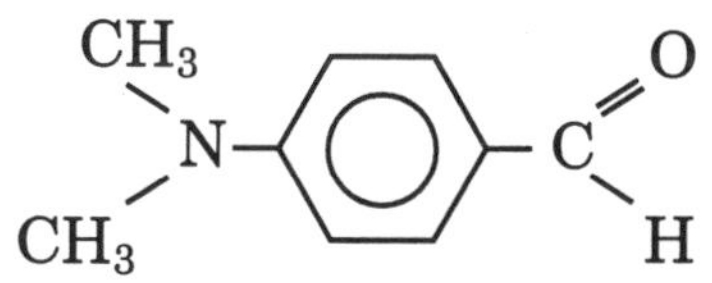

    c.

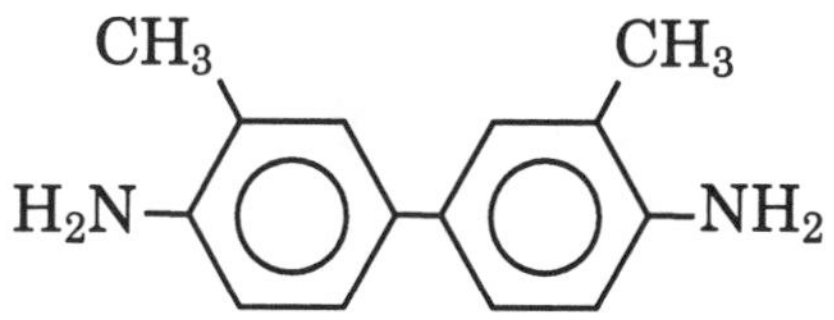

3. In determining glucose concentration in urine you use a Clinistix. Yet, the purple color of a positive test is the result of the interaction of *o*-toluidine indicator with atomic oxygen. How is atomic oxygen related to glucose?

4. A patient's urine shows a high specific gravity, 1.04. The pH is 7.8, and the Phenistix test indicates a purple color that is not characteristic of PKU. The patient has had a high fever for a few days and has been given aspirin. Do these tests indicate any specific disease, or are they symptomatic of recovering from a high fever? Explain.

NAME ______________________ SECTION __________ DATE __________

PARTNER ______________________ GRADE ______________________

Experiment 50

REPORT SHEET

Urine Samples

Test	Normal	Pathological A	Pathological B	Your own	Remarks
Glucose					
Ketone bodies					
Albumin					
Urobilinogen					
pH					
Phenylpyruvate					
Specific gravity					

POST-LAB QUESTIONS

1. Did you find any indication that your urine is not normal? If so, what may be the reason?

2. Why is the phenylpyruvate test mandatory with newborns in many states?

3. Urine samples came from three different persons. Each had a different diet for a week. One lived on a high protein diet; the second, on a high carbohydrate diet; and the third lived on a high fat (no carbohydrate) diet. How would their urine analyses differ from each other?

4. A patient's urine was tested with Clinistix, and the color was read 60 sec. after wetting the strip. It showed 1.0% glucose in the urine. Is the patient diabetic? Explain.

Appendix

List of Apparatus and Equipment in Student's Locker

Amount and description

(1) Beaker, 50 mL
(1) Beaker, 100 mL
(1) Beaker, 250 mL
(1) Beaker, 400 mL
(1) Beaker, 600 mL
(1) Clamp, test tube
(1) Cylinder, graduated by 0.1 mL, 10 mL
(1) Cylinder, graduated by 1 mL, 100 mL
(1) Dropper, medicine with rubber bulb
(1) Evaporating dish
(1) Flask, Erlenmeyer, 125 mL
(1) Flask, Erlenmeyer, 250 mL
(1) Flask, Erlenmeyer, 500 mL
(1) File, triangular
(1) Forceps
(1) Funnel, short stem
(1) Gauze, wire
(1) Spatula, stainless steel
(1) Sponge
(1) Striker (or box of matches)
(6) Test tubes, approximately 15 x 150 mm
(1) Test tube brush
(1) Thermometer, 150°C
(1) Tongs, crucible
(1) Wash bottle, plastic
(1) Watch glass

Appendix 2

List of Common Equipment and Materials in the Laboratory

Each laboratory should be equipped with hoods and safety-related items such as fire extinguisher, fire blankets, safety shower, and eye wash fountain. The equipment and materials listed here for 25 students should be made available in each laboratory.

Acid tray
Aspirators (splashgun type) on sink faucet
Balances, triple beam (or centigram) or top-loading
Barometer
Clamps, extension
Clamps, thermometer
Clamps, utility
Containers for solid chemical waste disposal
Containers for liquid organic waste disposal
Corks
Detergent for washing glass ware
Drying Oven
Filter paper
Glass rods, 4 and 6 mm OD
Glass tubing, 6 and 8 mm OD
Glycerol (glycerine) in dropper bottles
Hot plates
Ice maker
Paper towel dispensers
Pasteur pipets
Rings, support, iron, 76mm OD
Ring stands
Rubber tubing, pressure
Rubber tubing, latex (0.25 in. OD)
Water, deionized or distilled
Weighing dishes, polystyrene, disposable, 73 × 73 × 25 mm
Weighing paper

Appendix 3

Special Equipment and Chemicals

In the instructions below every time a solution is to be made up in "water" you must use *distilled water*.

Experiment 1 **Laboratory techniques: use of the laboratory gas burner; basic glassworking**

Special Equipment

(25)	Wing tops
(25)	Crucible tongs
(25)	Wire gauze
(50)	Glass tubing (6-mm OD), 25 cm segments
(25)	Solid glass rod, 25 cm segments

Experiment 2 **Laboratory measurements**

Special Equipment

(25)	50-mL graduated beakers
(25)	50-mL graduated Erlenmeyer flasks
(25)	Metersticks or rulers, with both English and metric scale
(25)	Hot plates
(25)	Centigram balances
(5)	Platform balances
(2)	Top-loading balances

Experiment 3 **Density determination**

Special Equipment

(4)	250-mL beakers (labeled for unknown metals)
(25)	Magnetic stir-bars, small (1/2 × 5/16 in.; must be small enough to fit into a 50-mL graduated cylinder)
(25)	Magnetic stirrers
(25)	Solid wood blocks, rectangular or cubic
(25)	Spectroline pipet fillers
(25)	10-mL volumetric pipets
(50)	Polyethylene plastic chips, 2–4 mm dia.

Chemicals

(1 L)	Acetone, reagent

Unknown metals

(100 g)	Aluminum, pellets or rod
(100 g)	Tin, pellets or cut strips

(100 g)	Zinc, pellets
(100 g)	Lead, shot

Experiment 4   The separation of the components of a mixture

Special Equipment

(2)	Top-loading balances (weigh to 0.001 g)
[(15)	Centigram balances (weigh to 0.01 g) as an alternative]
(25)	Evaporating dishes, porcelain, 6 cm dia.
(25)	Rubber policeman
(1 box)	Filter paper, 15 cm, fast flow
(25)	Mortar and pestle

Chemicals

(30 g)	Unknown mixture: mix 3.0 g naphthalene (10%), 15 g sodium chloride (50%), 12 g sea sand (40%)
(1 jar)	Boiling stones (silicon carbide chips if available)

Experiment 5   Resolution of a mixture by distillation

Special Equipment

(25)	Distillation kits with 19/22 standard taper joints: 100-mL round bottom flasks (2); distilling head; thermometer adapter; 110°C thermometer; condenser; vacuum adapter
(25)	Nickel wires

Chemicals

(1 jar)	Boiling chips
(1 jar)	Silicone grease
(2 L)	5% sodium chloride solution: dissolve 100 g NaCl in enough water to make 2 L
(100 mL)	0.5 M silver nitrate: dissolve 8.5 g $AgNO_3$ in enough water to make 100 mL solution
(50 mL)	Concentrated nitric acid, HNO_3

Experiment 6   Determination of the formula of a metal oxide

Special Equipment

(25)	Porcelain crucibles and covers
(25)	Clay triangles
(25)	Crucible tongs
(25)	Eye droppers

Chemicals

(25)	Magnesium ribbon, 12 cm strip
(500 mL)	6 M HCl: take 300 mL concentrated HCl and add to enough cold water to make 500 mL of solution. **Wear a face mask, rubber gloves and a rubber apron during the preparation. Do in the hood.**

Experiment 7   Classes of chemical reactions

Special Equipment

(1 box)   Wood splints

Chemicals

(25 pieces) Aluminum foil (2 × 0.5 in. each)
(25 pieces) Aluminum wire (1 cm each)
(25 pieces) Copper foil (2 × 0.5 in. each)
(25)   Pre 1982 copper penny (optional)
(20 g)   Ammonium carbonate, $(NH_4)_2CO_3$
(20 g)   Potassium iodide, KI
(40 g)   Potassium iodate, KIO_3
(20 g)   Calcium turnings, Ca
(20 g)   Iron filings, Fe
(20 g)   Mossy zinc, Zn
(20 g)   Lead shot, Pb

All the following solutions should be placed in dropper bottles:

(100 mL)   3 M hydrochloric acid, 3 M HCl: 20 mL concentrated HCl
            diluted with water to 100 mL
(100 mL)   6 M hydrochloric acid, 6 M HCl: 40 mL concentrated HCl
            diluted with water to 100 mL
(100 mL)   3 M sulfuric acid, 3 M H_2SO_4: 16.5 mL concentrated
            H_2SO_4 added to 60 mL ice cold water; stir slowly and
            dilute with water to 100 mL
(100 mL)   3 M sodium hydroxide, 3 M NaOH: dissolve 12 g NaOH
            per 100 mL solution

**In preparing the above solutions, rubber gloves, a rubber
apron, and a face shield should be worn. Do all preparations
in the hood.**

(100 mL)   0.1 M silver nitrate, 0.1 M $AgNO_3$: dissolve 1.7 g $AgNO_3$
            per 100 mL solution
(100 mL)   0.1 M sodium chloride, 0.1 M NaCl: dissolve 0.58 g NaCl
            per 100 mL solution
(100 mL)   0.1 M sodium nitrate, 0.1 M $NaNO_3$: dissolve 0.9 g
            $NaNO_3$ per 100 mL solution
(100 mL)   0.1 M sodium carbonate, 0.1 M Na_2CO_3: dissolve 1.2 g
            $Na_2CO_3 \cdot H_2O$ per 100 mL solution
(100 mL)   0.1 M potassium nitrate, 0.1 M KNO_3: dissolve 1.01 g
            KNO_3 per 100 mL solution
(100 mL)   0.1 M potassium chromate, 0.1 M K_2CrO_4: dissolve 2.0 g
            K_2CrO_4 per 100 mL solution
(100 mL)   0.1 M barium chloride, 0.1 M $BaCl_2$: dissolve 2.07 g
            $BaCl_2$ per 100 mL solution
(100 mL)   0.1 M copper(II) nitrate, 0.1 M $Cu(NO_3)_2$: dissolve 1.88 g
            $Cu(NO_3)_2$ per 100 mL solution
(100 mL)   0.1 M copper(II) chloride, 0.1 M $CuCl_2$: dissolve 1.34 g
            $CuCl_2$ per 100 mL solution
(100 mL)   0.1 M lead(II) nitrate, 0.1 M $Pb(NO_3)_2$: dissolve 3.3 g
            $Pb(NO_3)_2$ per 100 mL solution
(100 mL)   0.1 M iron(III) nitrate, 0.1 M $Fe(NO_3)_3$: dissolve 4.0 g
            $Fe(NO_3)_3 \cdot 9H_2O$ per 100 mL solution

Experiment 8   Chemical properties of consumer products

Special Equipment

(1 roll)	Copper wire
(2 vials)	Litmus papers, blue
(2 vials)	Litmus papers, red

Chemicals

All solutions should be placed in dropper bottles. In preparing all acid and base solutions, observe personal safety practices. Use a face shield, rubber gloves, and a rubber apron. Do preparations in the hood

(100 mL)   Commercial ammonia solution, NH_3 (3M ammonia solution can be substituted: 10 mL 30% NH_3 solution diluted to 100 mL with water)

(100 mL)   Commercial bleach containing sodium hypochlorite, NaOCl

(100 g)   Commercial baking soda, $NaHCO_3$

(100 g)   Commercial detergent containing sodium phosphate, Na_3PO_4

(100 g)   Garden fertilizer containing ammonium phosphate, $(NH_4)_3PO_4$

(100 g)   Epsom salt, $MgSO_4 \cdot 7H_2O$

(100 g)   Table salt, sodium chloride, NaCl

(100 g)   Ammonium chloride, NH_4Cl

(100 g)   Potassium iodide, KI

(100 mL)   Ammonium molybdate reagent: dissolve 1.6 g $(NH_4)_6 Mo_7O_{24} \cdot 4H_2O$ in 6 mL water. Put in ice bath.

Pour slowly 4 mL concentrated H_2SO_4 into solution and stir slowly. After cooling to room temperature, dilute to 100 mL with water.

(100 mL)   1 M barium chloride, 1 M $BaCl_2$: dissolve 24.4 g $BaCl_2 \cdot 2H_2O$ per 100 mL solution

(100 mL)   1% barium hydroxide, 1% $Ba(OH)_2$: dissolve 1 g $Ba(OH)_2 \cdot 8H_2O$ per 100 mL solution

(200 mL)   3 M nitric acid, 3 M HNO_3: 40 mL concentrated HNO_3 diluted to 200 mL solution with water

(200 mL)   6 M nitric acid, 6 M HNO_3: 80 mL concentrated HNO_3 diluted to 200 mL solution with water.

(100 mL)   0.1 M silver nitrate, 0.1 M $AgNO_3$: dissolve 1.7 g $AgNO_3$ per 100 mL solution

(100 mL)   6 M sodium hydroxide, 6 M NaOH: dissolve 24 g NaOH per 100 mL solution

(100 mL)   0.1 M sodium phosphate, 0.1 M Na_3PO_4: dissolve 3.7 g $Na_3PO_4 \cdot 12H_2O$ per 100 mL solution

(100 mL)   6 M sulfuric acid, 6 M H_2SO_4: pour 33 mL concentrated H_2SO_4 into 50 mL ice cold water. Stir slowly. Dilute to 100 mL volume.

(100 mL)   0.01% *p*-nitrobenzene azoresorcinol: dissolve 0.01 g *p*-nitrobenzene azoresorcinol in 100 mL 0.025 M NaOH

(100 mL)   Hexane, $CH_3CH_2CH_2CH_2CH_2CH_3$

Experiment 9   Calorimetry: the determination of the specific heat of a metal

Special Equipment

(25)	Wire loops for stirring
(25)	Rubber rings (cut from latex tubing)
(50)	Styrofoam cups (8 oz.)
(25)	Lids for styrofoam cups
(25)	Thermometers, 110°C
(25)	Stopwatches

Chemicals

(1 Kg)	Lead shot, no. 8
(1 Kg)	Aluminum metal, turnings or wire
(1 Kg)	Iron metal
(1 Kg)	Tin metal, granular (mossy)
(1 Kg)	Zinc shot

Experiment 10   Boyle's Law: the pressure-volume relationship of a gas

Special Equipment

(25)   Boyle's Law apparatus: the apparatus can be constructed as follows. A piece of glass tubing, 30 cm in length, 3-mm OD, is sealed at one end. A Pasteur disposable pipet is drawn out to form a capillary; the capillary needs to be only small enough to be inserted into the 3-mm OD glass tubing and long enough to reach half the length of the tubing (approx. 15 cm). Mercury is transferred with the pipet into the tubing; enough mercury is placed in the tube to give a 10 cm long column. The tube is attached to a 1-ft. ruler by means of rubber bands. The ruler should read with both English (to nearest 1/16 in.) and metric (to nearest mm) scales.

(25)   30°-60°-90° plastic triangles

Experiment 11   Charles' Law: the volume-temperature relationship of a gas

Special Equipment

(25)	Bunsen burner (or hot plate)
(50)	250-mL Erlenmeyer flask
(50)	800-mL beaker
(50)	Clamp
(25)	Glass tubing (5 to 8 cm length; 7-mm OD)
(25)	Marking pencil
(25)	One-hole rubber stopper (size no. 6)
(25)	Pre-made stopper assembly for 250-mL Erlenmeyer flask (optional alternative)
(25)	Ring stand
(25)	Ring support
(25)	Rubber tubing, latex (2-ft. length)
(25)	Thermometer, 110°C
(25)	Wire gauze

Chemicals

(1 jar)   Boiling stones

Experiment 12   Properties of gases: determination of the molecular weight of a volatile liquid

Special Equipment

(25)	Aluminum foil, 2.5 × 2.5 in.
(25)	Aluminum foil, 3 × 3 in.
(1 roll)	Copper wire
(25)	Beaker tongs
(25)	Crucible tongs
(25)	Rubber bands
(25)	Hot plates
(25)	Lead sinkers

Chemicals

(1 jar)   Boiling chips

The following liquids should be placed in dropper bottles.

(100 mL)	Pentane
(100 mL)	Acetone
(100 mL)	Methanol
(100 mL)	Hexane
(100 mL)	Ethanol
(100 mL)	2-Propanol

Experiment 13   Physical properties of chemicals: melting point, sublimation and boiling point

Special Equipment

(1 roll)	Aluminum foil
(1 bottle)	Boiling chips
(1)	Commercial melting point apparatus (if available)
(25)	Glass tubing, 20 cm segments
(25)	Hot plates
(100)	Melting point capillary tubes
(50)	Rubber rings (cut 0.25-in. rubber tubing into narrow segments)
(25)	Thermometer clamps
(25)	Thiele tube melting point apparatus

Chemicals

(20 g)	Acetamide
(20 g)	Acetanilide
(20 g)	Benzophenone
(20 g)	Benzoic acid
(20 g)	Biphenyl
(20 g)	Lauric acid
(20 g)	Naphthalene, pure
(50 g)	Naphthalene, impure: mix 47.5 g (95%) naphthalene and 2.5 g (5%) charcoal powder.
(20 g)	Stearic acid

The following liquids should be placed in dropper bottles.

(200 mL) Acetone
(200 mL) Cyclohexane
(200 mL) Ethyl acetate
(200 mL) Hexane
(200 mL) 2-Methyl-1-propanol (isopropyl alcohol)
(200 mL) Methanol (methyl alcohol)
(200 mL) 1-Propanol

Experiment 14 Entropy, a measure of disorder

Special Equipment

(12) Strips, 10 x 2 cm of polypropylene sheet; amorphous, high clarity sheets can be bought in any photographic store and are sold to protect photos. They may be 0.006 inch or 0.15 mm thick.
(25) Rubber bands, 3 mm width $\times$ 90 mm length
(25) Weights, approximately 300 g
(50) Bulldog clips
(50) Large paper clips
(5) Heat guns to be shared by the class

Experiment 15 Solubility and solutions

Special Equipment

(12) Electrical conductivity apparatus (one for each pair of students)
(25) Beaker tongs
(25) Hot plates

Chemicals

(2 lb) Granulated table sugar, sucrose
(10 g) Table salt, NaCl
(10 g) Naphthalene
(10 g) Iodine
(500 mL) Ethanol (ethyl alcohol)
(500 mL) Acetone
(500 mL) Petroleum ether (b.p. 30-60°C)
(500 mL) 1 M NaCl: dissolve 29 g of NaCl in water and bring to 500 mL volume
(500 mL) 0.1 M NaCl: take 50 mL of 1 M NaCl and add enough water to make 500 mL
(500 mL) 1 M sucrose: dissolve 166 g sucrose in water and bring to 500 mL volume
(500 mL) 0.1 M sucrose: take 50 mL of 1 M sucrose and add enough water to make 500 mL
(500 mL) 1 M HCl: add 41.5 mL concentrated HCl to 200 mL of ice water; add water to bring to 500 mL volume. **Use a face shield, rubber gloves, and a rubber apron during the preparation. Do in the hood.**
(500 mL) 0.1 M HCl: add 50 mL of 1 M HCl to enough water to make 500 mL **(use the same precautions as in the above preparation)**

| (500 mL) | Glacial acetic acid |
| (500 mL) | 0.1 M acetic acid: take 3 mL glacial acetic acid and add water to bring to 500 mL volume |

Experiment 16   Water of hydration

Special Equipment

(25)	Crucibles and covers
(25)	Clay triangles
(25)	Crucible tongs
(25)	Ring stands
(25g)	Calcium chloride, anhydrous, $CaCl_2$
(100g)	Copper(II) sulfate pentahydrate, $CuSO_4 \cdot 5H_2O$

Experiment 17   Colligative properties: freezing point depression and osmotic pressure

Special Equipment

(150)	Capillary melting point tubes
(50)	Narrow rubber bands
(25)	Test tubes (25 x 200 mm)
(25)	Hot plates
(5)	Student microscopes
(250)	Microscope slides
(250)	Cover slides
(5)	Razor blades or dissecting knives
(3)	Mortars
(3)	Pestles

Chemicals

(5 g)	Lauric acid
(5 g)	Benzoic acid
(100 mL)	0.1 M glucose solution: dissolve 1.8 g glucose in water and dilute to 100 mL solution
(100 mL)	0.5 M glucose solution: dissolve 9.0 g glucose in water and dilute to 100 mL solution
(250 mL)	0.89% NaCl solution: dissolve 2.23 g NaCl in water and dilute to 250 mL solution
(250 mL)	3.0% NaCl solution: dissolve 7.5 g NaCl in distilled water and dilute to 250 mL solution
(2 each)	Fresh carrot, scallion, celery
(50 mL)	Fresh (not more than 1 week old) whole bovine blood. Obtainable from a slaughterhouse. *Refrigerate.* **Even though bovine blood is not a source of HIV or hepatitis virus, prudence requires handling of the blood samples with care, using plastic gloves. The blood samples after the experiment should be collected in special jar. Both the blood and the gloves should be autoclaved before disposal.**

Experiment 18   Kinetic factors affecting rate of reactions

Special Equipment

(5)	Mortars
(5)	Pestles
(25)	10-mL graduated pipets
(25)	5-mL volumetric pipets

Chemicals

Solutions should be put into dropper bottles. In preparing the solutions, wear a face shield, rubber gloves and a rubber apron. Do in the hood.

(100 mL)   3 M H_2SO_4; dissolve 16.8 mL concentrated H_2SO_4 (95%) in 60 mL ice cold water. Stir gently and bring to 100 mL volume.

(500 mL)   6M HCl: add 258 mL concentrated HCl (36%) to 200 mL ice cold water. Mix and bring it to 500 mL volume.

(100 mL)   2 M H_3PO_4: add 13.6 mL concentrated H_3PO_4 (85%) to 50 mL ice cold water. Mix and bring it to 100 mL volume.

(100 mL)   6 M HNO_3: add 39.0 mL concentrated HNO_3 (69%) to 50 mL ice cold water. Mix and bring it to 100 mL volume.

(100 mL)   6 M acetic acid: add 34.4 mL glacial acetic acid (99-100%) to 50 mL water. Mix and bring it to 100 mL volume.

(500 mL)   0.1 M KIO_3: **Caution!** *This solution must be fresh. Prepare it on the day of the experiment.* Dissolve 10.7 g KIO_3 in 500 mL water.

(250 mL)   4% starch indicator: add 10 g soluble starch to 50 mL cold water. Stir it to make a paste. Bring 200 mL water to a boil in a 500-mL beaker. Pour the starch paste into the boiling water. Stir and cool to room temperature.

(500 mL)   0.01 M $NaHSO_3$: dissolve 0.52 g $NaHSO_3$ in 100 mL water. Add slowly 2 mL concentrated sulfuric acid. Stir and bring it to 500 mL volume.

(250 mL)   3% hydrogen peroxide: take 25 mL concentrated H_2O_2 (30%) and bring it to 250 mL volume with water.

(150)	Mg ribbons, 1 cm long
(25)	Zn ribbons, 1 cm long
(25)	Cu ribbons, 1 cm long
(25 g)	MnO_2

Experiment 19   Law of chemical equilibrium and Le Chatelier's principle

Special Equipment

(2 rolls)	Litmus paper, blue
(2 rolls)	Litmus paper, red

Chemicals

(50 mL)   0.1 M copper(II) sulfate: dissolve 1.1 g $CuSO_4$ (or 1.3 g $CuSO_4 \cdot 5H_2O$) in 50 mL water

(50 mL)	1 M ammonia: dilute 3.3 mL concentrated NH_3 (28%) with water to 50 mL volume. **In the preparation wear a face shield, rubber gloves, and a rubber apron. Do in the hood.**
(25 mL)	Concentrated HCl (36%)
(100 mL)	1 M hydrochloric acid: add 8.6 mL concentrated HCl (36%) to 50 mL ice water; add enough water to bring to volume. **In the preparation wear a face shield, rubber gloves and a rubber apron. Do in the hood.**
(150 mL)	0.1 M phosphate buffer: dissolve 1.74 g K_2HPO_4 in 100 mL water. Also dissolve 1.36 g K_2HPO_4 in 100 mL water. Mix 100 mL K_2HPO_4 with 50 mL of K_2HPO_4 solution.
(100 mL)	0.1 M potassium thiocyanate: dissolve 0.97 g KSCN in 100 mL water
(100 mL)	0.1 M iron(III) chloride: dissolve 2.7 g $FeCl_3 \cdot 6H_2O$ (or 1.6 g $FeCl_3$) in 100 mL water
(100 mL)	Saturated saline solution: add 290 g NaCl to warm (60°C) water. Stir until dissolved. Cool to room temperature.
(50 mL)	1.0 M cobalt chloride: dissolve 11.9 g $CoCl_2 \cdot 6H_2O$ in 50 mL water.

Experiment 20   pH and buffer solutions

Special Equipment

(5)	pH meters
(12 rolls)	pHydron paper (pH range 0 to 12)
(5 boxes)	Kimwipes
(5)	Wash bottles
(100)	10-mL graduate pipets
(25)	Spot plates

Chemicals

(250 mL)	0.1 M acetic acid, 0.1 M CH_3COOH: dissolve 1.4 mL glacial acetic acid in water to make 250 mL volume
(500 mL)	0.1 M sodium acetate, 0.1 M CH_3COONa: dissolve 6.8 g $CH_3COONa \cdot 3H_2O$ in water to make 500 mL volume
(1 L)	0.1 M hydrochloric acid, 0.1 M HCl: add 8.3 mL concentrated HCl (36%) to 100 mL ice water with stirring; dilute with water to 1 L. **Prepare in the hood; wear a face shield, rubber gloves, and a rubber apron.**
(1 L)	0.1 M sodium bicarbonate, 0.1 M $NaHCO_3$: dissolve 8.2 g $NaHCO_3$ in water to make 1 L volume
(500 mL)	saturated carbonic acid, H_2CO_3: use a bottle of Club Soda or Seltzer water; these solutions are approximately 0.1 M carbonic acid.

The following solutions should be placed in dropper bottles:

(100 mL)	0.1 M HCl prepared above
(100 mL)	0.1 M ammonia, 0.1 M NH_3: dilute 0.7 mL concentrated NH_3 (28%) with water to make 100 mL volume
(100 mL)	0.1 M sodium hydroxide, 0.1 M NaOH: dissolve 0.4 g NaOH in water to make 100 mL volume

Experiment 21   Analysis of vinegar by titration

Special Equipment

(25)   25-mL burets
(25)   Buret clamps
(25)   Ring stands
(25)   5-mL volumetric pipets
(25)   Small funnels

Chemicals

(500 mL)   Vinegar

(2L)   0.2 N NaOH standardized solution: dissolve 16.8 g NaOH in water to make 2 L volume. Standardize the solution as follows: place approximately 1 g potassium hydrogen phthalate, $KC_8H_5O_4$, in a tared weighing bottle. Weigh it to the nearest 0.001 g. Dissolve it in 20 mL water. Add a few drops of phenolphthalein indicator and titrate with the NaOH solution prepared above. The molarity and hence the normality of NaOH is calculated as follows: N = mass of phthalate/(0.2043 x mL NaOH used in titration). Write the calculated normality (molarity) of the NaOH on the bottle of the standardized NaOH solution.

(100 mL)   Phenolphthalein indicator: dissolve 0.1 g phenolphthalein in 60 mL 95% ethanol and bring it to 100 mL volume with water

Experiment 22   Analysis of antacid tablets

Special Equipment

(25)   25-mL burets
(25)   10-mL burets
(25)   Buret clamps
(25)   Ring stands
(5)   Balances to read to 0.001 g

Chemicals

(5 bottles)   Commercial antacids such as Alka-Seltzer, Gelusil, Maalox, Rolaids, Di-Gel, Tums, etc. Have at least two different kinds available.

(1 L)   0.2 N NaOH, sodium hydroxide, standardized: dissolve 8.4 g NaOH in 1 L water. Standardize as follows: accurately weigh to the nearest 0.001g approximately 1 g potassium hydrogen phthlate, $KC_8H_5O_4$, MW = 204.3 g/mole., and dissolve it in 20 mL water. Add a few drops of phenolphthalein and titrate the potassium hydrogen phthalate with the prepared NaOH solution. The normality (N) of the NaOH solution is calculated as follows: N = mass of phthalate/(0.2043 × mL NaOH used in the titration). Write the calculated normality on the bottle of the standardized NaOH solution.

(1 L)   0.2 N HCl, hydrochloric acid: add 16.6 mL concentrated HCl (36%) to 100 mL ice water; dilute with water to 1 L

volume. **(Prepare in the hood; wear a face shield, rubber gloves, and a rubber apron.)** Standardize the acid solution by titration against the standardized 0.2 N NaOH solution. Write the calculated normality on the bottle of the standardized HCl solution.

(100 mL)   Thymol blue indicator: dissolve 0.1 g thymol blue in 50 mL 95% ethanol and dilute with water to 100 mL volume. Put in a dropper bottle.

(100 mL)   Phenolphthalein indicator: dissolve 0.1 g phenolphthalein in 60 mL 95% ethanol and bring to 100 mL volume with water. Put in a dropper bottle.

Experiment 23   Structure in organic compounds: use of molecular models. I

Special Equipment

(Color of spheres may vary depending on the set; substitute as necessary.)

(50)   Black spheres—4 holes
(300)   Yellow spheres—1 hole
(50)   Colored spheres (e.g. green)—1 hole
(25)   Blue spheres—2 holes
(400)   Sticks
(25)   Protractors
(75)   Springs (Optional)

Experiment 24   Stereochemistry: use of molecular models. II

Special Equipment

Commercial molecular model kits vary in style, size, material composition and the color of the components. The set which works best in this exercise is the *Molecular Model Set for Organic Chemistry* available from Allyn and Bacon, Inc. (Newton, MA). Wood ball and stick models work as well. For 25 students, 25 of these sets should be provided. If you wish to make up your own kit, you would need the following for 25 students:

(25)   Cyclohexane model kits: each consisting of the following components:
          8 carbons—black, 4-hole
          18 hydrogens—white, 1-hole
          2 substituents—red, 1-hole
          24 connectors—bonds

(25)   Chiral model kits: each consisting of the following components:
          8 carbons—black, 4-hole
          32 substituents—8 red, 1-hole; 8 white, 1-hole; 8 blue, 1-hole; 8 green, 1-hole
          28 connectors—bonds

(5)   Small hand mirrors

Experiment 25   Identification of hydrocarbons

Special Equipment

(2 vials)   Litmus paper, blue
(250)   100 × 13 mm test tubes

Chemicals

(25 g)   Iron filings or powder

The following solutions should be placed in dropper bottles.

(100 mL)   Concentrated H_2SO_4
(100 mL)   Cyclohexene
(100 mL)   Hexane
(100 mL)   Ligroin (b.p. 90-110°C)
(100 mL)   Toluene
(100 mL)   1% Br_2 in cyclohexane **(wear a face shield, rubber gloves, and a rubber apron; prepare under hood):** mix 1.0 mL Br_2 with enough cyclohexane to make 100 mL. Prepare fresh solutions prior to use; keep in a dark-brown dropper bottle; do not store.
(100 mL)   1% aqueous $KMnO_4$: dissolve 1.0 g potassium permanganate in 50 mL distilled water by gently heating for 1 hr.; cool and filter; dilute to 100 mL. Store in a dark-brown dropper bottle.
(100 mL)   Unknown A = hexane
(100 mL)   Unknown B = cyclohexene
(100 mL)   Unknown C = toluene

Experiment 26   Column and paper chromatography; separation of plant pigments

Special Equipment

(50)   Melting point capillaries open at both ends
(25)   25-mL burets
(1 jar)   Glass wool
(25)   Filter papers (Whatman no.1), 20 x 10 cm
(3)   Heat lamp (optional)
(25)   Ruler with both English and metric scale
(1)   Stapler
(15)   Hot plates with or without water bath

Chemicals

(1 lb)   Tomato paste
(500 g)   Aluminum oxide (alumina)
(500 mL)   95% ethanol
(500 mL)   Petroleum ether, b.p. 30-60°C
(500 mL)   Eluting solvent: mix 450 mL petroleum ether with 10 mL toluene and 40 mL acetone.
(10 mL)   0.5% β-carotene solution: dissolve 50 mg in 10 mL petroleum ether. Wrap the vial in aluminum foil to protect from light and keep in refrigerator until used.

(150 mL)	Saturated bromine water: mix 5.5 g bromine with 150 mL water. **Prepare in hood; wear a face shield, rubber gloves, and a rubber apron.**
(500 mg)	Iodine crystals

Experiment 27 Identification of alcohols and phenols

Special Equipment

(125)	Corks (for test tubes 100 x 13 mm)
(125)	Corks (for test tubes 150 x 18 mm)
(25)	Hot plate
(5 rolls)	Indicator paper (pH 1 - 12)

Chemicals

The following solutions should be placed in dropper bottles.

(100 mL)	Acetone (reagent grade)
(100 mL)	1-Butanol
(100 mL)	2-Butanol
(100 mL)	2-Methyl-2-propanol (*t*-butyl alcohol)
(200 mL)	20% aqueous phenol: dissolve 80 g of phenol in 20 mL distilled water; dilute to 400 mL.
(100 mL)	Lucas reagent **(prepare under hood; wear a face shield, rubber gloves, and a rubber apron):** cool 100 mL of concentrated HCl in an ice bath; with stirring, add 150 g anhydrous $ZnCl_2$ to the cold acid.
(150 mL)	Chromic acid solution **(prepare under hood; wear a face shield, rubber gloves, and a rubber apron):** dissolve 30 g chromic oxide, CrO_3, in 30 mL concentrated H_2SO_4. Carefully add this solution to 90 mL water.
(100 mL)	2.5% ferric chloride solution: dissolve 2.5 g anhydrous $FeCl_3$ in 50 mL water; dilute to 100 mL.
(100 mL)	Iodine in KI solution: mix 20 g of KI and 10 g of I_2 in 100 mL water
(250 mL)	6 M sodium hydroxide, 6 M NaOH: dissolve 60 g NaOH in 100 mL water. Dilute to 250 mL with water.
(100 mL)	Unknown A = 1-butanol
(100 mL)	Unknown B = 2-butanol
(100 mL)	Unknown C = 2-methyl-2-propanol (*t*-butyl alcohol)
(100 mL)	Unknown D = 20% aqueous phenol

Experiment 28 Identification of aldehydes and ketones

Special Equipment

(250)	Corks (to fit 100 x 13 mm test tube)
(125)	Corks (to fit 150 x 18 mm test tube)
(1 box)	Filter paper (students will need to cut to size)
(25)	Hirsch funnels
(25)	Hot plates
(25)	Neoprene adapters (no. 2)
(25)	Rubber stopper assemblies: a no. 6 one-hole stopper fitted with glass tubing (15 cm in length x 7 mm OD)

(25)	50-mL side-arm filter flasks
(25)	250-mL side-arm filter flasks
(50)	Vacuum tubing, heavy-walled (2 ft. lengths)

Chemicals

| (50 g) | Hydroxylamine hydrochloride |
| (100 g) | Sodium acetate |

The following solutions should be placed in dropper bottles.

(100 mL)	Acetone (reagent grade)
(100 mL)	Benzaldehyde (freshly distilled)
(100 mL)	Bis(2-ethoxymethyl) ether
(100 mL)	Cyclohexanone
(500 mL)	Ethanol (absolute)
(500 mL)	Ethanol (95%)
(100 mL)	Isovaleraldehyde
(500 mL)	Methanol
(100 mL)	Pyridine
(150 mL)	Chromic acid reagent: dissolve 30 g chromic oxide (CrO_3) in 30 mL concentrated H_2SO_4. Add carefully to 90 mL water. **Wear a face shield, rubber gloves, and a rubber apron during the preparation. Do in the hood.**

Tollens' reagent

(100 mL)	Solution A: dissolve 9.0 g silver nitrate in 90 mL of water; dilute to 100 mL.
(100 mL)	Solution B: 10 g NaOH dissolved in enough water to make 100 mL
(100 mL)	10% ammonia water: 33.3 mL of concentrated (28%) NH_3 diluted to 100 mL
(100 mL)	6 M sodium hydroxide, 6 M NaOH: dissolve 24 g NaOH in enough water to make 100 mL
(500 mL)	Iodine-KI solution: mix 100 g of KI and a 50 g of iodine in enough distilled water to make 500 mL
(100 mL)	2,4-Dinitrophenylhydrazine reagent: dissolve 3.0 g of 2,4-dinitrophenylhydrazine in 15 mL concentrated H_2SO_4. In a beaker, mix together 10 mL water and 75 mL 95% ethanol. With vigorous stirring slowly add the 2,4-dinitrophenylhydrazine solution to the aqueous ethanol mixture. After thorough mixing, filter by gravity through a fluted filter paper. **Wear a face shield, rubber gloves, and a rubber apron during the preparation. Do in the hood.**
(100 mL)	Semicarbazide reagent: dissolve 22.2 g of semicarbazide hydrochloride in 100 mL of distilled water
(100 mL)	Unknown A = isovaleraldehyde
(100 mL)	Unknown B = benzaldehyde
(100 mL)	Unknown C = cyclohexanone
(100 mL)	Unknown D = acetone

Additional compounds for use as unknowns:

Aldehydes

(100 mL)   2-Butenal (crotonaldehyde)
(100 mL)   Octanal (caprylaldehyde)
(100 mL)   Pentanal (valeraldehyde)

Ketones

(100 mL)   Acetophenone
(100 mL)   Cyclopentanone
(100 mL)   2-Pentanone
(100 mL)   3-Pentanone

Experiment 29   Properties of carboxylic acids and esters

Special Equipment

(5 rolls)   pH paper (range 1 - 12)
(100)   Disposable Pasteur pipets
(5 vials)   Litmus paper, blue
(25)   Hot plates

Chemicals

(10 g)   Salicylic acid
(10 g)   Benzoic acid

The following solutions are placed in dropper bottles.

(75 mL)   Acetic acid
(50 mL)   Formic acid
(25 mL)   Benzyl alcohol
(50 mL)   Ethanol (ethyl alcohol)
(25 mL)   2-Methyl-1-propanol (isobutyl alcohol)
(25 mL)   3-Methyl-1-butanol (isopentyl alcohol)
(50 mL)   Methanol (methyl alcohol)
(25 mL)   Methyl salicylate
(250 mL)   6 M hydrochloric acid, 6 M HCl: take 150 mL of concentrated HCl and add to 50 mL of cold water; dilute with enough water to 250 mL. **Wear a face shield, rubber gloves, and a rubber apron during the preparation. Do in the hood.**
(100 mL)   3 M hydrochloric acid, 3 M HCL: take 50 mL 6 M HCl and bring to 100 mL; **follow the same precautions as above.**
(300 mL)   6 M sodium hydroxide, 6 M NaOH: dissolve 72 g NaOH in enough water to bring to 300 mL; **follow the same precautions as above.**
(150 mL)   2 M sodium hydroxide, 2 M NaOH: take 50 mL 6 M NaOH and bring to 150 mL; **follow the same precautions as above.**
(25 mL)   Concentrated sulfuric acid, H_2SO_4

Experiment 30   Properties of amines and amides

Special Equipment

(2 rolls)	pH paper (range 0 to 12)
(25)	Hot plates

Chemicals

(20 g)	Acetamide

The following solutions should be placed in dropper bottles.

(25 mL)	Triethylamine
(25 mL)	Aniline
(25 mL)	N,N-Dimethylaniline
(100 mL)	Diethyl ether (ether)
(100 mL)	6 M ammonia solution, 6 M NH_3: add 40 mL concentrated (28%) NH_3 to 50 mL water; then add enough water to 100 mL volume. **Do in the hood.**
(100 mL)	6 M hydrochloric acid, 6 M HCl: add 50 mL concentrated HCl to 40 mL ice cold water; then add enough water to 100 mL volume. **Wear a face shield, rubber gloves and a rubber apron when preparing. Do in the hood.**
(50 mL)	Concentrated hydrochloric acid, HCl
(250 mL)	6 M sulfuric acid, 6 M H_2SO_4: pour 82.5 mL concentrated H_2SO_4 into 125 mL ice cold water. Stir slowly. Then add enough water to 250 mL volume. **Wear a face shield, rubber gloves, and a rubber apron when preparing. Do in the hood.**
(250 mL)	6 M sodium hydroxide, 6 M NaOH: dissolve 60 g NaOH in 100 mL water. Then add enough water to 250 mL volume. **Do in the hood.**

Experiment 31   Polymerization reactions

Special Equipment

(25)	Hot plates
(25)	Cylindrical paper rolls or sticks
(25)	Bent wire approximately 10-cm long
(25)	10-mL pipets or syringes
(25)	Spectroline pipet fillers
(25)	Beaker tongs

Chemicals

(750 mL)	Styrene
(250 mL)	Xylene
(75 mL)	t-butyl peroxide benzoate (also called t-butyl benzoyl peroxide); store at 4°C.
(75 mL)	20% sodium hydroxide: dissolve 15 g NaOH in 75 mL water
(300 mL)	5% adipoyl chloride: dissolve 15 g adipoyl chloride in 300 mL cyclohexane
(300 mL)	5% hexamethylene diamine: dissolve 15 g hexamethylene diamine in 300 mL water
(200 mL)	80% formic acid: add 40 mL water to 160 mL formic acid

Experiment 32   Preparation of acetylsalicylic acid (aspirin)

Special Equipment

(25)	Büchner funnels (85-mm OD)
(25)	Filtervac or no. 2 neoprene adapters
(1 box)	Filter paper (7.0 cm, Whatman no. 2)
(25)	250-mL filter flasks
(25)	Hot plates

Chemicals

(1 jar)	Boiling chips
(25)	Commercial aspirin tablets
(100 mL)	Concentrated phosphoric acid, H_3PO_4 (in a dropper bottle)
(100 mL)	1% iron(III) chloride: dissolve 1 g $FeCl_3 \cdot 6H_2O$ in enough distilled water to make 100 mL (in a dropper bottle)
(100 mL)	Acetic anhydride, freshly opened bottle
(300 mL)	95% ethanol
(100 g)	Salicylic acid

Experiment 33   Measurement of the active ingredient in aspirin pills

Special Equipment

(1)	Drying oven at 110°C
(25)	Mortars, 100-mL capacity
(25)	Pestles
(1 box)	Filter paper, 7.0 cm diameter, Whatman no. 2
(1 box)	Microscope slides, 3 × 1 in., plain
(25)	25-mL beakers

Chemicals

(1.5 L)	95% ethanol
(300 g)	Commercial asprin tablets
(100 mL)	Hanus iodine solution: dissolve 1.2 g KI in 80 mL water. Add 0.25 g I_2. Stir until the iodine dissolves. Add enough water to make 100 mL volume. Store in dark dropper bottle.

Experiment 34   Isolation of caffeine from tea leaves

Special Equipment

(25)	Cold finger condensers (115 mm long x 15 mm OD)
(1 box)	Filter paper; 7.0 cm, fast flow (Whatman no.1)
(25)	Hot plates
(50)	Latex tubing, 2 ft. lengths
(1 vial)	Melting point capillaries
(25)	No. 2 neoprene adaptors
(25)	Rubber stopper (no. 6, 1-hole) with glass tubing inserted (10 cm length x 7 mm OD)
(25)	125-mL separatory funnels
(25)	25-mL side-arm filter flasks
(25)	250-mL side-arm filter flasks
(25)	Small sample vials
(1)	Stapler

| (50) | Vacuum tubing, 2 ft. lengths |
| (1 box) | Weighing paper |

Chemicals

(1 jar)	Boiling chips
(500 mL)	Methylene chloride, CH_2Cl_2
(25 g)	Sodium sulfate, anhydrous, Na_2SO_4
(50 g)	Sodium carbonate, anhydrous, Na_2CO_3
(50)	Tea bags

Experiment 35   Carbohydrates

Special Equipment

(50)	Medicine droppers
(125)	Microtest tubes or 25 depressions white spot plates
(2 rolls)	Litmus paper, red

Chemicals

(20 g)	Boiling chips
(400 mL)	Fehling's reagent (solutions A and B, from Fisher Scientific Co.)
(200 mL)	3 M NaOH: dissolve 24 g NaOH in 100 mL water and then add enough water to 200 mL volume
(200 mL)	2% starch solution: place 4 g soluble starch in a beaker. With vigorous stirring, add 10 mL water to form a thin paste. Boil 190 mL water in another beaker. Add the starch paste to the boiling water and stir until the solution becomes clear. Store in a dropper bottle.
(200 mL)	2% sucrose: dissolve 4 g sucrose in 200 mL water
(50 mL)	3 M sulfuric acid: add 8 mL concentrated H_2SO_4 to 30 mL ice cold water; pour the sulfuric acid slowly along the walls of the beaker, this way it will settle on the bottom without much mixing; stir slowly in order not to generate too much heat; when fully mixed bring the volume to 50 mL. **Wear a face shield, rubber gloves, and a rubber apron when preparing. Do in the hood.**
(100 mL)	2% fructose: dissolve 2 g fructose in 100 mL water. Store in a dropper bottle.
(100 mL)	2% glucose: dissolve 2 g glucose in 100 mL water. Store in a dropper bottle.
(100 mL)	2% lactose: dissolve 2 g lactose in 100 mL water. Store in a dropper bottle.
(100 mL)	0.01 M iodine in KI: dissolve 1.2 g KI in 80 mL water. Add 0.25 g I_2. Stir until the iodine dissolves. Dilute the solution to 100 mL volume. Store in a dark dropper bottle.

Experiment 36   Preparation and properties of a soap

Special Equipment

(25)	Büchner funnels (85 mm OD)
(25)	No. 7 one-hole rubber stoppers
(1 box)	Filter paper (Whatman no.2) 7.0 cm
(1 roll)	pHydrion paper (pH range 0 to 12)

Chemicals

(1 jar)	Boiling chips
(1 L)	95 % ethanol
(1 L)	Saturated sodium chloride (sat. NaCl): dissolve 360 g NaCl in 1 L water
(1 L)	25% sodium hydroxide (25% NaOH): dissolve 250 g NaOH in 1 L water
(1 L)	Vegetable oil
(100 mL)	5% iron(III) chloride (5% $FeCl_3$): dissolve 5 g $FeCl_3 \cdot 6H_2O$ in 100 mL water. Store in a dropper bottle.
(100 mL)	5% calcium chloride (5% $CaCl_2$): dissolve 5 g $CaCl_2 \cdot H_2O$ in 100 mL water. Store in a dropper bottle.
(100 mL)	Mineral oil. Store in a dropper bottle.
(100 mL)	5% magnesium chloride (5% $MgCl_2$): dissolve 5 g $MgCl_2$ in 100 mL water. Store in a dropper bottle.

Experiment 37 Preparation of hand cream

Special Equipment

(25)	Bunsen burners

Chemicals

(100 mL)	Triethanolamine
(40 mL)	Propylene glycol (1,2-propanediol)
(500 g)	Stearic acid
(40 g)	Methyl stearate (ethyl stearate may be substituted)
(400 g)	Lanolin
(400 g)	Mineral oil

Experiment 38 Extraction and identification of fatty acids from corn oil

Special equipment

(12)	Water baths
(2)	Heat lamps or hair dryers
(25)	15 × 6.5 cm silica gel TLC plates
(25)	Rulers, metric scale
(25)	Polyethylene, surgical gloves
(150)	Capillary tubes, open on both ends
(1)	Drying oven, 110°C

Chemicals

(50 g)	Corn oil
(5 mL)	Methyl palmitate solution: dissolve 25 mg methyl palmitate in 5 mL petroleum ether
(5 mL)	Methyl oleate solution: dissolve 25 mg methyl oleate in 5 mL petroleum ether
(5 mL)	Methyl linoleate solution: dissolve 25 mg methyl linoleate in 5 mL petroleum ether
(100 mL)	0.5 M KOH: dissolve 2.81 g KOH in 25 mL water and add 75 ml of 95% ethanol
(500 g)	Sodium sulfate, Na_2SO_4, anhydrous, granular

(100 mL)	Concentrated hydrochloric acid, HCl
(1 L)	Petroleum ether (b. p. 30-60°C)
(300 mL)	Methanol: perchloric acid mixture: mix 285 mL methanol with 15 mL $HClO_4 \cdot 2H_2O$ (73% perchloric acid)
(400 mL)	Hexane:diethyl ether mixture: mix 320 mL hexane with 80 mL diethyl ether
(10 g)	Iodine crystals, I_2

Experiment 39   Analysis of lipids

Special Equipment

| (25) | Hot plates |
| (25) | Cheesecloth 3×3 in. |

Chemicals

(3 g)	Cholesterol (ash free) 95-98% pure from Sigma Co.
(3 g)	Lecithin (prepared from dried egg yolk) 60% pure from Sigma Co.
(10 g)	Glycerol
(10 g)	Corn oil
(10 g)	Butter
(1)	Egg yolk obtained from one fresh egg before the lab period. Stir and mix.
(250 mL)	Molybdate solution: dissolve 0.8 g $(NH_4)_6Mo_7O_{24} \cdot 4H_2O$ in 30 mL water. Put in an ice bath. Pour slowly 20 mL concentrated sulfuric acid into the solution and stir slowly. After cooling to room temperature bring the volume to 250 mL. **Wear a face shield, rubber gloves, and a rubber apron during the preparation. Do in the hood.**
(50 mL)	0.1 M ascorbic acid solution: dissolve 0.88 g ascorbic acid (vitamin C) in water and bring it to 50 mL volume. This must be prepared fresh every week and stored at 4°C.
(250 mL)	6 M sodium hydroxide, 6 M NaOH: dissolve 60 g NaOH in water and bring the volume to 250 mL
(250 mL)	6 M nitric acid, 6 M HNO_3: into a 250-mL volumetric flask containing 100 mL ice cold water, pipet 63 ml concentrated nitric acid, HNO_3; add enough water to bring to 250 mL. **Wear a face shield, rubber gloves, and a rubber apron during the preparation. Do in the hood.**
(200 mL)	Chloroform
(75 mL)	Acetic anhydride
(50 mL)	Concentrated sulfuric acid, H_2SO_4
(75 g)	Potassium hydrogen sulfate, $KHSO_4$

Experiment 40   TLC separation of amino acids

Special Equipment

| (1) | Drying oven, 105-110°C |
| (2) | Heat lamps or hair dryers |

(50)	15 x 6.5 cm silica gel TLC plates (or chromatographic paper Whatman no. 1)
(25)	Rulers, metric scale
(25)	Polyethylene, surgical gloves
(150)	Capillary tubes, open on both ends
(1 roll)	Aluminum foil

Chemicals

(25 mL)	0.12% aspartic acid solution: dissolve 30 mg aspartic acid in 25 mL distilled water
(25 mL)	0.12% phenylalanine solution: dissolve 30 mg phenylalanine in 25 mL distilled water
(25 mL)	0.12% leucine solution: dissolve 30 mg leucine in 25 mL distilled water
(25 mL)	Aspartame solution: dissolve 150 mg Equal sweetener powder in 25 mL distilled water
(50 mL)	3 M HCl solution: place 10 mL distilled water into a 50 mL volumetric flask. Add slowly 12.46 mL of concentrated HCl and bring it to volume with distilled water. **Wear a face shield, rubber gloves, and a rubber apron when preparing. Do in the hood.**
(1 L)	Solvent mixture: mix 600 mL 1-butanol with 150 mL acetic acid and 250 mL distilled water
(1 can)	Ninhydrin spray reagent (0.2% ninhydrin in ethanol or acetone). Do not use any reagent older than 6 months.
(1 can)	Diet Coca-Cola
(4 packets)	Equal or NutraSweet, sweeteners

Experiment 41 Acid-base properties of amino acids

Special Equipment

(10)	pH meters or
(5 rolls)	pHydrion short-range papers, from each range: pH: 0.0 to 3.0; 3.0 to 5.5; 5.2 to 6.6; 6.0 to 8.0; 8.0 to 9.5 and 9.0 to 12.0
(25)	20-mL pipets
(25)	50-mL burets
(25)	Spectroline pipet fillers
(25)	Pasteur pipets

Chemicals

| (500 mL) | 0.25 M NaOH: dissolve 5.00 g NaOH in 100 mL water and then add enough water to 500 mL volume |
| (750 mL) | 0.1 M alanine solution: dissolve 6.68 g L-alanine in 500 mL; add sufficient 1 M HCl to bring the pH to 1.5. Add enough water to 750 mL volume. |

or

Do as above but use either 5.63 g glycine or 9.84 g leucine or 12.39 g phenylalanine or 8.79 g valine.

Experiment 42   Isolation and identification of casein

Special Equipment

(25)	Hot plates
(25)	600-mL beakers
(25)	Büchner funnels (O.D. 85 mm) in no. 7-hole rubber stopper
(7 boxes)	Whatman no. 2 filter paper, 7 cm
(25)	Rubber bands
(25)	Cheesecloths (6 × 6 in.)

Chemicals

(1 jar)	Boiling chips
(1 L)	95% ethanol
(1 L)	Diethyl ether:ethanol mixture (1:1)
(0.5 gal)	Regular milk
(500 mL)	Glacial acetic acid

The following solutions should be placed in dropper bottles:

(100 mL)	Concentrated nitric acid, HNO_3, (69%)
(100 mL)	2% albumin suspension: dissolve 2 g albumin in 100 mL water
(100 mL)	2% gelatin: dissolve 2 g gelatin in 100 mL water
(100 mL)	2% glycine: dissolve 2 g glycine in 100 mL water
(100 mL)	5% copper(II) sulfate: dissolve 5 g $CuSO_4$ (or 7.85 g $CuSO_4 \cdot 5 H_2O$) in 100 mL water
(100 mL)	5% lead(II) nitrate: dissolve 5 g $Pb(NO_3)_2$ in 100 mL water
(100 mL)	5% mercury(II) nitrate: dissolve 5 g $Hg(NO_3)_2$ in 100 mL water
(100 mL)	Ninhydrin reagent: dissolve 3 g ninhydrin in 100 mL acetone
(100 mL)	10% sodium hydroxide: dissolve 10 g NaOH in 100 mL water
(100 mL)	1% tyrosine: dissolve 1 g tyrosine in 100 mL water
(100 mL)	5% sodium nitrate: dissolve 5 g $NaNO_3$ in 100 mL water

Experiment 43   Isolation and identification of DNA from yeast

Special equipment

(12)	Mortars
(12)	Pestles
(6)	Desk top clinical centrifuges (swinging bucket rotor) (optional)

Chemicals

(100 g)	Baker's yeast, freshly purchased
(500 g)	Acid washed sand
(1 L)	Saline-CTAB isolation buffer: dissolve 20 g hexadecyltrimethylammonium bromide (CTAB, Sigma 45882), 2 mL 2-mercaptoethanol, 7.44 g ethylenediamine tetraacetate (EDTA, Sigma ED2SS), 8.77 g NaCl in 1 L Tris

buffer. The Tris buffer is prepared by dissolving 12.1 g Tris in 700 mL water; adjust the pH to 8 by titrating with 4 M HCl. Add enough water to bring the volume to 1 L.

(200 mL)   6 M sodium perchlorate solution, 6 M $NaClO_4$: dissolve 147 g $NaClO_4$ in 100 mL water and add enough water to bring the volume to 200 mL

(100 mL)   Citrate buffer: dissolve 0.88 g NaCl and 0.39 g sodium citrate in 100 mL water

(1 L)   Chloroform-isoamyl alcohol mixture: to 960 mL chloroform, add 40 mL isoamyl alcohol. Mix throughly.

(2 L)   Isopropyl alcohol (2-propanol)

(50 mL)   1% glucose solution: dissolve 0.5 g D-glucose in 50 mL water

(50 mL)   1% ribose solution: dissolve 0.5 g D-ribose in 50 mL water

(50 mL)   1% deoxyribose solution: dissolve 0.5 g 2-deoxy-D-ribose in 50 mL water

(200 mL)   95% ethanol

(500 mL)   Diphenylamine reagent. *This must be prepared shortly before lab use.* Dissolve 7.5 g reagent grade diphenylamine (Sigma D3409) in 50 mL glacial acetic acid. Add 7.5 mL concentrated sulfuric acid. Prior to use add 2.5 mL 1.6% acetaldehyde (made by dissolving 0.16 g acetaldehyde in 10 mL water). **Wear a face shield, rubber gloves, and a rubber apron when preparing. Do in the hood.**

Experiment 44   Viscosity and secondary structure of DNA

Equipment

(5)   Ostwald (or Cannon-Ubbelhode) capillary viscometers; 3-mL capacity, approximate capillary diameter 0.2 mm; **efflux time of water = 40-50 sec.**

(5)   Stopwatches (Wristwatches can also time the efflux with sufficient precision.)

(5)   Stands with utility clamps

(25)   Pasteur pipets

(10)   Spectroline pipet fillers

Chemicals

(500 mL)   Buffer solution: dissolve 4.4 g sodium chloride, NaCl, and 2.2 g sodium citrate, $Na_3C_6H_5O_7 \cdot 2H_2O$ in 450 mL distilled water. Adjust the pH with either 0.1 M HCl or 0.1 M NaOH to pH 7.0. Add enough water to bring to 500 mL volume.

(200 mL)   DNA solution: dissolve 20 mg of calf thymus Type I highly polymerized DNA (obtainable from Sigma as well as from other companies) in 200 mL buffer solution at pH 7.0. The purchased DNA powder should be kept in the freezer. The DNA solutions should be prepared fresh or maximum 2–3 hrs. in advance of the experiment. The solution should be kept at 4°C.; 1–2 hrs. before the

experiment, the solution should be allowed to come to room temperature. Label the solution as 1×10^{-2} g/dL concentration.

(100 mL)   1 M hydrochloric acid, 1 M HCl: add 8.3 mL concentrated HCl (36%) to 50 mL ice cold water; add enough water to bring to 100 mL volume. **Wear a face shield, rubber gloves, and a rubber apron during preparation. Do in the hood.**

(100 mL)   0.1 M hydrochloric acid: add 0.83 mL concentrated HCl (36%) to 50 mL ice cold water; add enough water bring to 100 mL volume. **Follow safety procedure described above.**

(100 mL)   1 M sodium hydroxide: dissolve 4 g NaOH in 50 mL water; add enough water to bring to 100 mL volume.

(100 mL)   0.1 M sodium hydroxide: dissolve 0.4 g NaOH in 50 mL water; add enough water to bring to 100 mL volume.

Experiment 45   Kinetics of urease catalyzed decomposition of urea

Special Equipment

(25)	5-mL pipets
(25)	10-mL graduated pipets
(25)	10-mL volumetric pipets
(25)	50-mL burets
(25)	Buret holders
(25)	Spectroline pipet fillers

Chemicals

(3.5 L)   0.05 M Tris buffer: dissolve 21.05 g Tris buffer in water (3 L). Adjust the pH to 7.2 with 1 M HCl solution; add sufficient water to make 3.5 L. Portions of buffer solution will be used to make urea and enzyme solutions.

(2.5 L)   0.3 M urea solution: dissolve 45 g urea in 2.5 L Tris buffer

(50 mL) 1×10^{-3} M phenylmercuric acetate: dissolve 16.5 mg phenylmercuric acetate in 40 mL water; add enough water to bring the volume to 50 mL.

CAUTION! Phenylmercuric acetate is a poison. Do not touch the chemical with your hands. Do not swallow the solution. Wear rubber gloves in the preparation.

(50 mL)   1% $HgCl_2$ solution: dissolve 0.5 g $HgCl_2$ in enough water to make 50 mL solution

(100 mL)   0.04% methyl red indicator: dissolve 40 mg methyl red in 100 mL distilled water

(500 mL)   Urease solution: prepare the enzyme solution on the week of the experiment and store at 4°C. Take 1.0 g urease, dissolve in 500 mL Tris buffer. (One can buy urease with 5 to 6 units activity, for example, from Nutritional Biochemicals, Cleveland, Ohio.) The activity of the enzyme printed on the label should be checked by the stockroom personnel or instructor.

| (1.0 L) | 0.05 N HCl: add 5 mL concentrated HCl to 100 mL ice cold water; add enough water to bring to 1.0 L volume. **Wear a face shield, rubber gloves, and a rubber apron during the preparation. Do in the hood.** |

Experiment 46   Isocitrate dehydrogenase

Special Equipment

| (15) | Spectrophotometers |

Chemicals

(40 mL)	Phosphate buffer at pH 7.0: mix together 25 mL 0.1 M KH_2PO_4 and 15 mL 0.1 M NaOH. To prepare 0.1 M NaOH, add 0.2 g NaOH to 20 mL water in a 50-mL volumetric flask; stir to dissolve; add enough water to bring to 50 mL volume. To prepare 0.1 M KH_2PO_4, add 0.68 g potassium dihydrogen phosphate to 40 mL water in a 50-mL volumetric flask; stir to dissolve; add enough water to bring to 50 mL volume.
(20 mL)	0.1 M $MgCl_2$: add 0.19 g magnesium chloride to 20 mL water; stir to dissolve.
(50 mL)	Isocitrate dehydrogenase: commercial preparations from porcine heart are obtainable from companies such as Sigma, etc. (EC 1.1.1.42) (activity about 8 units per mg of solid). Dissolve 10 mg of the enzyme in 50 mL water. This solution should be made fresh before the lab period and kept in a refrigerator until used.
(20 mL)	6.0 mM β-Nicotinamide adenine dinucleotide, β-$NADP^+$, solution: dissolve 92 mg $NADP^+$ in 20 mL water
(50 mL)	15 mM sodium isocitrate solution: dissolve 160 mg sodium isocitrate in 50 mL water

Experiment 47   Quantitative analysis of vitamin C in foods

Special equipment

(25)	50-mL burets
(25)	Buret clamps
(25)	Ring stands
(25)	Spectroline pipet fillers
(25)	10-mL volumetric pipets
(1 box)	Cotton

Chemicals

(500 g)	Celite, filter aid
(1 can)	Hi-C orange drink
(1 can)	Hi-C grapefruit drink
(1 can)	Hi-C apple drink
(2 L)	0.01 M iodine solution: add 32 g KI to 800 mL water; stir to dissolve. Add 5 g I_2; stir to dissolve. Add enough water to bring to 2 L volume. Store in dark bottle. **Caution! Iodine is poisonous if taken internally.**

| (100 mL) | 3 M HCl: add 25 mL concentrated HCl (36%) to 50 mL ice cold water; add enough water to bring to 100 mL volume. **Wear a face shield, rubber gloves, and a rubber apron during the preparation. Do in the hood.** |
| (100 mL) | 2% starch solution: place 2 g soluble starch in a 50-mL beaker. Add 10 mL water. Stir vigorously to form a paste. Boil 90 mL water in a second beaker. Add the starch paste to the boiling water. Stir until the solution becomes translucent. Cool to room temperature. |

Experiment 48 Analysis of vitamin A in margarine

Special equipment

(1)	UV spectrophotometer with suitable UV light source. Preferably it should be able to read down to 200 nm.
(1 pair)	Matched quartz cells, with 1 cm internal path length.
(2)	Long wavelength UV lamp. The lamp should provide radiation in the 300 nm range (for example, #UVSL-55; LW 240 from Ultraviolet Products Inc.)
(12)	500-mL separatory funnels
(12)	25-mL burets
(12)	Hot plates, each with a water bath
(12)	Beaker tongs

Chemicals

(0.5 lb)	Margarine
(400 mL)	50% KOH; weigh 200 g KOH and add 200 mL water, with constant stirring
(1 L)	95% ethanol
(150 mL)	Absolute ethanol
(2 L)	Petroleum ether, 30°-60°C
(3 L)	Diethyl ether
(350 g)	Alkaline aluminum oxide (alumina)

Experiment 49 Measurement of sulfur dioxide preservative in foods

Special Equipment

(1)	Blender
(175)	100-mL volumetric flasks
(25)	10-mL pipets
(100)	1-mL graduated pipets
(75)	10-mL graduated pipets
(5)	Spectrophotometers

Chemicals

(10 g)	Raisins
(200 mL)	0.5 N NaOH: dissolve 4.0 g NaOH in 50 mL water and add enough water to bring to 200 mL volume
(200 mL)	0.5 N H_2SO_4 solution: place 100 mL water in a beaker. Cool it in an ice bath. Add slowly from a graduated cylinder, 10.7 mL concentrated sulfuric acid. **Make sure you**

pour the concentrated acid slowly along the walls of the beaker. If you add it too fast the acid may splash and create severe burns. Wear a face shield, rubber gloves, and a rubber apron during this procedure. Do in the hood. Wait a few minutes. Slowly stir the solution with a glass rod and add enough water to bring to 200 mL.

(1.5 L)   0.015% formaldehyde solution: take 0.56 mL 40% formaldehyde solution and add it to 1.5 L water

(1 L)   Rosaniline reagent: place 100 mg p-rosaniline ·HCl (Allied Chem. Corp.) and 200 mL water in a 1-L volumetric flask. Add 80 mL concentrated HCl to 80 mL ice cold water in a 250-mL beaker. Stir. Add the hydrochloric acid solution to the 1-L volumetric flask, mix and add enough water to bring to volume. The rosaniline reagent must stand at least 12 hrs. before use. Follow the safety precautions given above in this preparation.

(1.5 L)   Mercurate reagent: use polyethylene gloves to protect your skin from touching mercurate reagent. Mercury compounds are toxic and if spills occur, wash them immediately with copious amounts of water. Dissolve 17.6 g NaCl and 40.7 g $HgCl_2$ in 1 L water and add enough water to bring to to 1.5 L.

(500 mL)   Standard sulfur dioxide stock solution: dissolve 85 mg $NaHSO_3$ in 500 mL water.

Experiment 50   Urine analysis

Special Equipment

(12)   Hydrometers (urinometers) from 1.00 to 1.40 specific gravity

(3 bottles)   Clinistix (50 reagent strips/bottle)

(3 bottles)   Urobilistix (50 reagent strips/bottle)

(3 bottles)   Phenistix (50 reagent strips/bottle)

(1 bottle)   Albustix (100 reagent strips/bottle)

(1 bottle)   Ketostix (100 reagent strip/bottle)

These are obtainable from Ames Co., Division Miles Lab. Inc., Elkhart, Indiana, 46515. Instead of the individual "Stix," you may purchase 4 bottles of multipurpose Labstix (100 reagent strips/bottle).

Chemicals

(500 mL)   Normal urine. This and all other urine samples must be kept at 4°C until 30 min. prior to the lab period. Alternatively, you may ask each student to provide fresh urine samples for analysis.

(500 mL)   "Pathological urine A": add 4 g glucose, 2 mL acetone and 2 g citric acid to 500 mL water

(500 mL)   "Pathological urine B": add 50 mg phenylpyruvate, and 500 mg sodium phoshate, Na_3PO_4, to 500 mL water

(250 mL)   1% glucose: dissolve 2.5 g glucose in 250 mL water

(200 mL)   0.25% glucose: dilute 50 mL of 1% glucose solution with water to 200 mL volume